Martin Pelikan, Kumara Sastry, Erick Cantú-Paz (Eds.)

Scalable Optimization via Probabilistic Modeling

Studies in Computational Intelligence, Volume 33

Editor-in-chief
Prof. Janusz Kacprzyk
Systems Research Institute
Polish Academy of Sciences
ul. Newelska 6
01-447 Warsaw
Poland
E-mail: kacprzyk@ibspan.waw.pl

Further volumes of this series can be found on our homepage: springer.com

Vol. 16. Yaochu Jin (Ed.)
Multi-Objective Machine Learning, 2006
ISBN 3-540-30676-5

Vol. 17. Te-Ming Huang, Vojislav Kecman, Ivica Kopriva
Kernel Based Algorithms for Mining Huge Data Sets, 2006
ISBN 3-540-31681-7

Vol. 18. Chang Wook Ahn
Advances in Evolutionary Algorithms, 2006
ISBN 3-540-31758-9

Vol. 19. Ajita Ichalkaranje, Nikhil Ichalkaranje, Lakhmi C. Jain (Eds.)
Intelligent Paradigms for Assistive and Preventive Healthcare, 2006
ISBN 3-540-31762-7

Vol. 20. Wojciech Penczek, Agata Półrola
Advances in Verification of Time Petri Nets and Timed Automata, 2006
ISBN 3-540-32869-6

Vol. 21. Cândida Ferreira
Gene Expression on Programming: Mathematical Modeling by an Artificial Intelligence, 2006
ISBN 3-540-32796-7

Vol. 22. N. Nedjah, E. Alba, L. de Macedo Mourelle (Eds.)
Parallel Evolutionary Computations, 2006
ISBN 3-540-32837-8

Vol. 23. M. Last, Z. Volkovich, A. Kandel (Eds.)
Algorithmic Techniques for Data Mining, 2006
ISBN 3-540-33880-2

Vol. 24. Alakananda Bhattacharya, Amit Konar, Ajit K. Mandal
Parallel and Distributed Logic Programming, 2006
ISBN 3-540-33458-0

Vol. 25. Zoltán Ésik, Carlos Martín-Vide, Victor Mitrana (Eds.)
Recent Advances in Formal Languages and Applications, 2006
ISBN 3-540-33460-2

Vol. 26. Nadia Nedjah, Luiza de Macedo Mourelle (Eds.)
Swarm Intelligent Systems, 2006
ISBN 3-540-33868-3

Vol. 27. Vassilis G. Kaburlasos
Towards a Unified Modeling and Knowledge-Representation based on Lattice Theory, 2006
ISBN 3-540-34169-2

Vol. 28. Brahim Chaib-draa, Jörg P. Müller (Eds.)
Multiagent based Supply Chain Management, 2006
ISBN 3-540-33875-6

Vol. 29. Sai Sumathi, S.N. Sivanandam
Introduction to Data Mining and its Applications, 2006
ISBN 3-540-34689-9

Vol. 30. Yukio Ohsawa, Shusaku Tsumoto (Eds.)
Chance Discoveries in Real World Decision Making, 2006
ISBN 3-540-34352-0

Vol. 31. Ajith Abraham, Crina Grosan, Vitorino Ramos (Eds.)
Stigmergic Optimization, 2006
ISBN 3-540-34689-9

Vol. 32. Akira Hirose
Complex-Valued Neural Networks, 2006
ISBN 3-540-33456-4

Vol. 33. Martin Pelikan, Kumara Sastry, Erick Cantú-Paz (Eds.)
Scalable Optimization via Probabilistic Modeling, 2006
ISBN 3-540-34953-7

Martin Pelikan
Kumara Sastry
Erick Cantú-Paz (Eds.)

Scalable Optimization via Probabilistic Modeling

From Algorithms to Applications

With 98 Figures and 26 Tables

Springer

Dr. Martin Pelikan
Department of Math. and
Computer Science
320 CCB
University of Missouri at St. Louis
One University Blvd.
St. Louis, MO 63121
USA
E-mail: pelikan@cs.umsl.edu

Kumara Sastry
Illinois Genetic Algorithms
Laboratory, 414 TB
Department of Industrial and
Enterprise Systems Engineering
104 S. Mathews Ave.
Urbana, IL 61801
USA
E-mail: ksastry@uiuc.edu

Dr. Erick Cantú-Paz
Yahoo! Inc.
701 First Avenue
Sunnyvale, CA 94809
USA
E-mail: cantupaz@acm.org

ISSN print edition: 1860-949X
ISSN electronic edition: 1860-9503
ISBN 978-3-642-07116-4 e-ISBN 978-3-540-34954-9

Springer is a part of Springer Science+Business Media
springer.com

Softcover reprint of the hardcover 1st edition 2006

Cover design: deblik, Berlin

Advance Praise for Scalable Optimization via Probabilistic Modeling

This book is an excellent compilation of carefully selected topics in estimation of distribution algorithms—search algorithms that combine ideas from evolutionary algorithms and machine learning. The book covers a broad spectrum of important subjects ranging from design of robust and scalable optimization algorithms to efficiency enhancements and applications of these algorithms. The book should be of interest to theoreticians and practitioners alike, and is a must-have resource for those interested in stochastic optimization in general, and genetic and evolutionary algorithms in particular.

John R. Koza,
Stanford University

This edited book portrays population-based optimization algorithms and applications, covering the entire gamut of optimization problems having single and multiple objectives, discrete and continuous variables, serial and parallel computations, and simple and complex function models. Anyone interested in population-based optimization methods, either knowingly or unknowingly, use some form of an estimation of distribution algorithm (EDA). This book is an eye-opener and a must-read text, covering easy-to-read yet erudite articles on established and emerging EDA methodologies from real experts in the field.

Kalyanmoy Deb,
Indian Institute of Technology Kanpur

This book is an excellent comprehensive resource on estimation of distribution algorithms. It can serve as the primary EDA resource for practitioner or researcher. The book includes chapters from all major contributors to EDA state-of-the-art and covers the spectrum from EDA design to applications. These algorithms strategically combine the advantages of genetic and evolutionary computation with the advantages of statistical, model building machine learning techniques. EDAs are useful to solve classes of difficult real-world problems in a robust and scalable manner.

Una-May O'Reilly,
Massachusetts Institute of Technology

Machine-learning methods continue to stir the public's imagination due to its futuristic implications. But, probability-based optimization methods can have great impact now on many scientific multiscale and engineering design problems, especially true with use of efficient and competent genetic algorithms (GA) which are the basis of the present volume. Even though efficient and competent GAs outperform standard techniques and prevent negative issues, such as solution stagnation, inherent in the older but more well-known GAs, they remain less known or embraced in the scientific and engineering communities. To that end, the editors have brought together a selection of experts that (1) introduce the current methodology and lexicography of the field with illustrative discussions and highly useful references, (2) exemplify these new techniques that dramatic improve performance in provable hard problems, and (3) provide real-world applications of these techniques, such as antenna design. As one who has strayed into the use of genetic algorithms and genetic programming for multiscale modeling in materials science, I can say it would have been personally more useful if this would have come out five years ago, but, for my students, it will be a boon.

Duane D. Johnson,
University of Illinois at Urbana-Champaign

Foreword

I'm not usually a fan of edited volumes. Too often they are an incoherent hodgepodge of remnants, renegades, or rejects foisted upon an unsuspecting reading public under a misleading or fraudulent title. The volume Scalable Optimization via Probabilistic Modeling: From Algorithms to Applications is a worthy addition to your library because it succeeds on exactly those dimensions where so many edited volumes fail.

For example, take the title, Scalable Optimization via Probabilistic Modeling: From Algorithms to Applications. You need not worry that you're going to pick up this book and find stray articles about anything else. This book focuses like a laser beam on one of the hottest topics in evolutionary computation over the last decade or so: estimation of distribution algorithms (EDAs). EDAs borrow evolutionary computation's population orientation and selectionism and throw out the genetics to give us a hybrid of substantial power, elegance, and extensibility.

The article sequencing in most edited volumes is hard to understand, but from the get go the editors of this volume have assembled a set of articles sequenced in a logical fashion. The book moves from design to efficiency enhancement and then concludes with relevant applications. The emphasis on efficiency enhancement is particularly important, because the data-mining perspective implicit in EDAs opens up the world of optimization to new methods of data-guided adaptation that can further speed solutions through the construction and utilization of effective surrogates, hybrids, and parallel and temporal decompositions.

The author selection in many edited volumes is catch as catch can, but the editors themselves are well decorated authors in this subject area, and they have reached out to some of the most important voices in the field.

Finally, edited volumes leave many loose ends hanging, but the coverage and coherence of this volume is outstanding. Although different authors have different perspectives, the overall impression one is left with is the correct one. That is, EDAs are an important current technique that is leading to breakthroughs in genetic and evolutionary computation and in optimization more generally. I'm putting Scalable Optimization via Probabilistic Modeling in a prominent place in my library, and I urge you to do so as well. This volume summarizes the state of the art at the same time it points to where that art is going. Buy it, read it, and take its lessons to heart.

Urbana, Illinois, USA,
May 2006

David E. Goldberg
Jerry S. Dobrovolny Distinguished Professor
Director, Illinois Genetic Algorithms Laboratory
University of Illinois at Urbana-Champaign
http://www-illigal.ge.uiuc.edu

Acknowledgments

We thank all the people who contributed towards the preparation of this book. Foremost, we are indebted to the authors of the chapters for their excellent contributions. We are extremely grateful to David E. Goldberg for writing the foreword for the book and encouraging us in this project. We would like to thank Janusz Kacprzyk, the editor of the Studies in Computational Intelligence series, for including this book as a volume in this series. We also thank Thomas Ditzinger from Springer for his help with the successful completion of this edited volume.

This work was sponsored by the Air Force Office of Scientific Research, Air Force Materiel Command, USAF, under grant FA9550-06-1-0096, the National Science Foundation under NSF CAREER grant ECS-0547013 and ITR grant DMR-03-25939 at Materials Computation Center. The work was also supported by the Research Award and the Research Board at the University of Missouri. The US Government is authorized to reproduce and distribute reprints for government purposes notwithstanding any copyright notation thereon.

The views and conclusions contained herein are those of the authors and should not be interpreted as necessarily representing the official policies or endorsements, either expressed or implied, of the Air Force Office of Scientific Research, the National Science Foundation, or the US Government.

St. Louis, Missouri, USA
Urbana, IL, USA
Sunnyvale, CA, USA
March 2006

Martin Pelikan
Kumara Sastry
Erick Cantú-Paz

List of Contributors

Hussein A. Abbass
University of New South Wales
University College, ADFA Canberra
ACT 2600, Australia
abbass@itee.adfa.edu.au

Uwe Aickelin
University of Nottingham
Nottingham, NG8 IBB, UK
uxa@cs.nott.ac.uk

Shumeet Baluja
Google Inc., 1600 Amphitheatre
Parkway Mountain View, CA 94043,
USA
shumeet@google.com

Peter A.N. Bosman
Centre for Mathematics and
Computer Science (CWI),
P.O. Box 94079, 1090 GB
Amsterdam, The Netherlands,
Peter.Bosman@cwi.nl

Martin Butz
University of Würzburg Röntgenring
11 Würzburg, Germany
mbutz@psychologie.uni-wuerzburg.de

Erick Cantú-Paz
Yahoo!, Inc. 701 First Avenue
Sunnyvale, CA 94809, USA
cantupaz@acm.org

Daryl Essam
University of New South Wales
University College, ADFA Canberra
ACT 2600, Australia
daryl@itee.adfa.edu.au

David E. Goldberg
University of Illinois at
Urbana-Champaign 104 S.
Mathews Avenue Urbana, IL 61801,
USA
deg@uiuc.edu

Georges R. Harik
Google Inc. 1600 Amphitheatre
Parkway Mountain View, CA 94043,
USA
georges@gmail.com

Alexander K. Hartmann
Universität Göttingen Friedrich-
Hund-Platz 1 37077 Göttingen,
Germany
hartmann@physik.uni-goettingen.de

Robin Höns
Fraunhofer Institute for
Autonomous Intelligent Systems
53754 Sankt Augustin,
Germany
robin.hoens@web.de

Jingpeng Li
University of Nottingham
Nottingham, NG8 IBB, UK
jpl@cs.nott.ac.uk

Xavier Llorà
University of Illinois at Urbana-Champaign 1205 W. Clark Street
Urbana, Illinois 61801, USA
xllora@ncsa.uiuc.edu

Fernando G. Lobo
University of Algarve DEEI-FCT,
Campus de Gambelas 8000-117 Faro,
Portugal
flobo@ualg.pt

Robert I. McKay
University of New South Wales
University College, ADFA Canberra
ACT 2600, Australia
rim@itee.adfa.edu.au

Heinz Mühlenbein
Fraunhofer Institute for Autonomous
Intelligent Systems 53754 Sankt
Augustin, Germany
heinz.muehlenbein@ais.fraunhofer.de

Jiri Ocenasek
Kimotion Technologies Leuvens-esteenweg 200 B-3370 Boutersem,
Belgium
jiri@ocenasek.com

Martin Pelikan
University of Missouri at St. Louis
One University Blvd. St. Louis, MO
63121, USA
pelikan@cs.umsl.edu

Scott Santarelli
Air Force Research Laboratory 80
Scott Drive, Hanscom AFB MA
01731, USA
Scott.Santarelli@hanscom.af.mil

Kumara Sastry
University of Illinois at Urbana-Champaign 104 S. Mathews Ave.
Urbana, IL 61801, USA
ksastry@uiuc.edu

Josef Schwarz
Brno University of Technology
Bozetechova 2 612 66 Brno, Czech
Republic
schwarz@fit.vutbr.cz

Yin Shan
University of New South Wales
University College, ADFA Canberra
ACT 2600, Australia
shanyin@itee.adfa.edu.au

Dirk Thierens
Utrecht University, P.O. Box 80.089,
3508 TB Utrecht, The Netherlands
Dirk.Thierens@cs.uu.nl

Tian-Li Yu
University of Illinois at Urbana-Champaign 104 S. Mathews
Avenue Urbana, Illinois 61801,
USA
tianliyu@uiuc.edu

Contents

1

Introduction

Martin Pelikan[1], Kumara Sastry[2], and Erick Cantú-Paz[3]

Summary. This chapter provides motivation for estimation of distribution algorithms and discusses the scope of this book. Additionally, the chapter provides a road map to the book and pointers to additional information.

1.1 Motivation for EDAs

Estimation of distribution algorithms (EDAs) [1, 5, 8, 11] address broad classes of optimization problems by learning explicit probabilistic models of promising solutions found so far and sampling the built models to generate new candidate solutions. By incorporating advanced machine learning techniques into genetic and evolutionary algorithms, EDAs can scalably solve many challenging problems, significantly outperforming standard genetic and evolutionary algorithms and other optimization techniques. In the recent decade, many impressive results have been produced in the design, theoretical investigation, and applications of EDAs.

An EDA evolves a population of candidate solutions to the given problem. Each iteration starts by evaluating the candidate solutions and selecting promising solutions so that solutions of higher quality are given more copies than solutions of lower quality. EDAs can use any standard selection method of genetic and evolutionary algorithms, such as binary tournament selection. Next, a probabilistic model is build for the selected solutions and new solutions are generated by sampling the built probabilistic model. New solutions are then incorporated into the original population using some replacement strategy, and the next iteration is executed unless the termination criteria have been met. EDAs usually differ in the representation of candidate solutions, the considered class of probabilistic models, or the procedures for learning and sampling probabilistic models. The pseudocode of an EDA follows:

M. Pelikan et al.: *Introduction*, Studies in Computational Intelligence (SCI) **33**, 1–10 (2006)
www.springerlink.com

```
Estimation of Distribution Algorithm (EDA)
  t := 0;
  generate initial population P(0);
  while (not done) {
    select population of promising solutions S(t);
    build probabilistic model P(t) for S(t);
    sample P(t) to generate O(t);
    incorporate O(t) into P(t);
    t := t+1;
  }
```

EDAs derive inspiration from two areas: genetic and evolutionary computation and machine learning. The remainder of this section discusses these two sources of inspiration.

1.1.1 Motivation from Genetic and Evolutionary Computation

EDAs borrow two important concepts from genetic and evolutionary computation:

1. Population-based search
2. Exploration by combining and perturbing promising solutions

Using a population of solutions as opposed to a single solution has several advantages; for example, it enables simultaneous exploration of multiple regions in the search space, it can help to alleviate the effects of noise in evaluation, and it allows the use of statistical and learning techniques to automatically identify problem regularities.

Exploration of the search space by combining and perturbing promising solutions can be effective because in most real-world problems, high quality solutions are expected to share features. By effective identification of important features and their juxtaposition, the global optimum can be identified even in problems where local operators fail because of exponentially many local optima and strong large-order interactions between problem variables.

1.1.2 Motivation from Machine Learning

EDAs use probabilistic models to guide exploration of the search space. Using probabilistic models enables the use of rigorous statistical modeling and sampling techniques to automatically discover and exploit problem regularities for effective exploration.

In most EDAs, probabilistic models are represented by graphical models [2, 4, 9], which combine graph theory, modularity and statistics to provide a flexible tool for learning and sampling probability distributions, and probabilistic inference. Graphical models provide EDAs with a powerful tool for identifying and exploiting problem decomposition, whereas evolutionary algorithms provide EDAs with robust operators for maintaining diverse populations of promising candidate solutions. Since most complex real-world systems are nearly decomposable and hierarchical [16], the combination of machine learning techniques and evolutionary algorithms should enable EDAs to solve broad classes of difficult real-world problems in a robust and scalable manner. This hypothesis was supported with a number of theoretical and empirical results [3, 6, 7, 10, 12–15].

1.2 Scope and Road Map

This book provides a selection of some of the important contributions to research and application of EDAs. There are three main areas covered in this book:

1. *Design of robust and scalable EDAs.* (Chaps. 2–6, 10 and 11)
2. *Efficiency enhancement of EDAs.* (Chaps. 7–9, 11, 13, and 15)
3. *Applications of EDAs.* (Chaps. 12–15)

The content of each chapter is discussed next.

Chapter 2. *The Factorized Distribution Algorithm and the Minimum Relative Entropy Principle* by Heinz Mühlenbein and Robin Höns.
In this chapter, Mühlenbein and Höns discuss major design issues of EDAs using an interdisciplinary framework: The *minimum relative entropy* (*MinRel*) approximation. They demonstrate the relation between the Factorized Distribution Algorithm (FDA) and the *MinRel* approximation. Mühlenbein and Höns propose an EDA derived from the *Bethe-Kikuchi* approach developed in statistical physics and present details of a concave–convex minimization algorithm to solve optimization problems. The two algorithms are compared using popular benchmark problems – 2-d grid problems, 2-d Ising spin glasses, and Kaufman's $n-k$ function – with instances of up to 900 variables.

Chapter 3. *Linkage Learning via Probabilistic Modeling in the Extended Compact Genetic Algorithm (ECGA)* by Georges R. Harik, Fernando G. Lobo, and Kumara Sastry.
The first-generation genetic algorithms (GAs) are not very successful in automatically identifying and exchanging structures consisting of several correlated genes. This problem, referred in the literature as the *linkage-learning* problem, has been the subject of extensive research over the last few decades. Harik et al. explore the relationship between the linkage-learning problem and that of learning probability distributions over multivariate spaces. They argue that the linkage-learning problem and learning probability distributions are equivalent. Using a simple yet effective approach to learning distributions, and by implication linkage,

Harik et al. propose a GA-like algorithm – the extended compact GA – that is potentially orders of magnitude faster and more accurate than simple GAs.

Chapter 4. *Hierarchical Bayesian Optimization Algorithm* by Martin Pelikan and David E. Goldberg.
Pelikan and Goldberg describe the hierarchical Bayesian optimization algorithm (hBOA) and its predecessor, the Bayesian optimization algorithm (BOA), and outline some of the important theoretical and empirical results in this line of research. The hierarchical Bayesian optimization algorithm (hBOA) solves nearly decomposable and hierarchical optimization problems scalably by combining concepts from evolutionary computation, machine learning and statistics. Since many complex real-world systems are nearly decomposable and hierarchical, hBOA is expected to provide scalable solutions for many complex real-world problems.

Chapter 5. *Numerical Optimization with Real–Valued Estimation-of-Distribution Algorithms* by Peter A.N. Bosman and Dirk Thierens.
In this chapter, Bosman and Thierens focus on the design of real-valued EDAs for the task of numerical optimization. Here, both the problem variables as well as their encoding are real values, and concordantly, the type of probability distribution to be used for estimation and sampling in the EDA is continuous. Bosman and Thierens indicate the main challenges in real-valued EDAs and review the existing literature to indicate the current EDA practice for real-valued numerical optimization. They draw conclusions about the feasibility of existing EDA approaches and provide an explanation for some observed deficiencies of continuous EDAs as well as possible improvements and future directions of research in this branch of EDAs.

Chapter 6. *A Survey of Probabilistic Model Building Genetic Programming* by Yin Shan, Robert I. McKay, Daryl Essam, and Hussein A. Abbass.
While the previous chapters address EDAs that mainly operate on variables encoded into fixed-length chromosomes, there has been growing interest in extending EDAs to operate on variable-length representations, especially for evolving computer programs. In this chapter, Shan et al. provide a critical and comprehensive review of EDAs for automated programming. They discuss important lessons learned from genetic programming (GP) for better design of probabilistic models for GP. Shan et al. also present key strengths and limitations of existing EDAs for GP.

Chapter 7. *Efficiency Enhancement of Estimation of Distribution Algorithms* by Kumara Sastry, Martin Pelikan, and David E. Goldberg.
Estimation of distributions have taken problems that were *intractable* with first generation GAs and rendered them *tractable*, whereas *efficiency-enhancement* take EDAs from *tractability* to *practicality*. That is, efficiency-enhancement techniques speedup the search process of estimation of distribution algorithms (EDAs) and thereby enable EDAs to solve hard problems in *practical* time. Sastry et al. provide a decomposition and

a review of different efficiency-enhancement techniques for EDAs. They illustrate a principled approach for designing efficiency enhancement techniques by developing an evaluation-relaxation scheme in the Bayesian optimization algorithm, and a time-continuation method in the extended compact genetic algorithm.

Chapter 8. *Design of Parallel Estimation of Distribution Algorithms* by Jiri Ocenasek, Erick Cantú-Paz, Martin Pelikan, and Josef Schwarz.
In this chapter, Ocenasek et al. focus on the parallelization of Estimation of Distribution Algorithms (EDAs) and present guidelines for designing efficient parallel EDAs that employ parallel fitness evaluation and parallel model building. They employ scalability analysis techniques to identify and parallelize the main performance bottlenecks to ensure that the achieved speedup grows almost linearly with the number of utilized processors. Ocenasek et al. demonstrate their proposed approach on the parallel Mixed Bayesian Optimization Algorithm (MBOA) and verify it with experiments on the problem of finding ground states of 2-d Ising spin glasses.

Chapter 9. *Incorporating a priori Knowledge in Probabilistic-Model Based Optimization* by Shumeet Baluja.
Complex dependency networks that can account for the interactions between parameters are often used in advanced EDAs; however, they may necessitate enormous amounts of sampling. In this chapter, Baluja demonstrates how a priori knowledge of parameter dependencies, even incomplete knowledge, can be incorporated to efficiently obtain accurate models that account for parameter interdependencies. This is achieved by effectively putting priors on the network structures that are created. These more accurate models yield improved results when used to guide the sample generation process for search. Baluja demonstrates the results on a variety of graph coloring problems, and examines the benefits of a priori knowledge as problem difficulty increases.

Chapter 10. *Multiobjective Estimation of Distribution Algorithms* by Martin Pelikan, Kumara Sastry, and David E. Goldberg.
Many real-world optimization problems contain multiple competing objectives and that is why the design of optimization techniques that can scalably discover an optimal tradeoff between given objectives (Pareto-optimal solutions) represents an important challenge. Pelikan et al. discuss EDAs that address this challenge. The primary focus is on scalability on discrete multiobjective decomposable problems and the multiobjective hierarchical BOA (mohBOA), but other approaches to multiobjective EDAs are also discussed.

Chapter 11. *Effective and Reliable Online Classification Combining XCS with EDA Mechanisms* by Martin Butz, Martin Pelikan, Xavier Llorà, and David E. Goldberg.
Learning Classifier Systems (LCSs), such as XCS and other accuracy-based classifier systems, evolve a distributed problem solution online.

During the learning process, rule quality is assessed iteratively using techniques based on gradient-descent, while the rule structure is evolved using selection and variation operators of evolutionary algorithms. While using standard variation operators suffices for solving some problems, it does not assure an effective evolutionary search in many difficult problems that contain strong interactions between features. Butz et al. describe how advanced EDAs can be integrated into XCS in order to ensure effective exploration even for problems in which features strongly interact and standard variation operators lead to poor XCS performance. In particular, they incorporate the model building and sampling techniques from BOA and ECGA into XCS and show that the two proposed algorithms ensure that the solution is found efficiently and reliably. The results thus suggest that the research on combining standard LCSs with advanced EDAs holds a big promise and represents an important area for future research on LCSs and EDAs.

Chapter 12. *Military Antenna Design Using a Simple Genetic Algorithm and hBOA* by Tian-Li Yu, Scott Santarelli, and David E. Goldberg.
In this chapter, Yu et al. describe the optimization of an antenna design problem via a simple genetic algorithm (SGA) and the hierarchical Bayesian optimization algorithm (hBOA). Three objective functions are designed in an effort to find a solution that meets the system requirements/specifications. Yu et al. show empirical results that indicate that the SGA and hBOA perform comparably when the objective function is "easy" (that is, traditional mask). When the objective function more accurately reflects the true objective of the problem (that is, "difficult"), however, hBOA consistently outperforms the SGA both computationally and the optimal antenna design obtained via hBOA also outperforms that obtained via the SGA.

Chapter 13. *Feature Subset Selection with Hybrids of Filters and Evolutionary Algorithms* by Erick Cantú-Paz.
The performance of classification algorithms is affected by the features used to describe the labeled examples presented to the inducers. Therefore, the problem of feature subset selection has received considerable attention. Approaches to this problem based on evolutionary algorithms typically use the wrapper method, treating the inducer as a black box that is used to evaluate candidate feature subsets. However, the evaluations might take a considerable time and the wrapper approach might be impractical for large data sets. Alternative filter methods use heuristics to select feature subsets from the data and are usually considered more scalable than wrappers to the dimensionality and volume of the data. In this chapter, Cantú-Paz describes hybrids of evolutionary algorithms and filter methods applied to the selection of feature subsets for classification problems. The proposed hybrids are compared against each of their components, two feature selection wrappers that are in wide use, and another filter-wrapper hybrid. Cantú-Paz investigates if the proposed evolutionary

hybrids present advantages over the other methods in terms of accuracy or speed. He uses decision tree and naive Bayes classifiers on public-domain and artificial data sets. The experimental results in this chapter suggest that the evolutionary hybrids usually find compact feature subsets that result in the most accurate classifiers, while beating the execution time of the other wrappers.

Chapter 14. *BOA for Nurse Scheduling* by Jingpeng Li and Uwe Aickelin. Li and Aickelin have shown that schedules can be built mimicking a human scheduler by using a set of rules that involve domain knowledge. Li and Aickelin present a Bayesian Optimization Algorithm (BOA) for the nurse scheduling problem that chooses such suitable scheduling rules from a set for each nurse's assignment. Based on the idea of using probabilistic models, the BOA builds a Bayesian network for the set of promising solutions and samples these networks to generate new candidate solutions. Computational results from 52 real data instances demonstrate the success of this approach. The authors also suggest that the learning mechanism in the proposed algorithm may be suitable for other scheduling problems.

Chapter 15. *Searching for Ground States of Ising Spin Glasses with Hierarchical BOA and Cluster Exact Approximation* by Martin Pelikan and Alexander K. Hartmann.
In this chapter, Pelikan and Hartmann apply the hierarchical Bayesian optimization algorithm (hBOA) to the problem of finding ground states of Ising spin glasses with $\pm J$ and Gaussian couplings in two and three dimensions. The authors compare the performance of hBOA to that of the simple genetic algorithm (GA) and the univariate marginal distribution algorithm (UMDA). The performance of all tested algorithms is improved by incorporating a deterministic hill climber based on single-bit flips. The results in the chapter show that hBOA significantly outperforms GA and UMDA on a broad spectrum of spin glass instances. The authors also describe and incorporate the cluster exact approximation (CEA) into hBOA and GA to improve their efficiency. The results show that CEA enables all tested algorithms to solve larger spin glass instances and that hBOA significantly outperforms other compared algorithms even in this case.

1.3 Additional Information

1.3.1 Conferences

Most EDA researchers present their results at the following conferences:

- *Congress on Evolutionary Computation (CEC)*; IEEE
- *Genetic and Evolutionary Computation Conference (GECCO)*; SIGEVO, ACM Special Interest Group for Genetic and Evolutionary Computation
- *Parallel Problem Solving from Nature*

1.3.2 Journals

The following journals publish majority of EDA articles:

- *Evolutionary Computation*; MIT Press
- *Genetic Programming and Evolvable Machines*; Springer
- *IEEE Transactions on Evolutionary Computation*; IEEE Press

A number of EDA articles can also be found in the following journals:

- *Computational Optimization and Applications (COAP)*; Kluwer
- *Information Sciences*; Elsevier
- *International Journal of Approximate Reasoning*; Elsevier
- *New Generation Computing*; Springer

1.3.3 World Wide Web

The following search engines can be used to search for many EDA papers:

- CiteSeer, Scientific Literature Digital Library
  `http://citeseer.ist.psu.edu/`
- Google Scholar
  `http://scholar.google.com/`

More papers can be found on personal and institutional Web pages of the researchers that contributed to this book or were cited in the references therein.

1.3.4 Online Source Code

Source code of various EDAs can be downloaded from the following sources:

- Extended Compact Genetic Algorithm, C++; F. G. Lobo, G. R. Harik
  Bayesian Optimization Algorithm, C++; M. Pelikan
  Bayesian Optimization Algorithm with Decision Graphs; M. Pelikan
  `http://www-illigal.ge.uiuc.edu/`
- Learning Factorized Distribution Algorithm (LFDA); H. Mühlenbein, T. Mahnig; `http://www.ais.fraunhofer.de/~muehlen/`
- Adaptive mixed Bayesian optimization algorithm (AMBOA); J. Ocenasek
  `http://jiri.ocenasek.com/`
- Real-coded Bayesian Optimization Algorithm; C.-W. Ahn
  `http://www.evolution.re.kr/`
- Probabilistic Incremental Program Evolution (PIPE); R. P. Salustowicz
  `http://www.idsia.ch/~rafal/`
- Naive Multi-objective Mixture-based Iterated Density-Estimation Evolutionary Algorithm (MIDEA), Normal IDEA-Induced Chromosome Elements Exchanger (ICE), Normal Iterated Density-Estimation Evolutionary Algorithm (IDEA); P. A. N. Bosman
  `http://homepages.cwi.nl/~bosman/`

– Java applets for several real-valued and permutation EDAs; S. Tsutsui http://www.hannan-u.ac.jp/~tsutsui/index-e.html

References

[1] Baluja, S. (1994). Population-based incremental learning: A method for integrating genetic search based function optimization and competitive learning. Tech. Rep. No. CMU-CS-94-163, Carnegie Mellon University, Pittsburgh, PA

[2] Buntine, W. L. (1991). Theory refinement of Bayesian networks. *Proceedings of the Uncertainty in Artificial Intelligence (UAI-91)*, pages 52–60

[3] Inza, I., Larrañaga, P., Etxeberria, R., and Sierra, B. (2000). Feature subset selection by Bayesian network-based optimization. *Artificial Intelligence*, 123(1–2):157–184

[4] Jordan, M. I., editor (1999). *Learning in Graphical Models.* MIT, Cambridge, MA

[5] Larrañaga, P. and Lozano, J. A., editors (2002). *Estimation of Distribution Algorithms: A New Tool for Evolutionary Computation.* Kluwer, Boston, MA

[6] Li, J. and Aickelin, U. (2003). A Bayesian optimization algorithm for the nurse scheduling problem. *Proceedings of the IEEE Congress on Evolutionary Computation 2003 (CEC-2003)*, pages 2149–2156

[7] Mühlenbein, H. and Mahnig, T. (1999). FDA – A scalable evolutionary algorithm for the optimization of additively decomposed functions. *Evolutionary Computation*, 7(4):353–376

[8] Mühlenbein, H. and Paaß, G. (1996). From recombination of genes to the estimation of distributions I. Binary parameters. *Parallel Problem Solving from Nature*, pages 178–187

[9] Pearl, J. (1988). *Probabilistic reasoning in intelligent systems: Networks of plausible inference.* Morgan Kaufmann, San Mateo, CA

[10] Pelikan, M. (2005). *Hierarchical Bayesian optimization algorithm: Toward a new generation of evolutionary algorithms.* Springer, Berlin Heidelberg New York

[11] Pelikan, M., Goldberg, D. E., and Lobo, F. (1999). A survey of optimization by building and using probabilistic models. IlliGAL Report No. 99018, University of Illinois at Urbana-Champaign, Illinois Genetic Algorithms Laboratory, Urbana, IL

[12] Pelikan, M., Sastry, K., and Goldberg, D. E. (2002). Scalability of the Bayesian optimization algorithm. *International Journal of Approximate Reasoning*, 31(3):221–258

[13] Rothlauf, F., Goldberg, D. E., and Heinzl, A. (2000). Bad codings and the utility of well-designed genetic algorithms. IlliGAL Report No. 200007, University of Illinois at Urbana-Champaign, Illinois Genetic Algorithms Laboratory, Urbana, IL

[14] Santarelli, S., Goldberg, D. E., and Yu, T.-L. (2004). Optimization of a constrained feed network for an antenna array using simple and competent genetic algorithm techniques. *Proceedings of the Workshop Military and Security Application of Evolutionary Computation (MSAEC-2004)*

[15] Sastry, K. (2001). Efficient atomic cluster optimization using a hybrid extended compact genetic algorithm with seeded population. IlliGAL Report No. 2001018, University of Illinois at Urbana-Champaign, Illinois Genetic Algorithms Laboratory, Urbana, IL

[16] Simon, H. A. (1968). *The Sciences of the Artificial.* The MIT, Cambridge, MA

2

The Factorized Distribution Algorithm and the Minimum Relative Entropy Principle

Heinz Mühlenbein and Robin Höns

Summary. Estimation of distribution algorithms (EDA) have been proposed as an extension of genetic algorithms. In this paper the major design issues of EDA's are discussed using an interdisciplinary framework, the *minimum relative entropy* (MinRel) approximation. We assume that the function to be optimized is additively decomposed (ADF). The interaction graph G_{ADF} of the ADF is used to create exact or approximate factorizations of the Boltzmann distribution. The relation between the Factorized distribution algorithm (FDA) and the MinRel approximation is shown. We present a new algorithm, derived from the *Bethe–Kikuchi* approach developed in statistical physics. It minimizes the relative entropy $\mathrm{KLD}(q|p_\beta)$ to the Boltzmann distribution p_β by solving a difficult constrained optimization problem. We present in detail the concave–convex minimization algorithm (CCCP) to solve the optimization problem. The two algorithms are compared using popular benchmark problems (2-D grid problems, 2-D Ising spin glasses, Kaufman's $n-k$ function.) We use instances up to 900 variables.

Key words: Estimation of distributions, Boltzmann distribution, factorization of distributions, maximum entropy principle, minimum relative entropy, maximum log-likelihood, Bethe–Kikuchi approximation

2.1 Introduction

The *Estimation of distribution algorithms* (EDA) family of population based search algorithms was introduced in [21] as an extension of genetic algorithms.[1] The following observations lead to this proposal. First, genetic algorithm have difficulties to optimize deceptive and nonseparable functions, and second, the search distributions implicitly generated by recombination and crossover do not exploit the correlation of the variables in samples of high fitness values.

[1] In [21] they have been named *conditional distribution algorithms.*

H. Mühlenbein and R. Höns: *The Factorized Distribution Algorithm and the Minimum Relative Entropy Principle*, Studies in Computational Intelligence (SCI) **33**, 11–37 (2006)
www.springerlink.com

EDA uses probability distributions derived from the function to be optimized to generate search points instead of crossover and mutation as done by genetic algorithms. The other parts of the algorithms are identical. In both cases a population of points is used and points with good fitness are selected either to estimate a search distribution or to be used for crossover and mutation.

The family of EDA algorithms can be understood and further developed without the background of genetic algorithms. The problem to estimate empirical distributions has been investigated independently in several scientific disciplines. In this paper we will show how results in statistics, belief networks and statistical physics can be used to understand and further develop EDA. In fact, an interdisciplinary research effort is well under way which cross-fertilizes the different disciplines.

Unfortunately each discipline uses a different language and deals with slightly different problems. In EDA we want to generate points with a high probability $p(\mathbf{x})$, in belief networks one computes a single marginal distribution $p(\mathbf{y}|\mathbf{z})$ for new evidence $\mathbf{z}$, and statistical physicists want to compute the free energy of a Boltzmann distribution. Thus the algorithms developed for belief networks concentrate on computing a single marginal distribution, whereas for EDA we want to sample $p(\mathbf{x})$ in areas of high fitness values, i.e., we are interested in a sampling method which generates points with a high value of $p(\mathbf{x})$. But all disciplines need fast algorithms to compute marginal distributions. The foundation of the theory is the same for all disciplines. It is based on graphical models and their decomposition.

Today two major branches of EDA can be distinguished. In the first branch the factorization of the distribution is computed from the structure of the function to be optimized, in the second one the structure is computed from the correlations of the data generated. The second branch has been derived from the theory of belief networks [11]. It computes a factorization from samples instead from the analytical expression of the fitness function. The underlying theory is the same for both branches. For large real life applications often a hybrid between these two approaches is most successful [17]. In this paper we investigate the first branch only.

We discuss the problem of computing approximations of distributions using factorizations is investigated with the framework of *minimum relative entropy* (MinRel). We distinguish exact factorizations and approximate factorizations. We shortly summarize the results for our well-known algorithm FDA. We present in detail a new algorithm BKDA. It is derived from an approach used in statistical physics to approximate the Boltzmann distribution. It is called the *Bethe–Kikuchi* approximation. In this approach the marginals from the unknown Boltzmann distribution are not computed from data, but from a difficult constrained optimization problem. This paper extends the theory first described in [15].

In Sect. 2.9, we summarize the functionality of our software system FDA. It can be downloaded from *http://www.ais.fraunhofer.de/~muehlen*. The different EDA algorithms are shortly numerically compared, using large benchmark optimization problems like 2-D Ising spin glasses, and Kaufman's $n-k$ function. We investigate problems with up to 900 variables, continuing the work in [17], where graph bipartitioning problems of 1,000 nodes have been solved.

2.2 Factorization of the Search Distribution

EDA has been derived from a search distribution point of view. We just recapitulate the major steps published in [17, 20]. We will use in this paper the following notation. Capital letters denote variables, lower cases instances of variables. If the distinction between variables and instances is not necessary, we will use lower case letters. Vectors are denoted by x, a single variable by $\mathbf{x}_i$. Let a function $f : \mathbf{X} \to \mathbb{R}_{\geq 0}$ be given. We consider the optimization problem

$$\mathbf{x}_{\text{opt}} = \operatorname{argmax} f(\mathbf{x}). \tag{2.1}$$

A promising search distribution for optimization is the Boltzmann distribution.

Definition 1 *For* $\beta \geq 0$ *define the* Boltzmann distribution[2] *of a function* $f(\mathbf{x})$ *as*

$$p_\beta(x) := \frac{e^{\beta f(\mathbf{x})}}{\sum_{\mathbf{y}} e^{\beta f(\mathbf{y})}} =: \frac{e^{\beta f(\mathbf{x})}}{Z_f(\beta)}. \tag{2.2}$$

where $Z_f(\beta)$ *is the partition function. To simplify the notation* β *and/or* f *might be omitted.*

The Boltzmann distribution concentrates with increasing β around the global optima of the function. Obviously, the distribution converges for $\beta \to \infty$ to a distribution where only the optima have a probability greater than 0 (see [18]). Therefore, if it were possible to sample efficiently from this distribution for arbitrary β, optimization would be an easy task. But the computation of the partition function needs an exponential effort for a problem of n variables. We have therefore proposed an algorithm which incrementally computes the Boltzmann distribution from empirical data using Boltzmann selection.

Definition 2 *Given a distribution* p *and a selection parameter* $\Delta\beta$, Boltzmann selection *calculates the distribution for selecting points according to*

$$p^s(\mathbf{x}) = \frac{p(\mathbf{x}) e^{\Delta\beta f(\mathbf{x})}}{\sum_y p(\mathbf{y}) e^{\Delta\beta f(\mathbf{y})}}. \tag{2.3}$$

---

[2] The Boltzmann distribution is usually defined as $e^{-\frac{E(\mathbf{x})}{T}}/Z$. The term $E(x)$ is called the energy and $T = 1/\beta$ the temperature. We use the inverse temperature β instead of the temperature.

The following theorem is easy to prove.

Theorem 3. *If* $p_\beta(\mathbf{x})$ *is a Boltzmann distribution, then* $p^s(\mathbf{x})$ *is a Boltzmann distribution with inverse temperature* $\beta(t+1) = \beta(t) + \Delta\beta(t)$.

Algorithm 1 describes the Boltzmann estimated distribution algorithm (BEDA).

Algorithm 1: BEDA – Boltzmann estimated distribution

---

1. $t \Leftarrow 1$. Generate N points according to the uniform distribution $p(\mathbf{x}, 0)$ with $\beta(0) = 0$.
2. **do** {
3. With a given $\Delta\beta(t) > 0$, let
$$p^s(\mathbf{x}, t) = \frac{p(\mathbf{x}, t)e^{\Delta\beta(t)f(\mathbf{x})}}{\sum_{\mathbf{y}} p(\mathbf{y}, t)e^{\Delta\beta(t)f(\mathbf{y})}}.$$
4. Generate N new points according to the distribution $p(\mathbf{x}, t+1) = p^s(\mathbf{x}, t)$.
5. $t \Leftarrow t + 1$.
6. } **until** (stopping criterion reached)

BEDA is a conceptional algorithm, because the calculation of the distribution $p^s(\mathbf{x}, t)$ requires a sum over exponentially many terms. In Sect. 2.2.1 we transform BEDA into a practical numerical algorithm.

2.2.1 Factorization of the Distribution

From now on we assume that the fitness function is additively decomposed.

Definition 4 *Let* $s_1, \ldots, s_m$ *be index sets,* $s_i \subseteq \{1, \ldots, n\}$. *Let* f_i *be functions depending only on variables* x_j *with* $j \in s_i$. *Then*

$$f(\mathbf{x}) = \sum_{i=1}^{m} f_i(\mathbf{x}_{s_i}) \tag{2.4}$$

is an additive decomposition of the fitness function (ADF). The ADF is k*-bounded if* $\max_i |s_i| \leq k$.

Definition 5 *Let an ADF be given. Then the interaction graph* G_{ADF} *is defined as follows: The vertices of* G_{ADF} *represent the variables of the ADF. Two vertices are connected by an arc iff the corresponding variables are contained in a common subfunction.*

Given an ADF we want to estimate the Boltzmann distribution (2.2) using a product of marginal distributions. The approximation has to fulfill two conditions:

- The approximation should only use marginals of low order.
- Sampling from the approximation should be easy.

A class of distributions fulfilling these conditions are the acyclic Bayesian network (acBN)

$$q(\mathbf{x}) = \prod_{i=1}^{n} q(x_i|\pi_i), \tag{2.5}$$

where π_i are called the parents of x_i. For acyclic Bayesian networks sampling can easily be done starting with the root x_1. Cyclic Bayesian networks are difficult to sample from.

Note that any distribution can be written in the form of an acyclic Bayesian network because of

$$p(\mathbf{x}) = p(x_1)p(x_2|x_1)p(x_3|x_1,x_2)\cdots p(x_n|x_1,\ldots,x_{n-1}). \tag{2.6}$$

But this factorization uses marginal distributions of size $O(n)$, thus sampling from the distribution is exponential in n. Therefore we are looking for factorizations where the size of the marginals is bounded independent of n.

For ADF's the following procedure can be used to create factorizations. We need the following sets:

Definition 6 *Given $s_1,\ldots,s_m$, we define for $i = 1,\ldots,m$ the sets d_i, b_i, and c_i:*

$$d_i := \bigcup_{j=1}^{i} s_j, \qquad b_i := s_i \setminus d_{i-1}, \qquad c_i := s_i \cap d_{i-1}. \tag{2.7}$$

We demand $d_m = \{1,\ldots,n\}$ and set $d_0 = \emptyset$. In the theory of decomposable graphs, d_i are called histories, *b_i* residuals *and c_i* separators *[13].*

The next definition is stated a bit informally.

Definition 7 *A set of marginal distributions $\tilde{q}(\mathbf{x}_{b_i},\mathbf{x}_{c_i})$ is called* consistent *if the marginal distributions fulfill the laws of probability, e.g.,*

$$\sum_{\mathbf{x}_{b_i},\mathbf{x}_{c_i}} \tilde{q}(\mathbf{x}_{b_i},\mathbf{x}_{c_i}) = 1, \tag{2.8}$$

$$\sum_{\mathbf{x}_{b_i}} \tilde{q}(\mathbf{x}_{b_i},\mathbf{x}_{c_i}) = \tilde{q}(\mathbf{x}_{c_i}). \tag{2.9}$$

Definition 8 *If $b_i \neq \emptyset$ we define a FDA factorization for a given ADF by*

$$q(\mathbf{x}) = \prod\nolimits_{i=1}^{m} \tilde{q}(\mathbf{x}_{b_i}|\mathbf{x}_{c_i}). \tag{2.10}$$

A FDA factorization is k-bounded *if the size of the sets $\{b_i,c_i\}$ is bounded by a constant k independent of n.*

Remark: *Any FDA factorization can easily be transformed into an acyclic Bayesian network which has the same largest clique size. The FDA factorization is only a more compact representation. Therefore the class of FDA factorizations is identical to the class of acyclic Bayesian networks. Sampling is done iteratively, starting with $\tilde{q}(x_{b_1})$.*

Proposition 9 *Let a consistent set of marginal distributions* $\tilde{q}(\mathbf{x}_{b_i}, \mathbf{x}_{c_i})$ *be given. Then the FDA factorization defines a valid distribution* ($\sum q(\mathbf{x}) = 1$). *Furthermore*

$$q(\mathbf{x}_{b_i}|\mathbf{x}_{c_i}) = \tilde{q}(\mathbf{x}_{b_i}|\mathbf{x}_{c_i}), \quad i = 1, \dots m \tag{2.11}$$

whereas in general

$$q(\mathbf{x}_{b_i}, \mathbf{x}_{c_i}) \neq \tilde{q}(\mathbf{x}_{b_i}, \mathbf{x}_{c_i}), \quad i = 1, \dots m. \tag{2.12}$$

The proof follows from the definition of marginal probabilities. The proof of (2.11) is somewhat technical, but straightforward. The inequality (2.12) is often overlooked. It means that sampling from the factorization does not reproduce the given marginals.

The next theorem was proven in [20]. It formulates a condition under which the *FDA* factorization reproduces the marginals.

Theorem 10 (Factorization Theorem). *Let* $f(\mathbf{x}) = \sum_{i=1}^{m} f_{s_i}(\mathbf{x})$ *be an additive decomposition. If*

$$\forall i = 1, \dots, m; \quad b_i \neq \emptyset \tag{2.13}$$

$$\forall i \geq 2 \, \exists j < i \;\; \textit{such that} \; c_i \subseteq s_j \tag{2.14}$$

then

$$p_\beta(\mathbf{x}) = \prod_{i=1}^{m} p_\beta(\mathbf{x}_{b_i}|\mathbf{x}_{c_i}) = \frac{\prod_{i=1}^{m} p_\beta(\mathbf{x}_{b_i}, \mathbf{x}_{c_i})}{\prod_{i=2}^{m} p_\beta(\mathbf{x}_{c_i})}. \tag{2.15}$$

The Factorization Theorem is not restricted to distributions and their factorization. It is connected to the decomposition of G_{ADF}. A general formulation and a historical survey of this important theorem can be found in [1].

Definition 11 *The constraint defined by equation* (2.14) *is called the* running intersection property (*RIP*).

The above theorem does not address the problem how to compute a good or even an exact factorization. The construction defined by (2.7) depends on the sequence $s_1, \dots, s_m$. For many sequences the RIP might not be fulfilled. But if the sequence is permutated, it might be possible that the RIP will be fulfilled. Furthermore, we can join two or more subfunctions, resulting in larger sets $\tilde{s}_i$. It might be that using these larger sets, the RIP is fulfilled. For an efficient FDA we are looking for k – *bounded* factorizations which fulfill the RIP.

Testing all the sequences is prohibitive. Actually, it turns out that the computation of exact factorizations is done better by investigating the corresponding interaction graph G_{ADF}. A well-known algorithm computes a *junction tree* of G_{ADF}(see [10]). From the junction tree a factorization can easily be obtained. This factorization fulfills the RIP. A short description of the algorithm can be found in [15]. The largest clique of the junction tree gives the largest marginal of the factorization. The decision problem if there exists an k-bounded junction tree is NP in general.

The space complexity of exact factorizations has been investigated in [6]. For many problems the size of the largest clique is $O(n)$, making a numerical application using this factorization prohibitive. Thus for real applications we are looking for good k-bounded factorizations which violate the RIP in a few cases only.

2.2.2 The Factorized Distribution Algorithm

We first describe our algorithm Factorized distribution algorithm (FDA) which runs with any FDA factorization.

Algorithm 2: *FDA* – Factorized distribution algorithm

```
Calculate b_i and c_i by the Subfunction Merger Algorithm.
t ⇐ 1. Generate an initial population with N individuals from
the uniform distribution.
do {
    Select M ≤ N individuals using Boltzmann selection.
    Estimate the conditional probabilities p(x_{b_i}|x_{c_i}, t) from the
    selected points.
    Generate new points according to p(x, t+1) =
    ∏_{i=1}^m p(x_{b_i}|x_{c_i}, t).
    t ⇐ t + 1.
} until (stopping criterion reached)
```

The computational complexity of FDA is $O(N * \sum_{i=1}^{m} 2^{|s_i|})$, where $|s_i|$ denotes the size of the marginals. For a k-bounded FDA factorization we obtain $O(N * m * 2^k)$, thus the complexity is linear in m. If the FDA factorization is exact, then the convergence proof of BEDA is valid for FDA too. But since FDA uses finite samples to estimate the conditional probabilities, convergence to the optimum will depend on the size of the sample. Sampling theory can be used to estimate the probability of convergence for a given sample size.

A factorization fulfilling the RIP is sufficient for convergence to the optimum, but not necessary. But such a factorization can be difficult to compute or may be not $k-bounded$. Therefore we use FDA mainly with approximate factorizations. A good approximate factorization should include all edges of the interaction graph.

We have implemented a general heuristic which automatically computes a FDA factorization. The heuristic uses mainly *merging* of subfunctions. Let us discuss the problem with a simple loop.

$$s_1 = \{1,2\}, s_2 = \{2,3\}, s_3 = \{3,4\}, \mathbf{s}_4 = \{1,4\}$$

All possible sequences end in $b_4 = \emptyset$. We can use the factorization $q(\mathbf{x}) = \tilde{q}(x_1, x_2)\tilde{q}(x_3|x_2)\tilde{q}(x_4|x_3)$ using s_1, s_2, s_3 only. But if the subfunctions f_3 and f_4 are merged then we obtain from our procedure.

$$q(\mathbf{x}) = \tilde{q}(x_1, x_2)\tilde{q}(x_3|x_2)\tilde{q}(x_4|x_3, x_1)$$

This factorization uses all edges from G_{ADF} but violates the RIP. Merging of subfunctions lead to larger marginals. Therefore a good heuristic has three conflicting goals: to minimize the number of RIP violations, to use all dependencies of G_{ADF}, and to find a k-bounded factorization with a small k.

Algorithm 3: Subfunction Merger

$\mathcal{S} \Leftarrow \{s_1, \ldots, s_m\}$
$j \Leftarrow 1$
while $\tilde{d}_j \neq \{1, \ldots, n\}$ **do** {
  Chose an $s_i \in \mathcal{S}$ to be added
  $\mathcal{S} \Leftarrow \mathcal{S} \setminus \{s_i\}$
  Let the indices of the new variables in s_i be $b_i = \{k_1, \ldots, k_l\}$
  for $\lambda = 1$ **to** l **do** {
    $\delta_\lambda \Leftarrow \{k \in \tilde{d}_{j-1} | (x_k, x_{k_\lambda}) \in G_{ADF}\}$
  }
  for $\lambda = 1$ **to** l **do** {
    if exists $\lambda' \neq \lambda$ with $\delta_\lambda \subseteq \delta_{\lambda'}$ and $k_{\lambda'}$ not marked superfluous
      $\delta_{\lambda'} \Leftarrow \delta_{\lambda'} \cup \{k_\lambda\}$
      Mark k_λ superfluous
  }
  for $\lambda = 1$ **to** l **do** {
    if not k_λ superfluous
      $\tilde{s}_j \Leftarrow \delta_\lambda \cup \{k_1, \ldots, k_\lambda\}$
      $j \Leftarrow j + 1$
  }
}

Algorithm 3 describes our heuristic. It is given a cut size, which bounds the size of the sets. Each new variable is included in a set together with the previous variables on which it depends. However, if another variable depends on a superset of variables, the two sets are merged. If the size is larger than the cut size, variables are randomly left out. After completing the merge phase, the algorithm calculates $\tilde{c}_j$, $\tilde{b}_j$, and $\tilde{d}_j$ analogous to the construction given by (2.7).

For 2-D grid problems exact factorizations are not k-bounded. If the ADF consists of subfunctions of two neigboring grid points only, our subfunction merger algorithm computes a factorization using marginals up to order 3. The factorization covers the whole interaction graph G_{ADF}. For a 3*3 grid the subfunction merger constructs the following factorization

$$q(\mathbf{x}) = \tilde{q}(x_4, x_5)\tilde{q}(x_3|x_4)\tilde{q}(x_2|x_5)\tilde{q}(x_1|x_2, 4)\tilde{q}(x_7|x_4)\tilde{q}(x_0|x_1, x_3) \times \tilde{q}(x_8|x_5, x_7)\tilde{q}(x_6|x_3, x_7). \tag{2.16}$$

Because grids are very common, we have in addition implemented a number of specialized factorizations for functions defined on grids. Our presentation of the subfunction merger algorithm has been very short. In the area of Bayesian networks, the problem has been investigated in [3].

FDA has experimentally proven to be very successful on a number of functions where standard genetic algorithms fail to find the global optimum. In [16] the scaling behavior for various test functions has been studied. For recent surveys the reader is referred to [15, 17, 19].

2.3 The Maximum Entropy Principle

In this section we investigate the problem of approximating an unknown distribution given some information in a theoretical framework.

Let $\mathbf{x} = (x_1, \ldots, x_n)$, $B = \{0,1\}^n$. Let $\phi_j : B \to \{0,1\}, j = 1, m$ be binary functions, often called features. Let a sample S be given, $\tilde{p}(\mathbf{x})$ the observed distribution. Let

$$E_{\tilde{p}}(\phi_j) = \sum_{x \in B} \tilde{p}(\mathbf{x})\phi_j(\mathbf{x}). \tag{2.17}$$

Note that ϕ_j can specify any marginal distribution, but also more general expectations.

Problem *We are looking for a distribution $p(\mathbf{x})$ which fulfills the constraints*

$$E_p(\phi_j) = E_{\tilde{p}}(\phi_j) \tag{2.18}$$

and is in some sense plausible.

If only a small number of features is given the problem is under-specified. Consequently, for incomplete specifications the missing information must be added by some automatic completion procedure. This is achieved by the *maximum entropy principle.* Let us recall

Definition 12 *The entropy of a distribution is defined by*

$$H(p) = -\sum_{x} p(\mathbf{x}) \ln(p(\mathbf{x})). \tag{2.19}$$

Maximum entropy principle (MaxEnt): *Let*

$$P = \{p | E_p(\phi_j) = E_{\tilde{p}}(\phi_j), \quad j = 1, \ldots, m\}. \tag{2.20}$$

Then the MaxEnt solution is given by

$$p^* = \operatorname{argmax}_{p \in P} H(p). \tag{2.21}$$

The MaxEnt formulates the *principle of indifference.* If no constraints are specified, the uniform random distribution will be the solution. MaxEnt has a long history in physics and probabilistic logic. The interested reader is referred to [8, 9].
The MaxEnt solution can be computed from the constrained optimization problem

$$p^* = \operatorname{argmax}_{p \in P} H(p), \tag{2.22}$$

$$\sum_x p(x) = 1, \tag{2.23}$$

$$\sum_x p(\mathbf{x})\phi_j(\mathbf{x}) = E_{\tilde{p}}(\phi_j). \tag{2.24}$$

This is a convex optimization problem with linear constraints. Therefore it has a unique solution. It can be found by introducing Lagrange multipliers.

$$L(p, \Lambda, \gamma) = -\sum_x p(\mathbf{x}) \ln p(\mathbf{x}) + \gamma(\sum_x p(x) - 1) \tag{2.25}$$

$$+\sum_{i=j}^{m} \lambda_j (E_p(\phi_j) - E_{\tilde{p}}(\phi_j)), \tag{2.26}$$

where $\Lambda = (\lambda_1, \dots, \lambda_m)$.
The maxima of L can be obtained by computing the derivatives of L. We compute

$$\frac{\partial L}{\partial p(\mathbf{x})} = -\ln p(\mathbf{x}) - 1 - \sum_{j=1}^{m} \lambda_j \phi_j(\mathbf{x}) + \gamma. \tag{2.27}$$

Setting the derivative to zero gives the parametric form of the solution

$$p^*(\mathbf{x}) = \exp(\gamma - 1) \exp \sum_{j=1}^{m} \lambda_j \phi_j(\mathbf{x}). \tag{2.28}$$

Definition 13 *Let Q be the space of distributions of the parametric form*

$$Q = \{q | q(\mathbf{x}) = \frac{1}{Z} \exp \sum_{j=1}^{m} \lambda_j \phi_j(\mathbf{x})\}. \tag{2.29}$$

In order to characterize the MaxEnt solution, the relative entropy between distributions has to be introduced.

Definition 14 *The relative entropy or Kullback–Leibler divergence between two distributions p and q is defined as (see [4])*

$$\mathrm{KLD}(p,q) = \sum_x p(\mathbf{x}) \ln \frac{p(\mathbf{x})}{q(\mathbf{x})}. \tag{2.30}$$

Note that $\mathrm{KLD}(p,q) \neq \mathrm{KLD}(q,p)$, i.e., KLD is not symmetric! If $q(\mathbf{x}) = 0$ and $p(\mathbf{x}) > 0$ we have $\mathrm{KLD}(p,q) = \infty$. This means that KLD gives large weights to values near zero. In all other aspects KLD is a distance measure. The following lemma holds [4].

Lemma 15 *For any two probability distributions p and q,* $\mathrm{KLD}(p,q) \geq 0$ *and* $\mathrm{KLD}(p,q) = 0$ *iff* $p = q$.

In our application KLD fulfills the Pythagorean property.

Lemma 16 (Pythagorean Property) *Let* $p \in P$, $q \in Q$, *and* $p^* \in P \cap Q$, *then*

$$\mathrm{KLD}(p,q) = \mathrm{KLD}(p,p^*) + \mathrm{KLD}(p^*,q). \tag{2.31}$$

The proof is straightforward. The following theorem follows easily from the lemma:

Theorem 17 (Maximum Entropy Solution). *If* $p^* \in P \cap Q$, *then*

$$p^*(\mathbf{x}) = \mathrm{argmax}_{p \in P}\, H(p). \tag{2.32}$$

Furthermore, p^* *is unique.*

The constrained optimization problem can be solved by standard mathematical algorithms. But also specialized algorithms have been invented, a popular one is the *Generalized iterative scaling algorithm (GIS)* [5]. Unfortunately the computational amount of the algorithm is exponential in general.

Obviously the MaxEnt approximation minimizes the relative entropy KLD (p,u) to the uniform distribution u. Thus MaxEnt is a special case of the MinRel principle. But there exists another justification of the MaxEnt solution, it is given by the *Maximum Log-Likelihood* principle.

Definition 18 *Let* $S = \{X_1, \ldots, X_N\}$ *be an empirical sample,* $\tilde{p}(\mathbf{x})$ *the empirical distribution. Let* $q(\mathbf{x})$ *be a distribution. Then the likelihood that q generates the data is given by*

$$\mathrm{LH}(q) = \prod_{i=1}^{N} q(X_i) = \prod_{x \in B} q(\mathbf{x})^{N\tilde{p}(\mathbf{x})}. \tag{2.33}$$

The log-likelihood is defined as

$$\mathrm{LogLH}(q) = \sum_{x \in B} N\tilde{p}(\mathbf{x}) \ln q(\mathbf{x}). \tag{2.34}$$

Theorem 19 (Maximum Log-Likelihood solution). *If $p^* \in P \cap Q$, then*

$$p^*(\mathbf{x}) = \text{argmax}_{q \in Q} \text{LogLH}(q). \qquad (2.35)$$

Furthermore, p^* is unique.

Proof. Let $\tilde{p}(\mathbf{x})$ be the observed distribution. Clearly $\tilde{p} \in P$. Suppose $q \in Q$ and $p^* \in P \cap Q$. We show that $\text{LogLH}(q) \leq \text{LogLH}(p^*)$. The Pythagorean property gives

$$\text{KLD}(\tilde{p}, q) = \text{KLD}(\tilde{p}, p^*) + \text{KLD}(p^*, q)$$

Therefore

$$\begin{aligned} \text{KLD}(\tilde{p}, q) &\geq \text{KLD}(\tilde{p}, p^*) \\ -H(\tilde{p}) - \frac{1}{N}\text{LogLH}(q) &\geq -H(\tilde{p}) - \frac{1}{N}\text{LogLH}(p^*) \\ \text{LogLH}(q) &\leq \text{LogLH}(p^*) \quad \square \end{aligned}$$

Thus the MaxEnt solution can be viewed under both the maximum entropy framework as well as the maximum log-likelihood framework. This means that p^* will fit the data as closely as possible while as the maximum entropy solution it will not assume facts beyond those given by the constraints.

We next investigate the relation of FDA factorizations and the MaxEnt solution.

Definition 20 *Given an ADF the MaxEnt problem is called* complete marginal *if all marginal distributions $\tilde{p}(x_{s_k})$ are given. The FDA factorization is called* complete, *if the graphical model of the factorization contains the interaction graph G_{ADF}.*

Theorem 21. *The MaxEnt solution of a complete marginal MaxEnt problem is the exact distribution. The MaxEnt solution of any complete FDA factorization is the exact distribution.*

Proof. Let a complete marginal MaxEnt problem be given. Then the features $\phi(\mathbf{x}_{s_i})$ are defined by $E_{\tilde{p}}\phi(\mathbf{x}_{s_i}) = \tilde{p}(\mathbf{x}_{s_i})$. We abbreviate the parameters in equation (2.29) by $\lambda(\mathbf{x}_{s_i})$. Now set $\lambda(\mathbf{x}_{s_i})\tilde{p}(\mathbf{x}_{s_i}) = \beta f(\mathbf{x}_{s_i})$. Thus the exact distribution is in the set Q. Obviously fulfills the exact distribution the marginalization constraints. Therefore the exact distribution is the MaxEnt solution. The proof for complete FDA factorizations works accordingly. $\square$

This theorem is the justification of the MaxEnt principle for FDA factorizations. If all relevant information of an ADF is given, the unique MaxEnt solution is the exact distribution. But the computation of the MaxEnt solution for a complete FDA factorization is *exponential* if it does not fulfill the RIP. Therefore FDA just samples from the factorization using the computed marginals. But if the RIP is violated the generated distribution might be different from the exact distribution, even if the factorization is complete. Thus in contrast to the MaxEnt solution, the FDA factorization with its simple sampling procedure might not converge to the optima of the function.

We next describe another approach to approximate the Boltzmann distribution. In this method the Kullback–Leibler divergence to the Boltzmann distribution is minimized without computing the marginal distributions from samples. Instead the values of the marginals are computed from a difficult constrained minimization problem.

2.4 Approximating the Boltzmann Distribution

The Boltzmann distribution plays an important role in statistical physics. Therefore a number of approximation techniques have been developed. We present an approach where an approximation q

$$q(\mathbf{x}) = \frac{1}{Z}\prod_{i=1}^{k}\tilde{q}(\mathbf{x}_k) \tag{2.36}$$

is computed which minimizes the relative entropy to the Boltzmann distribution. The method is described in [15] using the terminology of physics. We give here a short mathematical derivation. The relative entropy is given by

$$\begin{aligned}\mathrm{KLD}(q|p_\beta) &= \sum_x q(\mathbf{x})\ln q(\mathbf{x}) - \sum_x q(\mathbf{x})\ln p_\beta(\mathbf{x}) \\ &= -H(q) + \ln Z - \beta E_q(f).\end{aligned}$$

We again assume that the function is defined by an ADF. Then we easily obtain

$$E_q(f) = \sum_{i=1}^{m} q(\mathbf{x}_{s_i})f_i(x_{s_i}). \tag{2.37}$$

The expected average of the function can be computed using the marginals. More difficult problem is the computation of $H(q)$. We will restrict our discussion to *FDA* factorizations.

$$q(\mathbf{x}) = \frac{\prod_{i=1}^m \tilde{q}(\mathbf{x}_{s_i})}{\prod_{i=1}^m \tilde{q}(\mathbf{x}_{c_i})}. \tag{2.38}$$

For this factorization one computes

$$\begin{aligned}H(q) &= -\sum_{i=1}^{m} q(\mathbf{x}_{s_i})\ln\tilde{q}(\mathbf{x}_{s_i}) + \sum_{i=1}^{m} q(\mathbf{x}_{c_i})\ln\tilde{q}(\mathbf{x}_{c_i}) \\ &\approx -\sum_{i=1}^{m} q(\mathbf{x}_{s_i})\ln q(\mathbf{x}_{s_i}) + \sum_{i=1}^{m} q(\mathbf{x}_{c_i})\ln q(\mathbf{x}_{c_i}).\end{aligned} \tag{2.39}$$

Note that for FDA factorizations which do not fulfill the RIP we have $q(\mathbf{x}_{s_i}) \neq \tilde{q}(\mathbf{x}_{s_i})$ if we use FDA sampling. But in order to make the problem tractable we assume $q(\mathbf{x}_{s_i}) = \tilde{q}(\mathbf{x}_{s_i})$ Then minimizing KLD leads to the following constraint optimization problem.

Definition 22 (Bethe–Kikuchi approximation) *Compute the minimum of all FDA factorizations* $q(\mathbf{x})$ (2.38)

$$\begin{aligned}\operatorname{argmin}_q \mathrm{KLD}(q|p_\beta) &= \sum_{i=1}^{m} q(\mathbf{x}_{s_i}) \ln q(\mathbf{x}_{s_i}) \\ &\quad - \sum_{i=1}^{m} q(\mathbf{x}_{c_i}) \ln q(\mathbf{x}_{c_i}) - \beta \sum_{i=1}^{m} q(\mathbf{x}_{s_i}) f_i(x_{s_i}) \end{aligned} \tag{2.40}$$

subject to the constraints for all s_j *with* $c_i \subset s_j$

$$\sum_{x_{s_i}} q(\mathbf{x}_{s_i}) = 1, \tag{2.41}$$

$$\sum_{x_{s_j} \setminus x_{c_i}} q(\mathbf{x}_{s_j}) = q(\mathbf{x}_{c_i}). \tag{2.42}$$

Remark: The *minimization problem is not convex! There might exist many local minima. Furthermore, the exact distribution might not be a local minimum if the factorization violates the RIP.*

The constraints make the solution of the problem difficult. We again use the Lagrange function.

$$\begin{aligned} L(p, \Lambda, \Gamma) &= \mathrm{KLD}(q|p_\beta) + \sum_{i=1}^{m} \gamma_i \left(\sum_{x_{s_i}} q(\mathbf{x}_{s_i}) - 1 \right) \\ &\quad + \sum_{i=1}^{m} \sum_{x_{c_i}} \lambda(s_j, c_i) \left(\sum_{x_{s_j} \setminus x_{c_i}} q(\mathbf{x}_{s_j}) - q(\mathbf{x}_{c_i}) \right). \end{aligned} \tag{2.43}$$

The minima of L are determined be setting the derivatives of L zero. The independent variables are $q(\mathbf{x}_{s_i})$, $q(\mathbf{x}_{c_i})$, γ_i, and $\lambda(s_j, c_i)$. We obtain

$$\frac{\partial L}{\partial q(\mathbf{x}_{s_i})} = \ln q(\mathbf{x}_{s_i}) + 1 - \beta q(\mathbf{x}_{s_i}) f(\mathbf{x}_{s_i}) + \gamma_i + r(\Lambda). \tag{2.44}$$

Setting the derivative to zero, we obtain the parametric form

$$q(\mathbf{x}_{s_i}) = \mathrm{e}^{-1-\gamma_i} \mathrm{e}^{-r(\Lambda)} \mathrm{e}^{\beta f(x_{s_i})}. \tag{2.45}$$

Note that the parametric form is again exponential. The Lagrange factors Γ are easily computed from $\sum_{x_{s_i}} q(x_{s_i}) = 1$. The factors Λ have to be determined from a nonlinear system of equations. Before we describe an algorithm for solving it, we describe a simple special case, the mean-field approximation.

2.4.1 The Mean-Field Approximation

In the mean-field approximation uni-variate marginals only are used.

$$q(\mathbf{x}) = \prod_{i=1}^{n} q(x_i). \tag{2.46}$$

We obtain for its entropy and $E_q(f)$.

$$H(q) = -\sum_x \prod_{i=1}^{n} q(x_i) \sum_{j=1}^{n} \ln q(x_j) = -\sum_{i=1}^{n} \sum_{x_i} q(x_i) \ln q(x_i)$$
$$E_q(f) = \sum_x \prod_{i=1}^{n} q(x_i) f(\mathbf{x}) = \sum_{i=1}^{m} \prod_{j \in s_i}^{n} q(x_j) f(\mathbf{x}_{s_i}).$$

For the mean-field approximation the Kullback–Leibler divergence is convex, thus the minimum exists and is unique. The minimum is obtained by setting the derivative of KLD equal to zero, using the uni-variates as variables. We abbreviate $q_i = q(x_i = 1)$.

Theorem 23. *The uni-variate marginals of the mean-field approximation are given by the nonlinear equation*

$$q_i^* = \frac{1}{1 + \mathrm{e}^{\frac{\partial E_q}{\partial q_i}}}. \tag{2.47}$$

Proof. We compute the derivative

$$\frac{\partial \mathrm{KLD}}{\partial q_i} = \ln \frac{q_i}{1 - q_i} + \frac{\partial E_q}{\partial q_i} = 0. \tag{2.48}$$

The solution gives (2.47). □

Equation (2.47) can be solved by an iteration scheme.

2.5 Loopy Belief Models and Region Graphs

The computation of the Bethe–Kikuchi approximation is difficult. We decided to use a specialized algorithm, recently proposed in [26]. It is based on the concept of a *region graph.* A region graph is a loopy graphical model. It is strongly related to partially ordered sets (posets) or Hasse diagrams. Similar or identical structures have been presented in [2, 14, 24]. This section follows largely the notation of [26].

2.5.1 Regions

The region graph was introduced in [26] using a different graphical model, the *factor graph.* The factor graph is a more detailed way to describe an additive decomposition. We decided not to use the factor graph to show the connection of region graphs to junction trees.

Definition 24 *Let $S = \{s_1, \dots, s_m\}$ be the index set of an additive decomposition for a fitness function f, such that*

$$f(\mathbf{x}) = \sum_{s_i \in S} f_i(\mathbf{x}_{s_i}). \tag{2.49}$$

A ***region*** *$R = (s_R, I_R)$ is a set of variable indices $s_R \subseteq \{1, \dots, n\}$ and a set of subfunction indices $I_R \subseteq \{1, \dots, m\}$, such that*

$$\forall i \in I_R : s_i \subseteq s_R. \tag{2.50}$$

The variables contained in the region are indexed by s_R, whereas I_R contains the indices of the subfunctions which are contained in the region. It is asserted by (2.50) *that all variables needed for the contained subfunctions are in s_R.*

Our goal is to approximate the Boltzmann distribution with the energy $E(\mathbf{x}) = -f(\mathbf{x})$ by minimizing the relative entropy. For a region we define a local energy.

Definition 25 *For a region R, define the* ***region energy***

$$E_R(\mathbf{x}_{s_R}) := -\sum_{i \in I_R} f_i(\mathbf{x}_{s_i}). \tag{2.51}$$

Region energies are defined only for those regions which contain the variables of at least one subfunctions. We will try to compute the marginals q_R on R from the Bethe–Kikuchi minimization problem. In [26] the marginals are called the *beliefs* on R. This is the terminology of Bayesian networks.

2.5.2 Region Graph

Definition 26 *A* ***region graph*** *is a graph $G = (\mathcal{R}, E_\mathcal{R})$, where $\mathcal{R}$ is a set of regions and $E_\mathcal{R}$ is a set of directed edges. An edge $(R_\mathrm{p}, R_\mathrm{c}) \in E_\mathcal{R}$ is only allowed if $s_{R_\mathrm{c}} \subset s_{R_\mathrm{p}}$. If $(R_\mathrm{p}, R_\mathrm{c}) \in E_\mathcal{R}$, we call R_p a parent of R_c and R_c child of R_p.*

Since $E_\mathcal{R}$ imposes a partial ordering on the set of regions, in [14] the same structure was called a partially ordered set or *poset*.

Lemma 27 *A region graph is directed acyclic.*

Proof. This follows immediately from the requirement that edges are only allowed from supersets to subsets. □

A junction tree can be turned into a region graph by creating a region for every cluster and every separator and adding edges from each node to each neighboring separator.

The global distribution of a junction tree is the product of all distributions on the clusters divided by the distributions of all the separators (see [15]). We generalize this factorization by introducing counting numbers of the regions.

Definition 28 *The* ***counting number*** c_R *of a region* R *is defined recursively as*

$$c_R = 1 - \sum_{R' \in A(R)} c_{R'}, \tag{2.52}$$

where $A(R)$ *is the set of all ancestors of* R.

This is well-defined, because the region graph is cycle-free. The maximal regions (without ancestors) have counting number 1. From there, the counting numbers can be calculated from the top to the bottom of the graph.

2.5.3 Region Graph and Junction Tree

The junction tree property has an equivalent on the region graph, called the region graph condition.

Definition 29 *We call a region graph* ***valid*** *if it fulfills the* ***region graph condition****, which states that*

1. *For all variable indices* $i \in \{1, \dots, n\}$ *the set* $\mathcal{R}_{X,i} := \{R \in \mathcal{R} | i \in s_R\}$ *of all regions* R *that contain* X_i *form a connected subgraph with*

$$\sum_{R \in \mathcal{R}_{X,i}} c_R = 1, \tag{2.53}$$

and

2. *For all subfunction indices* $i \in \{1, \dots, m\}$ *the set* $\mathcal{R}_{f,i} := \{R \in \mathcal{R} | i \in I_R\}$ *of all regions* R *that contain* f_i *form a connected subgraph with*

$$\sum_{R \in \mathcal{R}_{f,i}} c_R = 1. \tag{2.54}$$

The connectivity of the subgraph, like the junction property, prevents that in different parts of the graph contradictory beliefs can evolve. The condition on the counting numbers makes sure that every variable and every subfunction is counted exactly once.

The Kikuchi approximation of the Boltzmann distribution is defined as follows (see also [23]).

Definition 30 *The **Kikuchi approximation** of the Boltzmann distribution for a region graph is*

$$k(\mathbf{x}) = \prod_{R \in \mathcal{R}} q_R(\mathbf{x}_{s_R})^{c_R} \tag{2.55}$$

*In general, it is not normalized and therefore no probability distribution. The **normalized Kikuchi approximation***

$$p_k(\mathbf{x}) = \frac{k(\mathbf{x})}{\sum_{\mathbf{y}} k(\mathbf{y})} \tag{2.56}$$

is a probability distribution.

If the region graph is derived from a junction tree, with q_R being the marginal distributions on the clusters and separators, $k(\mathbf{x})$ is a valid distribution, since its definition coincides with the junction tree distribution. It has been proven that cycle-free region graphs reproduce the exact distribution [25]. Thus they give the same result a junction tree.

The problem of sampling from a Kikuchi approximation is discussed later. We next describe a local iteration algorithm, based on the region graph and message passing between regions. The iteration algorithm minimizes the Kullback–Leibler divergence.

2.6 The Concave Convex Procedure

The Concave Convex Procedure (CCCP) [27] is a variant of Generalized Belief Propagation (GBP) proposed in [25]. It is based on the observation that the Lagrangian consists of a convex and a negative convex (concave) term. The CCCP algorithm alternates between updates of the convex and the concave term.

2.6.1 The Convex and Concave Lagrangian

We now derive the CCCP update procedure, following [27]. The algorithm is fairly complex. A detailed description can be found in the dissertation [7]. The Lagrangian to be minimized is given by (2.43)

$$\begin{aligned} L = & \sum_{R \in \mathcal{R}} c_R \left(\sum_{\mathbf{x}_{s_R}} q_R(\mathbf{x}_{s_R}) \beta E(\mathbf{x}_{s_R}) + \sum_{\mathbf{x}_{s_R}} q_R(\mathbf{x}_{s_R}) \log q_R(\mathbf{x}_{s_R}) \right) \\ & + \sum_{R \in \mathcal{R}} \gamma_R \left(1 - \sum_{\mathbf{x}_{s_R}} q_R(\mathbf{x}_{s_R}) \right) \\ & + \sum_{(P,R) \in E_{\mathcal{R}}} \sum_{\mathbf{x}_{s_R}} \lambda_{PR}(\mathbf{x}_{s_R}) \left(\sum_{\mathbf{x}_{s_P \setminus s_R}} q_P(\mathbf{x}_{s_P}) - q_R(\mathbf{x}_{s_R}) \right). \end{aligned} \tag{2.57}$$

The basic idea of CCCP is to split L in a convex and a concave part. The problematical part is the entropy term: For regions with $c_R > 0$, the entropy term is convex, for regions with $c_R < 0$ it is concave. The average energy and the constraints are linear in the q_R, so it does not matter where we put them.

To avoid an arbitrary separation into convex and concave regions, we set

$$c_{\max} = \max_R c_R \tag{2.58}$$

and use this definition to split up L into a convex part

$$\begin{aligned} L_{\text{vex}} = & \sum_{R \in \mathcal{R}} c_{\max} \left(\sum_{\mathbf{x}_{s_R}} q_R(\mathbf{x}_{s_R}) \beta E_R(\mathbf{x}_{s_R}) + \sum_{\mathbf{x}_{s_R}} q_R(\mathbf{x}_{s_R}) \ln q_R(\mathbf{x}_{s_R}) \right) \\ & + \sum_{R \in \mathcal{R}} \gamma_R \left(1 - \sum_{\mathbf{x}_{s_R}} q_R(\mathbf{x}_{s_R}) \right) \\ & + \sum_{(P,R) \in E_{\mathcal{R}}} \sum_{\mathbf{x}_{s_R}} \lambda_{PR}(\mathbf{x}_{s_R}) \left(\sum_{\mathbf{x}_{s_P \setminus s_R}} q_P(\mathbf{x}_{s_P}) - q_R(\mathbf{x}_{s_R}) \right) \end{aligned} \tag{2.59}$$

and a concave part

$$L_{\text{ave}} = \sum_{R \in \mathcal{R}} (c_R - c_{\max}) \left(\sum_{\mathbf{x}_{s_R}} q_R(\mathbf{x}_{s_R}) E_R(\mathbf{x}_{s_R}) + \sum_{\mathbf{x}_{s_R}} q_R(\mathbf{x}_{s_R}) \ln q_R(\mathbf{x}_{s_R}) \right) \tag{2.60}$$

Obviously $L = L_{\text{vex}} + L_{\text{ave}}$.

2.6.2 The Outer and Inner Loops

CCCP consists of an *inner loop* in which the messages are updated until convergence, and an *outer loop* in which the current estimates of the marginals are updated. For the inner loop we use the iteration index τ and for the outer the index ξ.

The Outer Loop

For the outer loop we make the ansatz

$$\nabla L_{\text{vex}}^{\xi+1} + \nabla L_{\text{ave}}^{\xi} = 0, \tag{2.61}$$

where ∇L denotes the vector of the partial derivatives of L with respect to the marginals $q_R(\mathbf{x}_{s_R})$. The derivatives are given by

$$\frac{\partial L_{\text{vex}}}{\partial q_R(\mathbf{x}_{s_R})} = c_{\max}\left(\beta E_R(\mathbf{x}_{s_R}) + \ln q_R(\mathbf{x}_{s_R}) + 1\right) - \gamma_R - \sum_{P|(P,R)\in E_{\mathcal{R}}} \lambda_{PR}(\mathbf{x}_{s_R}) + \sum_{C|(R,C)\in E_{\mathcal{R}}} \lambda_{RC}(\mathbf{x}_{s_C}) \quad (2.62)$$

and

$$\frac{\partial L_{\text{ave}}}{\partial q_R(\mathbf{x}_{s_R})} = (c_R - c_{\max})\left(\beta E_R(\mathbf{x}_{s_R}) + \ln q_R(\mathbf{x}_{s_R}) + 1\right). \quad (2.63)$$

Inserting (2.62) and (2.63) into (2.61) gives

$$c_{\max}\left(\beta E_R(\mathbf{x}_{s_R}) + \ln q_R^{\xi+1}(\mathbf{x}_{s_R}) + 1\right) - \gamma_R - \sum_{P|(P,R)\in E_{\mathcal{R}}} \lambda_{PR}(\mathbf{x}_{s_R}) + \sum_{C|(R,C)\in E_{\mathcal{R}}} \lambda_{RC}(\mathbf{x}_{s_C}) + (c_R - c_{\max})\left(\beta E_R(\mathbf{x}_{s_R}) + \ln q_R^{\xi}(\mathbf{x}_{s_R}) + 1\right) = 0. \quad (2.64)$$

Solving this for $q_R^{\xi+1}(\mathbf{x}_{s_R})$ gives the update equations for the marginals in the outer loop:

$$q_R^{\xi+1}(\mathbf{x}_{s_R}) = q_R^{\xi}(\mathbf{x}_{s_R})^{\frac{c_{\max}-c_R}{c_{\max}}} \exp\left[-\frac{c_R}{c_{\max}}\beta E_R(\mathbf{x}_{s_R})\right] \times \exp\left[\frac{\gamma_R - c_R}{c_{\max}} + \frac{1}{c_{\max}}\left(\sum_{P|(P,R)\in E_{\mathcal{R}}} \lambda_{PR}(\mathbf{x}_{s_R})\right)\right] \times \exp -\frac{1}{c_{\max}}\left[\sum_{C|(R,C)\in E_{\mathcal{R}}} \lambda_{RC}(\mathbf{x}_{s_C})\right]. \quad (2.65)$$

For the regions with $c_R = c_{\max}$ the previous marginal $q_R^{\xi}(\mathbf{x}_{s_R})$ cancels out. We next introduce messages (see [25])

$$m_{PC}(\mathbf{x}_{s_C}) := \mathrm{e}^{\frac{1}{c_{\max}}\lambda_{PC}(\mathbf{x}_{s_C})} \quad (2.66)$$

and choose γ_R appropriately for normalization, which changes the update equation to

$$q_R^{\xi+1}(\mathbf{x}_{s_R}) \propto q_R^{\xi}(\mathbf{x}_{s_R})^{\frac{c_{\max}-c_R}{c_{\max}}} \mathrm{e}^{-\frac{c_R}{c_{\max}}\beta E_R(\mathbf{x}_{s_R})} \frac{\prod_{P|(P,R)\in E_{\mathcal{R}}} m_{PR}(\mathbf{x}_{s_R})}{\prod_{C|(R,C)\in E_{\mathcal{R}}} m_{RC}(\mathbf{x}_{s_C})}. \quad (2.67)$$

The Inner Loop

The inner loop update equation for the messages can be derived by inserting (2.67) into the consistency equation

$$\sum_{\mathbf{x}_{s_P \setminus s_R}} q_P(\mathbf{x}_{s_P}) = q_R(\mathbf{x}_{s_R}). \quad (2.68)$$

This gives

$$\sum_{\mathbf{x}_{s_P\setminus s_R}} q_P^t(\mathbf{x}_{s_P})^{\frac{c_{\max}-c_P}{c_{\max}}} e^{-\frac{c_P}{c_{\max}}\beta E_P(\mathbf{x}_{s_P})} \frac{\prod_{Q|(Q,P)\in E_{\mathcal{R}}} m_{QP}(\mathbf{x}_{s_P})}{\prod_{C|(P,C)\in E_{\mathcal{R}}} m_{PC}(\mathbf{x}_{s_C})}$$

$$= q_R^t(\mathbf{x}_{s_R})^{\frac{c_{\max}-c_R}{c_{\max}}} e^{-\frac{c_R}{c_{\max}}\beta E_R(\mathbf{x}_{s_R})} \frac{\prod_{Q|(Q,R)\in E_{\mathcal{R}}} m_{QR}(\mathbf{x}_{s_R})}{\prod_{C|(R,C)\in E_{\mathcal{R}}} m_{RC}(\mathbf{x}_{s_C})}. \quad (2.69)$$

The message $m_{PR}(\mathbf{x}_{s_R})$ is independent of the summation variables $\mathbf{x}_{s_P\setminus s_R}$, so it can be extracted from the sum. It appears in the denominator on the left side of (2.69) and in the numerator on the right side. This allows to solve the equation for this message.

With the abbreviations

$$g_R(\mathbf{x}_{s_R}) := q_R^{\xi}(\mathbf{x}_{s_R})^{\frac{c_{\max}-c_R}{c_{\max}}} e^{-\frac{c_R}{c_{\max}}\beta E_R(\mathbf{x}_{s_R})}, \quad (2.70)$$

$$h_R(\mathbf{x}_{s_R}) := \frac{\prod_{Q|(Q,R)\in E_{\mathcal{R}}} m_{QR}^{\tau}(\mathbf{x}_{s_R})}{\prod_{C|(R,C)\in E_{\mathcal{R}}} m_{RC}^{\tau}(\mathbf{x}_{s_C})} \quad (2.71)$$

we arrive at the inner loop update equation

$$m_{PR}^{\tau,\mathrm{upd}}(\mathbf{x}_{s_R}) = m_{PR}^{\tau}(\mathbf{x}_{s_R})\sqrt{\frac{\sum_{\mathbf{x}_{s_P\setminus s_R}} g_P(\mathbf{x}_{s_P})h_P(\mathbf{x}_{s_P})}{g_R(\mathbf{x}_{s_R})h_R(\mathbf{x}_{s_R})}}. \quad (2.72)$$

In order to make the iteration more robust, damping is applied Linear damping [26, 27] calculates the messages as a linear combination between the old and update messages:

$$m_{P\to R}^{\tau}(\mathbf{x}_R) = (1-\alpha)m_{P\to R}^{\tau-1}(\mathbf{x}_R) + \alpha m_{P\to R}^{\tau,\mathrm{upd}}(\mathbf{x}_R). \quad (2.73)$$

2.6.3 FDA Factorizations and Region Graphs

The Kikuchi factorization and the concept of region graph has also been used for an EDA algorithm by [23]. But the marginals are not determined from minimization of the Kullback–Leibler divergence, they are estimated from samples. Thus instead of the FDA factorization the Kikuchi factorization is used. But here a difficult problem appears, namely to sample from the Kikuchi approximation. The factorization is a loopy graphical model, it contains cycles. Therefore Gibbs sampling has been used in [23], which gives a valid distribution but is computational very expensive.

In contrast, we have implemented the full Bethe–Kikuchi method. In order to circumvent the sampling problem, we decided to use FDA factorizations only for the Kikuchi method. Given an arbitrary FDA factorization, we use the marginals used for the factorization to create a region graph. This is always possible. Then the Bethe–Kikuchi approximation is computed using this region graph. After the computation of the marginals the FDA factorization is used again for sampling.

2.7 The FDA Software

The FDA software allows to optimize arbitrary fitness functions of binary variables using various evolutionary algorithms like the simple genetic algorithm (GA), the univariate marginal algorithm (UMDA), the FDA, the learning factorized distribution algorithm (LFDA), the Bethe–Kikuchi Approximation (BKDA) and the Iterated Kernighan–Lin algorithm (IKL). Details about these algorithms and a *free download of the FDA software* can be found at the web site *http://www.ais.fraunhofer.de/~muehlen.*
The following list summarizes the implemented algorithms:

- $-ag$ chooses the simple genetic algorithm GA. It allows one-point or two-point crossover and mutation.
- $-au$ chooses the UMDA algorithm. This algorithm estimates the univariate marginal distributions from the population and then samples the next generation with them.
- $-af$ chooses the FDA algorithm. The FDA algorithm expects the fitness function to be additively decomposable. To this end, the fitness functions are given as a sum of "local functions" on a subset of the variables. If there is no such structure given, UMDA is used instead. $-af$ should be combined with one of the implemented automatic factorization algorithms described under $-j$.
- $-al$ selects the LFDA algorithm. This algorithm estimates the structure from the data using the BIC measure. The edge that increases the BIC most is added to the graphical model, until there is no more improvement possible. There are two subparameters to this algorithm: "m" gives a maximal number of parents that a variable can have, and "a" gives the weight of the structure penalty. The default values are "-alm8a0.5."
- $-ab$ is our implementation of the Bethe–Kikuchi method BKDA. It is recommended for experienced users only. The user is prompted for a value of beta used for the Boltzmann distribution. It should be run for a single generation only.
- $-am$ is an iterated random algorithm. It creates the next population using mutation from the maximum of the current population. It requires an argument, which is the number of random bit flips which is performed on the copies of the maximum in order to generate the starting points of the next generation. It should be run together with a local optimizing algorithm like the Kernighan–Lin algorithm. In FDA this option should be used together with the $-k$ switch with a population size of 2 till 4.
- $-j$ selects the heuristic to compute the factorization for FDA. $-jm$ selects the subfunction join algorithm. The user should bound the size of the marginal distribution by specifying a number, e.g., $-jm9$. In addition a number of specialized factorizations for certain functions can be selected.
- $-k$ turns on the local optimizer Kernighan–Lin algorithm [12]. It can be used together with all the other algorithms, excluding BKDA of course.

2.7.1 Local Hill Climbing

The use of a powerful local hill climbing algorithm changes the character of the search dramatically. We have implemented a general Kernighan–Lin algorithm [12]. It is not a simple local hill climber, but it has a sophisticated backtracking procedure. Our implementation scales with $O(n^2)$. The algorithm can run together with GA, UMDA, FDA and LFDA using a small population size. It can also run as a simple iteration, the Iterated Kernighan–Lin algorithm *IKL* with a population size of two. New and promising start configurations are provided by randomly changing a certain percentage of the best solution obtained so far. The performance of *IKL* strongly depends on using a good value for this percentage.

For the graph bipartitioning problem a very fast version of Kernighan–Lin has been implemented. It uses hash tables and a cut for backtracking. This implementation scales approximately linearly with n. It allows the optimization of graphs with more than 1000 variables with pop sizes up to 50.

2.8 Numerical Results

The EDA family of algorithms seems to be mature, at least for binary problems. It is time to demonstrate the state-of-the-art with large instances of popular benchmark problems. In [17] large graph bipartitioning problems have been solved. Large problems have been also solved in [22]. We will continue this work here. We will use problems on 2-D grids and Kauffman's $n-k$ function. The number of variables will be up to 900. Kauffman's function is an example of an ADF with random connections, the 2-D grid problems are important problems with regular connections.

2-D Spin glass:

$$f(\mathbf{x}) = \sum_{i,j} f_{i,j}(s_i, s_j). \tag{2.74}$$

s_j is one of the four neighbors of s_i, $s_i \in \{-1, 1\}$. The function values are randomly drawn from a Gaussian distribution in the interval $[-1, +1]$.

2-D grid on plaquettes:

$$f(\mathbf{x}) = \sum_{i,j} f_{i,j}(x_{i,j}, x_{i+1,j}, x_{i,j+1}, x_{i+1,k+1}). \tag{2.75}$$

The indices define a plaquette on the 2-D grid. The function values for each subfunction are randomly drawn from a Gaussian distribution in the interval $[-1, +1]$.

Kauffman random $n-3$:

$$f(\mathbf{x}) = \sum_{i=1}^{n} f_i(x_i, x_j, x_k). \tag{2.76}$$

The indices i, j, k are randomly chosen. The function values are drawn from a Gaussian distribution in $[0, 1]$.

The following table can be easily generated by any user of the FDA software. We do not have the space to compare all the possible FDA algorithms. The reader is encouraged to do tests himself.

The results confirm our theory. The standard FDA algorithm with large population sizes ($N = 30,000$) performs very good on all instances. It should be no surprise that the population size has to be large. For the 2-D grid problems we used the special factorization $-jg5$. It uses marginals of size 5, even for the Ising problems. For the Kauffman function the subfunction merger algorithm $-jm$ creates marginals up to size 12 for $n = 400$ and size 15 for $n = 625$. It needs a large sample size to compute good estimates of these marginals. We remind the reader that for the BKDA algorithm the samples are computed only once, after computing the marginals. Surprisingly the BKDA algorithm runs slightly faster than FDA. Note that β has to be very large for the Kauffman function. This indicates that the Kauffman function has many local maxima. Still larger values of β do not improve the results for BKDA, because the convergence becomes a problem. Given the many assumptions used for BKDA we find the performance surprisingly good. But it seems to be not a breakthrough.

Table 2.1. Comparison of FDA and BKDA on large problems

problem	size	alg.	sample.	β	best value
Ising	400	FDA	30,000	-	297.259
Ising	400	BKDA	10,000	30	297.259
Ising	625	FDA	30,000	-	466.460
Ising	625	BKDA	10,000	30	463.622
Plaqu.	400	FDA	10,000	-	207.565
Plaque.	400	BKDA	10,000	30	207.565
Plaque.	625	FDA	30,000	-	320.069
Plaque.	625	BKDA	10,000	30	320.132
Plauqe.	900	FDA	30,000	-	459.274
Plaque.	900	BKDA	10,000	30	454.237
$n-3$	400	FDA	10,000	-	0.7535
$n-3$	400	BKDA	10,000	12,000	0.7520
$n-3$	625	FDA	30,000	-	0.7501
$n-3$	625	BKDA	10,000	15,000	0.7436

2.9 Conclusion and Outlook

The efficient estimation and sampling of distributions is a common problem in several scientific disciplines. Unfortunately each discipline uses a different language to formulate its algorithms. We have identified two principles used for the approximation – minimizing the Kullback–Leibler divergence $\mathrm{KLD}(q, u)$ to the uniform distribution u or minimizing $\mathrm{KLD}(q, p_\beta)$ to the Boltzmann distribution p_β.

We have shown that the basic theory is the same for the two algorithms. This theory deals with the decomposition of graphical models and the computation of approximate factorizations. If the interaction graph G_{ADF} allows an exact factorization fulfilling the RIP, then both methods compute the exact distribution.

We have discussed two EDA algorithms in detail. The standard FDA algorithm computes a factorization from the graph representing the structure. If the corresponding graphical model does not fulfill the assumptions of the factorization theorem the exact distribution is only approximated. Factorizations which cover as much as possible from the interaction graph G_{ADF} are obtained by merging of subfunctions. The marginals of the standard FDA algorithm are computed from sampling the FDA factorization.

The Bethe–Kikuchi algorithm BKDA computes the marginals from a difficult constrained minimization problem. Because sampling from the original Bethe–Kikuchi factorization is difficult, we have extended the original approach. We use the FDA factorization which contains no loops. From this factorization the marginals are computed using the Bethe–Kikuchi minimization.

Our results show that for binary problems the EDA algorithms perform as good or even better than other heuristics used for optimization. At this stage our algorithm is not yet optimized from a numerical point of view, nevertheless is already competitive to more specialized algorithms.

In our opinion too many EDA researchers still investigate 1-D problems. Our theory (and also practice) shows that these problems can be solved exactly in polynomial time if the junction tree factorization is used. They pose no problem for optimization at all.

The interested reader can download our software from the WWW site *http://www.ais.fraunhofer.de/~muehlen.*

References

[1] S. M. Aji and R. J. McEliece. The generalized distributive law. *IEEE Transactions of Information Theory*, 46(2):325–343, March 2000

[2] S. M. Aji and R. J. McEliece. The generalized distributive law and free energy minimization. In *Proceedings of the 39th Annual Allerton*

Conference on Communication, Control, and Computing, pages 672–681, 2001

[3] R. G. Almond. *Graphical Belief Modelling.* Chapman and Hall, London, 1995

[4] T. M. Cover and J. Thomas. *Elements of Information Theory.* Wiley, New York, 1989

[5] J. Darroch and D. Ratcliff. Generalized iterative scaling for log-linear models. *Ann.Math. Statistics*, 43:1470–1480, 1972

[6] Y. Gao and J. Culberson. Space complexity of estimation of distribution algorithms. *Evolutionary Computation*, 13(1):125–143, 2005

[7] R. Höns. *Estimation of Distribution Algorithms and Minimum Relative Entropy.* PhD thesis, University of Bonn, 2005

[8] E. T. Jaynes. Information theory and statistical mechanics. *Physical Review*, 6:620–643, 1957

[9] E. T. Jaynes. Where do we stand on maximum entropy? In R. D. Levine and M. Tribus, editors, *The Maximum Entropy Formalism.* MIT Press, Cambridge, 1978

[10] F. V. Jensen and F. Jensen. Optimal junction trees. In *Proceedings of the 10th Conference on Uncertainty in Artificial Intelligence*, pages 360–366, Seattle, 1994

[11] M. I. Jordan, editor. *Learning in Graphical Models.* MIT Press, Cambrigde, 1999

[12] B. W. Kernighan and S. Lin. An efficient heuristic for partitioning graphs. *Bells Systems Technical Journal*, 2:291–307, 1970

[13] S. L. Lauritzen. *Graphical Models.* Clarendon Press, Oxford, 1996

[14] R. J. McEliece and M. Yildirim. Belief propagation on partially ordered sets. In *Proceedings of the 15th Internatonal Symposium on Mathematical Theory of Networks and Systems (MTNS 2002)*, 2002

[15] H. Mühlenbein and R. Höns. The estimation of distributions and the minimum relative entropy principle. *Evolutionary Computation*, 13(1): 1–27, 2005

[16] H. Mühlenbein and T. Mahnig. FDA - a scalable evolutionary algorithm for the optimization of additively decomposed functions. *Evolutionary Computation*, 7(4):353–376, 1999

[17] H. Mühlenbein and T. Mahnig. Evolutionary optimization and the estimation of search distributions with applications to graph bipartitioning. *Journal of Approximate Reasoning*, 31(3):157–192, 2002

[18] H. Mühlenbein and T. Mahnig. Mathematical analysis of evolutionary algorithms. In C. C. Ribeiro and P. Hansen, editors, *Essays and Surveys in Metaheuristics*, Operations Research/Computer Science Interface Series, pages 525–556. Kluwer, Norwell, 2002

[19] H. Mühlenbein and T. Mahnig. Evolutionary algorithms and the Boltzmann distribution. In K. D. Jong, R. Poli, and J. C. Rowe, editors, *Foundations of Genetic Algorithms 7*, pages 525–556. Morgan Kaufmann Publishers, San Francisco, 2003

[20] H. Mühlenbein, T. Mahnig, and A. Ochoa. Schemata, distributions and graphical models in evolutionary optimization. *Journal of Heuristics*, 5(2):213–247, 1999

[21] H. Mühlenbein and G. Paaß. From recombination of genes to the estimation of distributions I. binary parameters. In H.-M. Voigt, W. Ebeling, I. Rechenberg, and H.-P. Schwefel, editors, *Lecture Notes in Computer Science 1141: Parallel Problem Solving from Nature - PPSN IV)*, pages 178–187, Berlin, 1996. Springer-Verlag

[22] M. Pelikan and D. Goldberg. Hierarchical BOA solves Ising spin glasses and MAXSAT. In *Genetic and Evolutionary Computation Conference 2003*, volume 2724 of *Lecture Notes in Computer Science*, pages 1271–1282. Springer, Berlin Heidelberg New York 2003. Also IlliGAL Report No. 2003001

[23] R. Santana. Estimation of distribution algorithms with Kikuchi approximations. *Evolutionary Computation*, 13(1):67–97, 2005

[24] Y. W. Teh and M. Welling. On improving the efficiency of the iterative proportional fitting procedure. In *Proceedings of the International Workshop on Artificial Intelligence and Statistics*, volume 9, 2003

[25] C. Yanover and Y. Weiss. Finding the M most probable configurations using loopy belief propagation. In S. Thrun, L. Saul, and B. Schölkopf, editors, *Advances in Neural Information Processing Systems 16*. MIT Press, Cambridge, MA, 2004

[26] J. S. Yedidia, W. T. Freeman, and Y. Weiss. Constructing free energy approximations and generalized belief propagation algorithms. Technical Report 2004-040, Mitsubishi Electric Research Laboratories, May 2004

[27] A. L. Yuille. CCCP algorithms to minimize the Bethe and Kikuchi free energies: Convergent alternatives to belief propagation. *Neural Computation*, 14(7):1691–1722, 2002

3

Linkage Learning via Probabilistic Modeling in the Extended Compact Genetic Algorithm (ECGA)

Georges R. Harik, Fernando G. Lobo, and Kumara Sastry

Summary. For a long time, genetic algorithms (GAs) were not very successful in automatically identifying and exchanging structures consisting of several correlated genes. This problem, referred in the literature as the linkage-learning problem, has been the subject of extensive research for many years. This chapter explores the relationship between the linkage-learning problem and that of learning probability distributions over multi-variate spaces. Herein, it is argued that these problems are equivalent. Using a simple but effective approach to learning distributions, and by implication linkage, this chapter reveals the existence of GA-like algorithms that are potentially orders of magnitude faster and more accurate than the simple GA.

3.1 Introduction

Linkage learning in genetic algorithms (GAs) is the identification of building blocks to be conserved under crossover. Theoretical studies have shown that if an effective linkage-learning GA were developed, it would hold significant advantages over the simple GA [1]. Therefore, the task of developing such an algorithm has drawn significant attention. Past approaches to developing such an algorithm have focused on evolving a problem's chromosomal representation along with its solution [2–4]. This has proven to be a difficult undertaking. This chapter reinterprets and solves the linkage-learning problem in the context of probabilistic optimization.

Recently, a number of algorithms have been developed that replace the GA's population and crossover operator with a probabilistic representation and generation method [5–8]. Studies have shown a close correspondence between these algorithms and equivalent simple GAs [7]. This chapter shows how a variant of these algorithms, that pays close attention to the probabilistic modeling of the population successfully tackles the linkage-learning problem.

We will begin by briefly reviewing the workings of the simple GA, as well as those of the related probability-based algorithms. We will then explore

G.R. Harik et al.: *Linkage Learning via Probabilistic Modeling in the Extended Compact Genetic Algorithm (ECGA)*, Studies in Computational Intelligence (SCI) **33**, 39–61 (2006)
www.springerlink.com

the close relationship between these two approaches to optimization. The remainder of this chapter is then concerned with the ultimate consequences of this relationship. The argument to be presented will consist of two separate assertions:

- That learning a good probability distribution is equivalent to learning linkage.
- That one "good" distribution can be found by searching for a jointly small representation of two components: (1) the compressed representation of the population under the given distribution and (2) the distribution's representation given the problem encoding.

Ultimately, this argument must stand on the legs of empirical observations, as in its essence it is but the application of Occam's Razor. The last part of this chapter presents a probabilistic algorithm, the extended compact GA (ECGA), designed to learn linkage through learning good probability distributions. It then demonstrates the advantage that this approach provides over the simple GA, on a theoretical problem that has traditionally been used to test other linkage-learning approaches. Finally, this chapter explores the consequences of the proposed probabilistic algorithm.

3.2 The Simplified Simple GA

A GA [9, 10] is a simulation of the genetic state of a population of individuals – their genetic state being their combined chromosomes. It typically includes those forces of genetics deemed most influential in nature, such as natural selection, mutation, and crossover (mating). In this chapter, we will restrict ourselves to one facet of the GA: its use as a problem solver, or optimization algorithm. Natural evolution typically leads to a set of individuals that are well suited to their environment. By controlling the computational nature of such an environment, the GA can be made to evolve chromosomes (structures) that are well suited to any given task.

An optimization is a search over a set of structures, to find the "best" structure under some given criteria. This paradigm maps over readily to implementation in a GA. Each structure is represented by its blueprint, or chromosome, in the GA's population. The GA's population is thus the current set of structures the algorithm has found to be most interesting or useful. At each point in time, it represents the current "state" of the search. The genetic operators of natural selection, crossover, and mutation then generate the next state of the search from the current one. The GA's goal is to reach a final state (population) that contains a good solution (structure) to the problem at hand.

In order to simplify this exposition, we will assume that the structures the GA will optimize over are the set of binary strings of fixed length L. A binary string is simply a consecutive sequence of characters each of which is a

0 or a 1. This restriction makes it easier to visualize and understand the GA and its operators. However, the theory developed in this chapter will remain applicable to the much wider domain of optimization over finite-dimensional spaces. We will for the same reason also consider only selecto-recombinative GAs, thus ignoring for the moment the effects of mutation.

In the course of optimization, the GA's population repeatedly undergoes processing by the two genetic operators of crossover and selection, until it converges. Convergence here means that only one type of chromosome remains in the population – hopefully the best or a good solution. The two operators have orthogonal goals. The crossover operator generates new chromosomes by mixing parts from other pairs of chromosomes. It roughly corresponds to mating and reproduction in nature.

The selection operator weeds out those chromosomes that are unsuited to their environment – that is, those that have a poor score under the current optimization. Again, for the purpose of simplicity, we focus on block selection [11] and uniform crossover [12] as representatives of possibly more general selection and crossover operators.

In block selection, a large fraction of the weakest chromosomes in the population are thrown out, and the stronger chromosomes are given their place. Strength here is measured according to the chosen optimization problem. Operationally, the optimization problem is represented by a fitness function that maps structures over to real numbers. The strongest structures are then those with the highest fitness score. Block selection is controlled by one parameter, S, which specifies that only the best fraction $1/S$ of the population is to be retained after the action of selection. Figure 3.1 shows the effects of selection with $S = 2$ on a population of size 8.

Under uniform crossover, the population is paired up, and each pair of chromosomes generates two children, which replace their parents in the population. A child is created from the two parents by randomly inheriting the value of each position (dimension) from one of its two parents, while its sibling

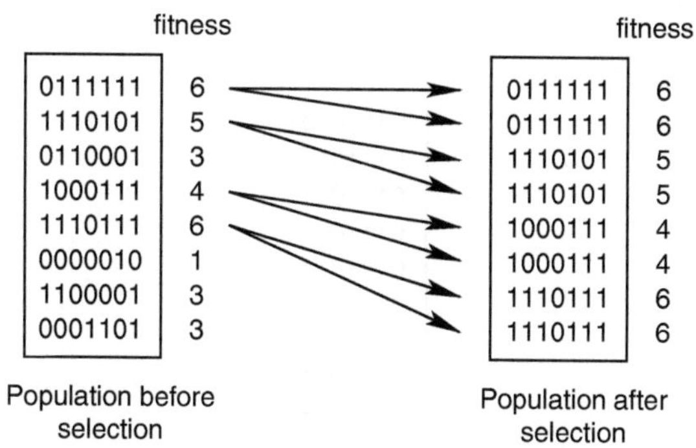

Fig. 3.1. With $S = 2$, each chromosome in the top half of the population gets two copies in the next generation

gets the value at that position from the other parent. In the parlance of GAs, each (position, value) combination is referred to as a gene. Figure 3.2 shows the possible effects of uniform crossover on two very dissimilar chromosomes.

Optimization by selecto-recombinative GAs thus consists of a random initialization of the population, followed by repeated applications of crossover and selection. This optimization is typically stopped when the population has converged, although a number of other stopping criteria are also possible. Figure 3.3 shows one particular population that has converged to the value 01110111. In a number of problems, GAs have been shown to consistently outperform standard optimization techniques. The reason why the GA does well is widely agreed to be a consequence of its effective propagation of substructures that are correlated with high fitnesses.

As an example, let us consider a simple optimization problem of maximizing 1s (onemax), where the fitness of each string is the number of 1s it contains. Figure 3.4 shows the possible evolution of a population under one generation of selection and crossover. Note how the ratio of 1s in the new population is higher than in the old, and that the 1s are well distributed. This is because selection increases the number of 1s, and crossover mixes them together in an attempt to combine all the 1s into a single chromosome. In this case, each 1 gene is correlated with a high fitness, and the GA has successfully exploited this information.

Although the reason for the GA's success is widely agreed upon, the nature of the structures it exchanges, and whose correlation with fitness it maintains, is under vigorous debate. The crux of the problem is the complexity of these structures. Two mutually exclusive possibilities vie for the explanation of the

Parents before crossover		Parents after crossover
1111111	→	10011000
0000000		01100111

Fig. 3.2. Note how all the genes are conserved in a crossover. Each parental gene ends up in one of the two children

```
01110111
01110111
01110111
01110111
01110111
01110111
01110111
01110111
```

Fig. 3.3. A converged population only has one type of chromosome in it, and can therefore not search for more structures

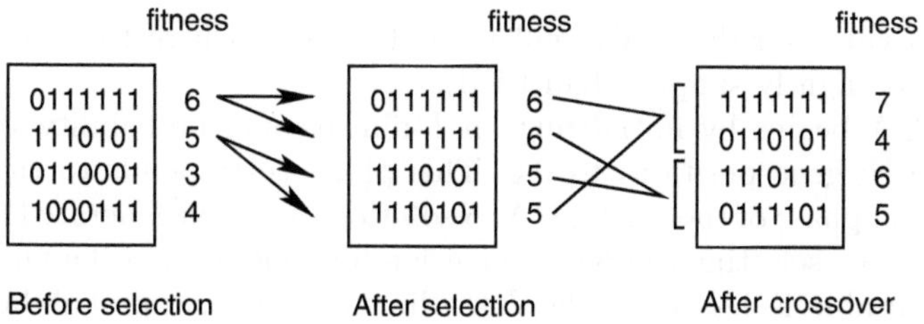

Fig. 3.4. Selection followed by crossover leads to a new chromosome (the first) that is more fit than any previous one

GA's success: (1) that single genes are the only structures that the GA can effectively deal with; and (2) that the GA can process more complex structures, consisting of several genes, which are referred to as building blocks.

The study of linkage learning is the study of how to make the GA process structures more complex than single genes, in the absence of information about which genes are related. In this context, related roughly means that the genes would have to exist in tandem to provide a fitness boost, but each gene alone would not provide a detectable fitness signal. As of yet, the most advanced of such methods have only been marginally successful in justifying the computational effort necessary to undertake them [3]. The remainder of this chapter addresses this issue by developing a computationally justifiable algorithm that learns linkage. First, however, we take a necessary detour to explore a set of probabilistic algorithms that are closely related to the GA.

3.3 Order-1 Probabilistic Optimization Algorithms

The population of the GA represents information about the parts of the search space that the GA has seen before. The crossover and selection operators tell the GA how to exploit this information to generate new, and potentially good, solutions. Along the course of time, researchers noticed that crossover tended to decorrelate the individual dimensions (or genes) in the solution structures, while selection tended to change the makeup of the population by rewarding the more successful genes. Thus were born a number of algorithms that replaced the population, crossover, and selection with a number of actions on marginal probability distributions on each of the representation's genes.

The idea behind these algorithms rested on representing the current state of the search as the fraction of each dimension (or gene) in the population that had a value of one. Using only this information, a new population could be generated that mimicked the effect of many consecutive crossovers. By altering these probabilities according to how well certain genes did against the competition, these algorithms could also mimic the effect of selection. The compact GA (cGA) [7] and PBIL [6] are two examples of these simplistic (but effective) algorithms. We will restrict ourselves here to looking at the cGA as it is slightly simpler than PBIL.

The cGA begins by initializing an L-dimensional probability vector $P[\,]$ (one for each gene position) to 0.5. This phase corresponds to the random initialization phase of the simple GA. S solutions are then generated by polling this vector, i.e., selecting the Kth dimension (or gene) to be 1 if a unit uniform random variable is less than the Kth dimension of the probability vector, $P[K]$. The gene positions of the fittest of these S solutions are rewarded in pairwise competitions with each of the less fit solutions. $P[K]$ is increased if the fittest has a 1 in the Kth position, and the less fit solution does not. $P[K]$ is likewise decreased if the fittest has a 0 in the Kth gene, and the less fit solution does not. The amount of increase or decrease in parameterized by a value E.

For instance, take the maximizing 1s problem, and assume $L = 4$, $S = 2$ and $E = 0.25$. Figure 3.5 shows one iteration taking place under this algorithm. Of the two generated chromosomes 0111 (with a fitness of 3) is fitter than 1010 (with a fitness of 2). The original probability vector is random, $P[\,] = [0.5, 0.5, 0.5, 0.5]$. In the first gene, the 0 is part of a fitter chromosome than the 1. Therefore $P[0]$ is decreased by 0.25. Note that the index of the first gene in the vector is taken to be 0, not 1 due to a programming convention. In the second gene, the opposite is true, therefore $P[1]$ is increased by 0.25. The third gene is the same in both chromosomes, so $P[2]$ is unchanged. $P[3]$ is again increased by 0.25. This leaves us with the new probability vector $P[\,] = [0.25, 0.75, 0.5, 0.75]$. This process continues until the $P[\,]$ vector implies a single solution, that is all its values are zeroes or ones. At this point, the cGA has converged.

One might see a close correlation between this algorithm and the simple GA, but still find it difficult to guess at the extent of this relationship. It has been shown that the simple GA using a population size N and a selection rate

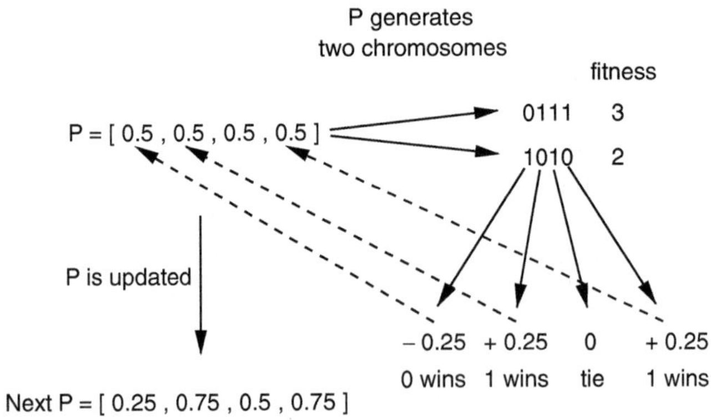

Fig. 3.5. The cGA evaluation step consists of generation, followed by an examination that changes the probability distribution

of S under tournament selection (which is a close cousin to block selection), and uniform crossover, can be mimicked very closely by the cGA generating S solutions and using $E = 1/N$. These algorithms are referred to as order-1 probabilistic algorithms as they maintain the population's distribution as a product of the marginal distributions of each of the separate genes; genes being considered order-1 or trivial building blocks.

3.4 Probabilistic Optimization and Linkage Learning

The correspondence between the operation of the cGA and the simple GA hints at a deeper connection. This connection is that the GA's population can be interpreted as representing a probability distribution over the set of future solutions to be explored. The GA's population consists of chromosomes that have been favored by evolution and are thus in some sense good. The distribution that this population represents tells the algorithm where to find other good solutions.

In that sense, the role of crossover is to generate new chromosomes that are very much like the ones found in the current population. This role can also be played by a more direct representation of the distribution itself. This is precisely what the cGA and PBIL do. Similarly, changes in the makeup of the population due to selection can be reflected in alterations of the probability distribution itself.

The probability distribution chosen to model the population can be crucial to the algorithm's success. In fact, the choice of a good distribution is equivalent to linkage learning. We take a moment now to explore this statement in the context of a problem that is difficult for the simple GA to solve without proper linkage learning.

3.4.1 Linkage Learning and Deceptive Problems

We begin by defining a "deceptive" version of the counting ones problem. Here the fitness of a string is the number of 1s it contains, unless it is all 0s, in which case its fitness is $L + 1$ (L recall is the problem length). The reason this is called a deceptive problem is that the GA gets rewarded incrementally for each 1 it adds to the problem, but the best solution consists of all 0s.

The initial conception of this problem is a needle in a haystack, which no optimization algorithm can be reasonably expected to solve. To transform it into one that requires linkage learning, we combine multiple copies of deceptive subproblems into one larger problem. For example, a 40-dimensional problem can be formed by grouping together each four dimensions into a deceptive subproblem. This problem will thus utilize 10 of the deceptive subproblems defined above. The fitness of a string will be the sum of the subproblem fitnesses, where each subproblem is defined over a separate group of four dimensions. Figure 3.6 shows how a sample 40-bit string is evaluated in this problem. This problem is an order-4 deceptive problem, and is typical of the kinds of problems used to test linkage learning algorithms.

A GA that learns linkage will operate by recombining each of the optimal solutions to the 4-bit subproblems into one optimal 40-bit string consisting of all 0s [2, 3]. A GA not learning linkage will tend to gravitate towards a suboptimal solution consisting of some 0000s, and some 1111s. This is what the simple GA will do [1]. In swapping genes between parents, it will often break up good combinations, such as 0000, by crossing them over with slightly worse combinations, such as 1111. The difficulty in learning linkage in this situation is that the four genes defining each subproblem don't have to be adjacent. To learn linkage, a GA must correctly pick out each set of four related genes. Even in this small problem, the number of such combinations is astronomical.

Figure 3.7 shows a possible size 8 population representing a set of solutions to this partially deceptive problem. This illustration depicts only one subproblem (four genes) of each chromosome. In Figure 3.7 the GA has found several good solutions with marginal fitness contributions of 4 and 5 over these four genes. The uniform crossover shown destroys the correlations among the genes that lead to a high fitness; and the average fitness in the population decreases after crossover! Similarly, the order-1 probability representing this population is $P[] = [0.5, 0.5, 0.5, 0.5]$. Yet, generating new solutions using this distribution leads to poor solutions.

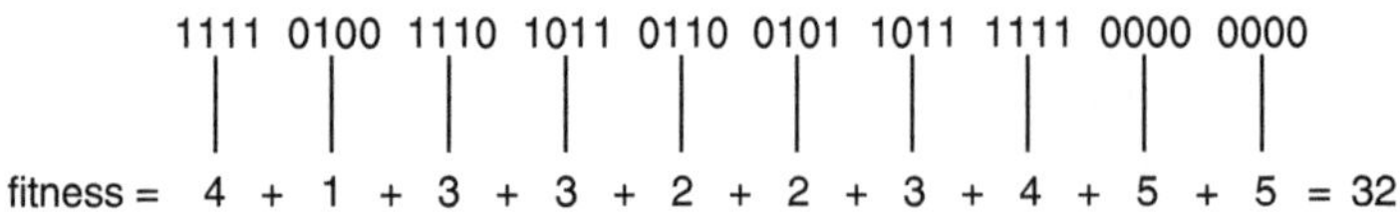

Fig. 3.6. A large partially deceptive problem can be formed by concatenating a number of smaller fully deceptive subproblems

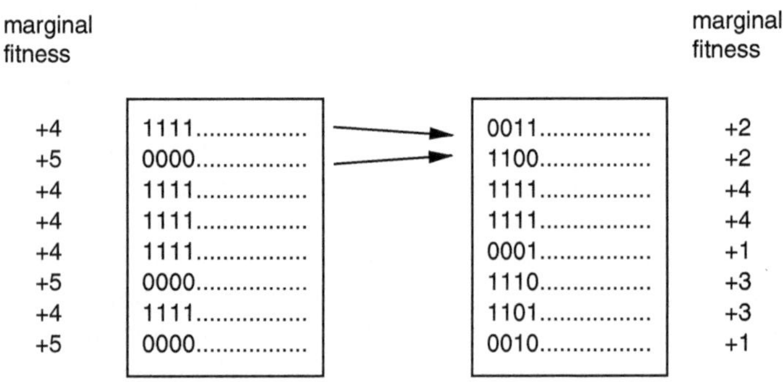

Fig. 3.7. The first four genes of a population before and after a crossover that does not recognize building block boundaries

The correspondence that holds between the cGA and the simple GA rears an ugly side to its head here. Both are unable to deal with this partially deceptive problem, in which linkage learning is crucial. That is, both uniform crossover, and order-1 probabilistic generation, fail to produce new chromosomes that are as good as the ones already in the population! Similarly, the solution to this problem holds dually in the realm of GAs and probabilistic algorithms.

In the GA, the crossover operator needs to understand that these four genes are related, and not break up the combinations they represent. A building block crossover can be developed for this purpose that only swaps whole solutions to subproblems, instead of single genes. In order to do this, however, the algorithm must guess correctly at which genes are related – it must learn linkage.

In probabilistic algorithms, the probability distribution needs to recognize that these four genes are related, and represent the joint probability of these four genes having the 16 possible configurations they can hold; as opposed to the marginal distributions over each of the four genes independently. Such an algorithm would model the original population in Fig. 3.7 using $P[0000] = 3/8$ and $P[1111] = 5/8$, and each set of four genes would maintain their correlation from one generation to the next.

What we have just seen is that linkage learning is a skill that is easily transferable into the domain of probabilistic algorithms. The remainder of this chapter shows that the reverse is also true: that an operational and computationally feasible search for good distributions can fulfill the traditional task of linkage learning.

3.4.2 What Makes a Good Probability Model?

The cGA and PBIL define what it means to generate new solutions that are like the current one. It is to poll the marginal distributions of each dimension or gene position, considering each of the gene positions to be independent. More complex algorithms have been developed that match some of the order-2 behavior of the population [8]. These algorithms act by investigating pairwise inter-gene correlations and generating a distribution that is very close to polling from the population. The closeness measure most easily used is an information-theoretic measure of probability distribution distances [13]. Modeling more complex, and more precise, higher-order behavior has been suggested, but the validity of doing so has been questioned [5].

Pursuing this last train of thought to its ultimate conclusion reveals the flaw in its prescription. We can directly model the order-L behavior of polling the population, by only generating new members through random selection of chromosomes that exist in the population already. This behavior will rapidly lead to the algorithm's convergence, while exploring no new structures. Thus, more accurate modeling of the population's distribution is not always a desirable course of action.

Probabilistic algorithms that use order-2 correlations have sometimes been found to be vastly superior to those using order-1 probabilities. Yet the argument above indicates that this trend cannot continue indefinitely up to order-L modeling of the population. At some point, this progression must stop. This puzzling combination seems to imply that more complicated models of the population are useful, but only up to a point.

These ruminations hint at a resolution to the problem of picking an appropriate distribution to model the population. The solution comes from realizing that the probability distribution to be used will represent a model of what makes the current population good; and that the population is simply a finite sample from this distribution. Fundamentally, the task of identifying a probability model to be used is then the induction of models that are likely to have generated the observed population.

It is well known that unbiased search for such models is futile. Thus we have no choice but to select a bias in this search space. The one we choose is that given all other things are equal, simpler distributions are better than complex ones. Simplicity here can be defined in terms of the representational complexity of the distribution, given the original problem encoding. All things are, however, rarely equal, and there remains a tradeoff between simplicity and accuracy. Our aim will therefore be to find a simple model that nonetheless is good at explaining the current population.

3.4.3 Minimum Description Length Models

Motivated by the above requirement, we venture forth a hypothesis on the nature of good distributions:

> By reliance on Occam's Razor, good distributions are those under which the representation of the distribution using the current encoding, along with the representation of the population compressed under that distribution, is minimal.

This definition is a minimum description length bias on the model search for distributions [14]. It directly penalizes complex models by minimizing over model size. In addition to doing so, it penalizes inaccurate models, because information theory tells us that these are unlikely to be of much use in the compression of the population [13].

3.4.4 MDL Restrictions on Marginal Product Models

We take a moment now to explore this hypothesis. The basis for compression is the availability of a probability distribution over the space of structures to be compressed. Given any particular distribution, we can calculate how many bits it takes to represent a given message. In our case, the message is the population, and the distribution is the one to be evaluated. This hypothesis reformulates the problem of finding a good distribution as a new optimization problem – that of finding the distribution model that minimizes the combined model and population representation.

For the remainder of this chapter we focus on a simple class of probability models: those formed as the product of marginal distributions on a partition of the genes – marginal product models (MPMs). These models are similar to those of the cGA and PBIL, excepting for the fact that they can represent probability distributions over more than one gene at a time. We choose these models for two reasons: (1) they make the exposition simpler; and (2) the structure of such a model can directly be translated into a linkage map, with the partition used defining precisely which genes should be tightly linked.

To make MPMs more concrete, Table 3.1 shows one possible model over a 4-dimensional problem. The partition chosen is [0,3] [1] [2], which means the distribution represents genes 1 and 2 independently, but genes 0 and 3 jointly. The probability distribution over [0,3] has a positive distribution for only the values 00 and 11. This means that at no time can the population generated by this distribution contain a 1 in the first position and a 0 in the fourth position. So, chromosomes of 1001 and 0100 are legal, but 0001 and 1010 are not! This form of restriction is not possible if the probabilities of genes 0 and 3 are represented independently. Obviously, this form of distribution is more powerful than that allowed by the cGA (or PBIL).

Let us now try to represent a population of N chromosomes, half of which are 0000, and half of which are 1111. Representing the population as is (simple bit listing), requires $4N$-bits of storage. On the other hand, an MPM of genes [0,1,2,3] could first represent the probability of a structure being any of the 16 possible binary structures over those four positions. This probability distribution would indicate that only 1111 and 0000 have a positive probability of being represented. Subsequently, the representation of each structure can be 0 for 0000 and 1 for 1111. This encoding uses only 1-bit per structure, and thus only N-bits for the whole population.

By recognizing that these 4-bits are correlated, and representing their distribution in an MPM, we have cut down to a fourth the amount of space required to store the population. Even when the probabilities are not so abrupt, and every string has a positive probability of being represented, the entropy

Table 3.1. A marginal probability model over four genes

[0,3]	[1]	[2]
00 : 0.5	0 : 0.5	0 : 0.6
01 : 0.0	1 : 0.5	1 : 0.4
10 : 0.0		
11 : 0.5		

of a distribution gives us the average number of bits it takes to represent structures randomly pulled from that distribution. By calibrating an MPM's probabilities to match those of the population, this number can be used to estimate that distribution's compression of the population. Furthermore, the calibrated MPM can easily be seen to be the one that compresses the population the most – that is, incorrectly modeling the population cannot possibly help. In fact, since the order of chromosomes is deemed unimportant in the representation, a randomization of the ordering, followed by sampling a chromosome then projecting onto any gene subset from the MPM will be identical to polling the MPM at that subset. Thus, no distribution over that MPM subset can do better than that incorporating the population's frequency counts [13].

The use of an MPM to represent the population consists of two parts: (1) choosing the partition structure of the MPM; and (2) calibrating the MPM by pulling the frequency counts of each of the subsets of the MPM directly from the population. The efficacy of a particular distribution is defined as the sum representation size of the model itself and the population compressed under the model. At this point, there is little recourse but to explore the actual equations defining this criterion.

3.4.5 The Combined Complexity Criterion

Let the Ith partition subset of an MPM be of size S_I, where the sum of the S_I is L. Each subset of size S requires $2^S - 1$ frequency counts to completely define its marginal distribution. Each of the frequency counts is of size $\log_2(N+1)$, where N is the population size. Therefore the total model representation size, or complexity is

$$\text{Model complexity} = \log_2(N+1) \sum_I (2^{S_I} - 1). \tag{3.1}$$

Now, the Ith subset represents S_I genes. Let M_I be the marginal distribution over this subset. The entropy of this distribution, Entropy(M_I), is defined as: $\sum -p_k \log_2(p_k)$, where p_k is the probability of observing outcome k. This number is the average number of bits it takes to represent these S_I genes in the population. This number will never be greater than S_I. This is how the population is compressed, by representing the Ith subset's genes only after the Ith marginal distribution has been represented! Therefore, the total compressed population complexity is

$$\text{Compressed population complexity} = N \sum_I \text{Entropy}(M_I). \tag{3.2}$$

Armed with these two definitions, we can evaluate the efficacy of any given MPM structure. By MPM structure, we mean the MPM partitioning, without the actual probabilities for each of the marginal distributions. These probabilities are determined by the population, and the required condition that the compression be optimal. First, we calibrate the MPM structure using the population's frequency counts to form a full MPM model. Second, we add the model complexity and compressed population complexity to get a combined complexity number

$$\begin{aligned}\text{Combined complexity} &= \text{Model complexity}\\ &\quad + \text{Compressed population complexity}. \end{aligned} \tag{3.3}$$

Section 3.5 describes a simple algorithm for searching for partitions of the gene space for a good MPM distribution over the population, given this operational criterion of finding a distribution such that its compressed complexity is suitably small.

3.5 The ECGA

In this section, we combine the above heuristic with a greedy search algorithm to invent an efficient probabilistic optimization algorithm. The proposed algorithm is very simple:

1. Generate a random population of size N
2. Undergo tournament selection at a rate S
3. Model the population using a greedy MPM search
4. If the model has converged, stop
5. Generate a new population using the given model
6. Return to step 2

This algorithm may also be stopped at any time, using the best found solution so far as its result. Most of the algorithm is self-explanatory, but we focus on two of its features. First, the algorithm requires both a population, and selection. Because we are remodeling the population at each generation, the structure of the models may not be stable. Therefore, selection cannot be replaced by a simple update as in the cGA. For the same reason, a concrete population is required also. Only the crossover step is replaced by probabilistic polling in this algorithm. Second, we have yet to describe the greedy MPM search.

The greedy MPM search begins each generation by postulating that all of the genes are independent – that is, that the MPM $[0][1]\cdots[L-2][L-1]$ is best. What it will then do is perform a steepest ascent search, where at each step, the algorithm attempts to merge all pairs of subsets into larger subsets. It judges such mergers again on the basis of their combined complexity. If the best such combination leads to a decrease in combined complexity, then that merger is carried out. This process continues until no further pairs of subsets can be merged. The resulting MPM is then the one that is used for that generation.

Let us illustrate the greedy MPM search with an example. Consider the population depicted in Fig. 3.8. The MPM search starts by assuming that the model [0] [1] [2] [3] is best and computes its combined complexity.

Model	Combined complexity
[0] [1] [2] [3]	44

The next step is to merge all possible pair of subsets of the current model and compute their combined complexity measure. Doing so yields:

Model	Combined complexity
[0, 1] [2] [3]	46.7
[0, 2] [1] [3]	39.8
[0, 3] [1] [2]	46.7
[0] [1, 2] [3]	46.7
[0] [1, 3] [2]	45.6
[0] [1] [2, 3]	46.7

The combined complexity of [0] [1] [2] [3] can be improved. [0, 2] [1] [3] is the model that gives the best improvement, thus the search proceeds from [0, 2] [1] [3]. Again, all possible pair of groups are merged:

Model	Combined complexity
[0, 2, 1] [3]	48.6
[0, 2, 3] [1]	48.6
[0, 2] [1, 2]	41.4

At this point the model search stops because it is not possible to improve upon the combined complexity measure. Subsequently, the model [0, 2][1][3] would be used to generate a new population of individuals. This particular model uncovers an important pattern in the population. It says that there is a strong correlation between genes 0 and 2 and that they should be processed accordingly.

A new MPM search is thus carried out each generation. Significant optimizations can and have been taken in the implementation here, such as caching

1000
1101
0111
1100
0010
0111
1000
1001

Fig. 3.8. A population after selection has been performed

delta values for all pair combinations at each step. For those interested, computer code of the algorithm is available elsewhere [15].

This combined greedy search algorithm along with the minimum description length search criteria, applied to the task of optimization, will henceforth be referred to as the extended compact GA (ECGA).

3.5.1 Experimental Results

Most of this chapter has concerned itself with the theoretical justification of the ECGA. This section shows how the ECGA can significantly speed the solution of problems that are partially deceptive. In the creation of partially deceptive functions, we will rely on the composition of small deceptive problems, like the 4-bit problem defined previously. The subproblems we will use are trap functions, whose fitness relies solely on the number of 1s present in a chromosome. These functions have been used extensively in the testing of linkage learning algorithms, and solving them has proven to be quite challenging in the absence of prior linkage information.

We will begin by exploring the relationship between the population size used and the proportion of subproblems solved correctly by both the ECGA and the simple GA using uniform crossover. By adding in information about the algorithms' running time, we can show comparisons of the number of function evaluations both algorithms need to achieve a comparable level of optimization. Without further ado then, we proceed to the experimental results.

3.5.2 Deceptive Trap Functions

In this section, ten copies of the 4-bit trap subproblem, are concatenated to form a difficult 40-bit problem, as in Fig. 3.6. This is the problem used to compare the ECGA with the simple GA. Each set of four neighboring genes, [0–3] [4–7] and so on, thus formed one subfunction to be optimized. But neither the simple GA nor the ECGA were told which genes were related, or that the related groups were contiguous, or for that matter the size of the subproblems.

Both the ECGA and the simple GA with uniform crossover were run on this problem 10 times, with a selection rate of 16 (which is higher than the ECGA needs, but which the simple GA requires), gathering the average number of subfunctions solved per population size. This measure is especially significant, as the performance of GAs and other optimization algorithms is typically judged by the number of objective function evaluations they undertake – and this number is the population size times the number of generations.

Table 3.2 shows the population size versus the average number of subfunctions solved for the simple GA, and the average number of function evaluations taken to do so. Table 3.3 does the same for the ECGA.

The differences between the simple GA and the ECGA are large, and in the favor of the ECGA. To consistently solve nine building blocks, the simple GA needs a population size of 100 thousand and over 3.8 million function evaluations! To do the same, the ECGA needs a population size of 500 and

Table 3.2. Simple GA complexity on deceptive subproblems

Population size	Subfunctions solved	Objective evaluations
100	3.9	740
500	5.1	5,100
1,000	6.1	15,600
5,000	6.8	1,00,000
10,000	7.3	2,48,000
20,000	8.0	6,14,000
50,000	7.9	15,60,000
1,00,000	8.8	37,90,000

Table 3.3. ECGA complexity on deceptive subproblems

Population size	Subfunctions solved	Objective evaluations
100	4.0	750
200	5.2	1,460
300	7.1	2,610
500	9.3	4,000
600	9.9	5,040
1,000	10.0	7,300

4 thousand function evaluations. On this small 40-bit problem, the ECGA is 1,000 times faster than the simple GA. This speedup is due to the careful attention paid to probabilistic modeling in the ECGA. This speedup should theoretically also become much greater when solving larger problems. The following shows the successive MPM structures used in one successful run of the ECGA:

```
GENERATION 0:  [0-3] [4 7 27] [5-6 15] [8-11] [12-14 24] [16-19]
               [20-23] [25] [26 32-35] [28-31] [36-39]

GENERATION 1:  [0-3 25] [4-7 15] [8-11] [12-14] [16-19] [20-23]
               [24 27] [26 32-35] [28-31] [36-39]

GENERATION 2:  [0-3] [4-7] [8-11] [12-15] [16-19] [20-23] [24-27]
               [28-31] [32-35] [36-39]
```

Note that this structure changes from generation to generation. The ECGA makes a few mistakes in the first pair of generations. The final result arrived at in generation 2, however, completely discerns the subfunction structure of the given problem – without being given this information ahead of time. By simply searching for MPM-structures that optimally compress the population, the ECGA has completely dissected the subproblem structure of the 40-bit problem! It is no surprise that armed with this information, the ECGA proceeded to optimize this problem much faster than the simple GA.

3.5.3 Scalability on Deceptive Trap Functions

In this section, experiments are performed to find out how the ECGA scales up when the number of deceptive subproblems increase. For that purpose, the number of deceptive subproblems ranges from 2 to 20 and the minimal number of function evaluations required to solve all but one subproblem is recorded, that is, the optimal solution with an error of $\alpha = 1/m$, where m denotes the number of subproblems.

A bisection method [16] is used over the population size to search for the minimal sufficient population size to achieve a target solution. The results for the minimal sufficient population size are averaged over 30 bisection runs. In each bisection run, the number of subproblems solved with a given population size is averaged over another 30 runs. Thus, the results for the number of function evaluations and the number of generations spent are averaged over 900 (30×30) independent runs. For all experiments, tournament selection without replacement is used with size $S = 8$. Both 4- and 5-bit deceptive subproblems were tested. The results are presented in Fig. 3.9.

The results show that the population size scales approximately as $O(m \log m)$ and the number of objective function evaluations scales as $O(m^{1.5} \log m)$, giving a good match with analytical models that have been developed elsewhere [18, 19]. These results contrast sharply with the exponential scalability of the simple GA to solve the same problems [1]. Good scalability is also obtained when exponentially scaled deceptive subproblems are tested with the ECGA [20].

3.5.4 The Role of Selection

The role of selection is a curious one in the ECGA and not at all immediately clear. If the proper level of selection is not maintained in the above runs however, the ECGA fails, and never models the problem structure correctly (the same is true but an often ignored aspect of the simple GA). Taking a second, we consider the dual roles of probabilistic generation and selection in the ECGA. The role of generation (the ECGA's crossover equivalent) is to create more chromosomes that are like the ones in the present population. These chromosomes will have no correlations across the MPM structure boundaries. That is, if the MPM structure says that genes 1 and 2 are independent, they will actually be independent in the generated population. Given that the algorithm begins by assuming that all genes are independent, one might wonder where dependencies come from at all.

The answer to that question is that selection recorrelates genes if a specific combination of theirs is correlated with high fitness. In the partially deceptive problems we have experimented on, selection correlates the group of 0s defined over a common subproblem. This correlation is what the ECGA detects. If the level of selection is very low, this correlation will never be generated, and the ECGA will never detect that the genes are related. Thus, a low level of selection can cause the ECGA to fail.

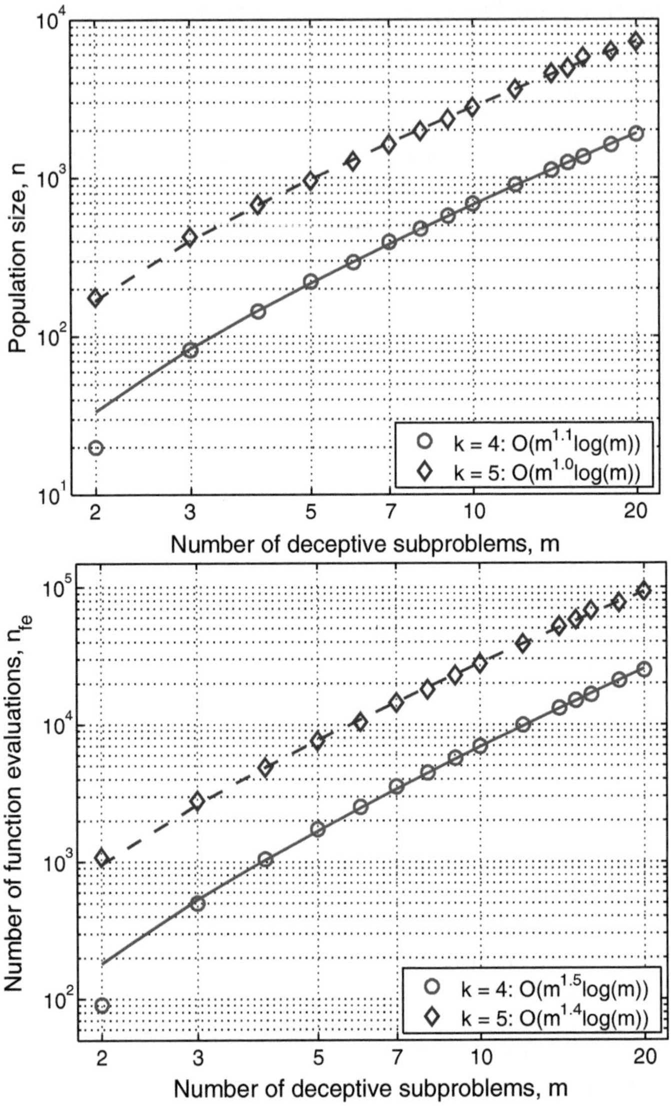

Fig. 3.9. Population size and number of function evaluations required by the ECGA to successfully solve all but one deceptive subproblems. Both 4- and 5-bit deceptive subproblems were tested. The number of subproblems m range from 2 to 20. The results for population size are averaged over 30 runs, while the number of function evaluations is averaged over 900 independent runs. Copied with permission from [17]

In practical terms, this issue can be resolved by not generating a complete new population in each generation, and instead, only replace a fraction of the selected individuals by new ones sampled from the MPM model. This mechanism is the equivalent of implementing a crossover probability in simple GAs.

It has been shown that if half of the selected individuals are kept from the previous generation (the equivalent of a crossover probability of 0.5), then a fixed level of selection pressure, not too high and not too low, is a robust setting for the ECGA [21, 22].

3.5.5 Practical Applications

The ECGA provides a huge advantage over the simple GA when optimizing deceptive subproblems. The real purpose of the ECGA, however, is not to solve artificial problems, but rather practical real-world problems. Since its introduction [23], the ECGA has been applied with success in a variety of application areas, including forest management [24], quantum chemistry [25], stock trading [26], and also as an improved learning mechanism in learning classifier systems [27].

3.6 Summary, Conclusions, and Future Work

This chapter began by reviewing the simple GA and a related set of probabilistic algorithms. Past work has equated some of these algorithms with the simple GA, and an exploration of this relationship has pointed to the existence of more general probabilistically based GA-like algorithms.

This chapter demonstrated that proper probabilistic modeling in these algorithms is in effect the long-sought solution to the linkage-learning problem. It has also introduced an operational complexity criterion for distinguishing between good models and bad models. Experimental results have shown that by focusing on learning marginal probability models, the ECGA can solve some difficult problems orders of magnitude faster than simple GAs not using linkage information.

This chapter has revealed a strong connection between linkage learning and proper probabilistic modeling. This is however, only the tip of the iceberg in as much as effective optimization is concerned. The goal of linkage learning, while ambitious, is only a small part of the more general goal of representation learning. In representation learning, the actual optimization problem is transformed into a different space, and optimized in that new space. The optimization and design of biological entities – which transforms fitness functions defined on 3-dimensional molecules into ones over a genetic representation – is proof that such techniques can be effective. It is this author's belief that even such a search for transformations can be placed into the framework of complexity based modeling.

Several questions and difficulties remain to be addressed by future work on probability based optimization in general, and MPM-like approaches in particular. A few of the more promising or addressable issues in this area are:

- The ECGA is simple to parallelize, by replacing the migration step of standard parallel GAs by one of probability model exchange. This has the

potential to greatly reduce the amount of communication and synchronization needed over that of the parallel simple GA. Preliminary work in this direction has been done with the compact GA [28] but needs to be extended for the case of more general probabilistic models.
- The MPM model search is potentially computationally expensive. In some problems, this poses the risk of overwhelming the function evaluation time with the search's own computational overhead. One recourse in these cases is to use simpler probability models, such as those used by the cGA or MIMIC. Another possible alternative is to implement this fixed search algorithm in hardware, or to look for heuristic approximations to the MPM algorithm. For example, the MPM search could be biased somewhat to the encoder's original linkage specification.
- Another approach to reducing the complexity of the MPM model search is to sample from the population when building the correct MPM structure.
- On the other spectrum of function evaluation complexity, it is possible that more time for population analysis might be available. In these cases, Bayesian network learning, although more complex, yield even more powerful algorithms than the one described in this chapter [29, 30]. In some cases, such as optimization in euclidean spaces, direct probabilistic modeling might also offer more accurate methodologies than modeling using arbitrary binary encodings over the selfsame space.
- The learning mechanism of the ECGA has been used to design a mutation-like operator for conducting local search using the linkage information learned [17], and its combination with the regular ECGA has also been suggested [20]. The hybridization issue deserves additional research. It is quite possible that by searching for mixtures of different models, one might be able to optimally decide between the application of different GA operators. That is, the probabilistic approach could settle once and for all, on a problem by problem basis, the efficacy question of crossover versus mutation.
- The probability modeling framework suggested here deals particularly well with multi-dimensional data, but does not trivially extend to optimization over more complex structures such as permutations, or programs. This issue deserves serious consideration.

References

[1] Thierens, D., Goldberg, D.E.: Mixing in genetic algorithms. In *Proceedings of the Fifth International Conference on Genetic Algorithms* (1993) 38–45

[2] Goldberg, D.E., Korb, B., Deb, K.: Messy genetic algorithms: Motivation, analysis and first results. *Complex Systems* **3** (1989) 493–530

[3] Harik, G.R.: *Learning gene linkage to efficiently solve problems of bounded difficulty using genetic algorithms.* PhD thesis, University of Michigan, Ann Arbor (1997) Also IlliGAL Report No. 97005
[4] Kargupta, H.: *SEARCH, polynomial complexity, and the fast messy genetic algorithm.* PhD thesis, University of Illinois at Urbana-Champaign, Urbana, IL (1995) Also IlliGAL Report No. 95008
[5] Mühlenbein, H., Mahnig, T., Rodriguez, A.O.: Schemata, distributions and graphical models in evolutionary optimization. *Journal of Heuristics* **5** (1999) 215–247
[6] Baluja, S.: *Population-based incremental learning: A method for integrating genetic search based function optimization and competitive learning.* Technical Report CMU-CS-94-163, Carnegie Mellon University, Pittsburgh, PA (1994)
[7] Harik, G.R., Lobo, F.G., Goldberg, D.E.: The compact genetic algorithm. In *Proceedings of the International Conference on Evolutionary Computation* 1998 (ICEC '98), IEEE New York (1998) 523–528
[8] De Bonet, J.S., Isbell, C.L., Viola, P.: MIMIC: Finding optima by estimating probability densities. In Mozer, M.C., et al. (Eds.): *Advances in Neural Information Processing Systems.* Vol. 9, MIT, Cambridge (1997) 424
[9] Goldberg, D.E.: *Genetic Algorithms in Search, Optimization, and Machine Learning.* Addison-Wesley, Reading, MA (1989)
[10] Holland, J.H.: *Adaptation in Natural and Artificial Systems.* University of Michigan Press, Ann Arbor, MI (1975)
[11] Mühlenbein, H., Schlierkamp-Voosen, D.: Predictive models for the breeder genetic algorithm: I. continuous parameter optimization. *Evolutionary Computation* **1** (1993) 25–49
[12] Syswerda, G.: Uniform crossover in genetic algorithms. In Schaffer, J.D., (Ed.): *Proceedings of the Third International Conference on Genetic Algorithms*, Morgan Kaufman, San Mateo, CA (1989) 2–9
[13] Cover, T.M., Thomas, J.A.: *Elements of Information Theory.* Wiley, New York (1991)
[14] Mitchell, T.: *Machine Learning.* McGraw Hill, New York (1997)
[15] Lobo, F.G., Harik, G.R.: *Extended compact genetic algorithm in C++.* IlliGAL Report No. 99016, Illinois Genetic Algorithms Laboratory, University of Illinois at Urbana-Champaign, Urbana, IL (1999)
[16] Sastry, K.: *Evaluation–relaxation schemes for genetic and evolutionary algorithms.* Master's thesis, University of Illinois at Urbana-Champaign, Urbana, IL (2001) Also IlliGAL Report No. 2002004
[17] Sastry, K., Goldberg, D.E.: Designing competent mutation operator via probabilistic model building of neighborhoods. In Deb, K., et al. (Eds.): *Proceedings of the Genetic and Evolutionary Computation Conference GECCO-2004,* Springer, Berlin Heidelberg New York (2004) 114–125 Part II, LNCS 3103

[18] Pelikan, M., Goldberg, D.E., Cantú-Paz, E.: Bayesian optimization algorithm, population sizing, and time to convergence. In Whitley, D., et al. (Eds.): *Proceedings of the Genetic and Evolutionary Computation Conference GECCO-2000*, Morgan Kaufmann, San Francisco, CA (2000) 275–282

[19] Pelikan, M., Sastry, K., Goldberg, D.E.: Scalability of the Bayesian optimization algorithm. *International Journal of Approximate Reasoning* **31** (2003) 221–258

[20] Lima, C.F., Sastry, K., Goldberg, D.E., Lobo, F.G.: Combining competent crossover and mutation operators: A probabilistic model building approach. In Beyer, H.G., et al. (Eds.): *Proceedings of the ACM SIGEVO Genetic and Evolutionary Computation Conference GECCO-2005*, ACM, NY, USA (2005) 735–742

[21] Harik, G.R., Lobo, F.G.: A parameter-less genetic algorithm. In Banzhaf, W., et al. (Eds.): *Proceedings of the Genetic and Evolutionary Computation Conference GECCO-99,* Morgan Kaufmann, San Francisco, CA (1999) 258–265

[22] Lobo, F.G.: *The parameter-less genetic algorithm: Rational and automated parameter selection for simplified genetic algorithm operation.* PhD thesis, Universidade Nova de Lisboa, Portugal (2000) Also IlliGAL Report No. 2000030

[23] Harik, G.R.: *Linkage learning via probabilistic modeling in the ECGA.* IlliGAL Report No. 99010, Illinois Genetic Algorithms Laboratory, University of Illinois at Urbadna-Champaign, Urbana, IL (1999)

[24] Ducheyne, E.I., De Wulf, R.R., De Baets, B.: Using linkage learning fo forest management planning. In Cantú-Paz, E., (Ed.): *Late Breaking Papers at the Genetic and Evolutionary Computation Conference (GECCO-2002)*, AAAI, New York (2002) 109–114

[25] Sastry, K.: Efficient cluster optimization using extended compact genetic algorithm with seeded population. In *Workshop Proceedings of the Genetic and Evolutionary Computation Conference*, San Francisco, CA, USA (2001) 222–225

[26] Lipiński, P.: *Evolutionaty data-mining methods in discovering stock market expertise from financial time series.* PhD thesis, Université Louis Pasteur and University of Wroclaw, Strasbourg and Wroclaw (2004)

[27] Butz, M.V., Pelikan, M., Llora, X., Goldberg, D.E.: *Automated global structure extraction for effective local building block processing in XCS.* IlliGAL Report No. 2005011, Illinois Genetic Algorithms Laboratory, University of Illinois at Urbana-Champaign, Urbana, IL (2005)

[28] Lobo, F.G., Lima, C.F., Mártires, H.: Massive parallelization of the compact genetic algorithm. In Ribeiro, B., et al. (Eds.): *Adaptive and Natural Computing Algorithms.* Springer Computer Series, Springer, Berlin Heidelberg New York (2005) 530–533

[29] Pelikan, M., Goldberg, D.E., Cantú-Paz, E.: BOA: The Bayesian Optimization Algorithm. In Banzhaf, W., et al. (Eds.): *Proceedings of the*

Genetic and Evolutionary Computation Conference GECCO-99, Morgan Kaufmann, San Francisco, CA (1999) 525–532

[30] Pelikan, M.: *Hierarchical Bayesian optimization algorithm: Toward a new generation of evolutionary algorithms.* Springer, Berlin Heidelberg New York (2005)

4

Hierarchical Bayesian Optimization Algorithm

Martin Pelikan and David E. Goldberg

Summary. The hierarchical Bayesian optimization algorithm (hBOA) solves nearly decomposable and hierarchical optimization problems scalably by combining concepts from evolutionary computation, machine learning and statistics. Since many complex real-world systems are nearly decomposable and hierarchical, hBOA is expected to provide scalable solutions for many complex real-world problems. This chapter describes hBOA and its predecessor, the Bayesian optimization algorithm (BOA), and outlines some of the most important theoretical and empirical results in this line of research.

Key words: Hierarchical problem solving, Evolutionary computation, Bayesian networks, Optimization, Scalability, Decomposable problems

4.1 Introduction

The design and application of robust and scalable optimization techniques that are applicable to broad classes of important real-world problems is among the most important challenges of computational optimization. Estimation of distribution algorithm (EDAs) [3, 27, 36, 50] address this challenge by combining concepts from evolutionary computation, machine learning and statistics. The Bayesian optimization algorithm (BOA) [48] and the hierarchical BOA (hBOA) [45] are among the most advanced EDAs. BOA and hBOA combine Bayesian networks and evolutionary algorithms to solve nearly decomposable and hierarchical optimization problems in which candidate solutions can be represented by fixed-length discrete vectors.

It has been theoretically and empirically shown that BOA and hBOA can solve nearly decomposable and hierarchical problems scalably and reliably [43, 55]. A number of applications have been successfully tackled using BOA and hBOA, including the applications to military antenna design [64], telecommunication network design [63], Ising spin glasses [46] and nurse scheduling [29].

M. Pelikan and D.E. Goldberg: *Hierarchical Bayesian Optimization Algorithm*, Studies in Computational Intelligence (SCI) **33**, 63–90 (2006)
www.springerlink.com

This chapter describes BOA and hBOA and provides an overview of the most important theoretical and empirical results in this line of research. The chapter also presents a number of experimental results, which confirm that BOA and hBOA can solve difficult nearly decomposable and hierarchical problems scalably and reliably.

The chapter starts by describing the basic BOA procedure and discussing the methods for learning and sampling Bayesian networks in Sect. 11.2. Section 4.3 describes hBOA and the differences between BOA and hBOA. Section 4.4 presents several sets of experimental results and discusses the results in the context of the existing theory. Section 4.5 provides an overview of related work. Section 4.6 discusses important topics for future work on BOA and hBOA. Section 4.7 provides information on obtaining BOA and hBOA implementations. Finally, Sect. 4.8 summarizes and concludes the chapter.

4.2 Bayesian Optimization Algorithm

This section describes the Bayesian optimization algorithm (BOA) [48]. The section starts with an outline of the basic BOA procedure. Next, the section provides a brief introduction to learning and sampling Bayesian networks. Finally, the section discusses the scalability of BOA on decomposable problems of bounded difficulty.

4.2.1 Basic BOA Procedure

BOA works with a population of candidate solutions to the given problem. Candidate solutions are represented by fixed-length strings over a finite alphabet; usually, fixed-length binary strings are used. The first population is generated randomly according to the uniform distribution over all solutions.

Each iteration starts by selecting promising solutions from the original population using any standard selection method of evolutionary algorithms, such as tournament or truncation selection. For example, truncation selection selects the best half of the current population. Then, a Bayesian network is built for the selected solutions and the built network is sampled to generate new candidate solutions. New candidate solutions are then incorporated into the original population using a replacement strategy. A number of replacement strategies may be used; for example, the new candidate solutions can replace the entire original population. The next iteration is then executed unless some predefined termination criteria have been met. For example, the run can be terminated after a given number of iterations or when all bits have nearly converged. The pseudocode of BOA is shown in Fig. 4.1.

4.2.2 Bayesian Networks

Bayesian networks [24, 32] combine graph theory, probability theory and statistics to provide a flexible and practical tool for probabilistic modeling and

```
Bayesian optimization algorithm (BOA)
  t := 0;
  generate initial population P(0);
  while (not done) {
    select population of promising solutions S(t);
    build Bayesian network B(t) for S(t);
    sample B(t) to generate O(t);
    incorporate O(t) into P(t);
    t := t+1;
  };
```

Fig. 4.1. Pseudocode of the Bayesian optimization algorithm (BOA)

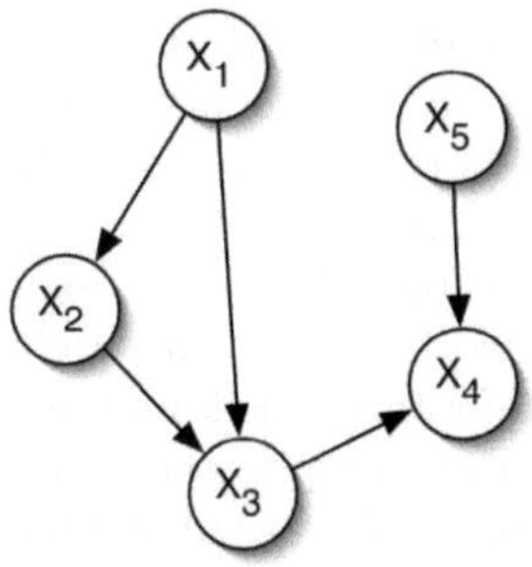

Fig. 4.2. A Bayesian network structure for five random variables X_1, X_2, X_3, X_4, X_5. In this network, for example, X_1 has no parents (i.e., $\Pi_1 = \emptyset$) whereas the set of parents of X_3 contains both X_1 and X_2 (i.e., $\Pi_3 = \{X_1, X_2\}$). To complete the definition of this network, a conditional probability table for each X_i with the condition Π_i would have to be specified

inference. BOA and hBOA use Bayesian networks to model promising solutions found so far and sample new candidate solutions. A Bayesian network consists of two components:

(1) *Structure*, which is defined by an acyclic directed graph with one node per variable and the edges corresponding to conditional dependencies between the variables (see Fig. 4.2 for example)
(2) *Parameters*, which consist of the conditional probabilities of each variable given the variables that this variable depends on

Mathematically, a Bayesian network with n nodes encodes a joint probability distribution of n random variables $X_1, X_2, \ldots, X_n$:

$$p(X_1, X_2, \ldots, X_n) = \prod_{i=1}^{n} p(X_i | \Pi_i), \tag{4.1}$$

where Π_i is the set of variables from which there exists an edge into X_i (members of Π_i are called parents of X_i).

In addition to encoding direct conditional dependencies, a Bayesian network may also encode a number of conditional independence assumptions. More specifically, a Bayesian network encodes the assumption that each variable X_i is conditionally independent of its predecessors in an ancestral ordering of variables given Π_i, where the ancestral ordering orders the variables so that variables in Π_i always precede X_i. Since Bayesian networks are acyclic, there always exists at least one ancestral ordering of the variables.

Like most other EDAs, BOA and hBOA consider each string position a random variable. Therefore, each string position is represented by one node in the Bayesian network and a candidate solution of n bits is an instantiation of n random variables. Ideally, learning a Bayesian network for the set of promising solutions in BOA should capture the structure of the underlying problem by identifying conditional dependencies and independencies between variables. Sampling the learned network exploits the identified dependencies and independencies by combining and juxtaposing partial solutions so that no important dependencies are broken. As is discussed in Sect. 4.2.5, if all or most independence assumptions are discovered and the problem is decomposable into subproblems of bounded order, exploration guided by building and sampling Bayesian networks significantly reduces problem dimensionality and ensures that the global optimum is found in a quadratic or subquadratic number of function evaluations despite the exponential number of potential solutions [15, 43, 55]. The quadratic or subquadratic growth of function evaluations until reliable convergence is retained even for problems where the number of local optima grows exponentially with problem size [15, 43].

The remainder of this section describes methods for learning and sampling Bayesian networks and the scalability of BOA on decomposable problems.

4.2.3 Learning Bayesian Networks from Data

For successful application of BOA, it is important that BOA can automatically learn a Bayesian network that encodes dependencies and independencies that properly decompose the problem. To learn an appropriate Bayesian network, BOA uses the set of promising solutions selected by the selection operator, which transforms dependencies between decision variables in the problem into statistical dependencies [43, 55]. There are two subtasks of learning a Bayesian network: (1) learn the structure and (2) learn the parameters.

Learning the parameters is straightforward. For each variable X_i with parents Π_i, we need to estimate the probabilities $p(X_i = x_i|\Pi_i = \pi_i)$ for all potential combinations of values x_i of X_i and π_i of Π_i. This can be done using the maximum likelihood estimator:

$$p(X_i = x_i|\Pi_i = \pi_i) = \frac{m(x_i, \pi_i)}{m(\pi_i)}, \tag{4.2}$$

where $m(x_i, \pi_i)$ denotes the number of instances with $X_i = x_i$ and $\Pi_i = \pi_i$, and $m(\pi_i)$ denotes the number of instances with $\Pi_i = \pi_i$.

Learning the structure is a much more difficult problem. Usually, the problem of learning the Bayesian network structure is split into two components: (1) a scoring metric and (2) a search procedure. The scoring metric provides a measure for evaluating network structures with respect to the given data D (the selected set of promising solutions). The search procedure searches the space of potential network structures to identify the structure with the best value of the scoring metric.

Two types of scoring metrics are commonly used to evaluate Bayesian network structures:

(1) Bayesian metrics [9, 22]
(2) Minimum description length metrics [59–61].

This work uses two basic scoring metrics: (1) The Bayesian information criterion (BIC) [67] and (2) the Bayesian–Dirichlet metric with likelihood equivalence (BDe) [9, 22].

BIC is a two-part minimum description length metric, where one part represents model accuracy, whereas the other part represents model complexity measured by the number of bits required to store model parameters. For simplicity, let us assume that the solutions are binary strings of fixed length n. BIC assigns the network structure B a score [67]

$$\mathrm{BIC}(B) = \sum_{i=1}^{n} \left(-H(X_i|\Pi_i)N - 2^{|\Pi_i|}\frac{\log_2(N)}{2} \right), \tag{4.3}$$

where $H(X_i|\Pi_i)$ is the conditional entropy of X_i given its parents Π_i; n is the number of variables; and N is the population size (the size of the training data set). The conditional entropy $H(X_i|\Pi_i)$ is given by

$$H(X_i|\Pi_i) = -\sum_{x_i,\pi_i} p(x_i,\pi_i)\log_2 p(x_i|\pi_i), \tag{4.4}$$

where $p(x_i,\pi_i)$ is the marginal probability of $X_i = x_i$ and $\Pi_i = \pi_i$; and $p(x_i|\pi_i)$ is the conditional probability of $X_i = x_i$ given that $\Pi_i = \pi_i$.

BDe is a Bayesian metric that approximates the marginal likelihood of the network structure given the data and is given by [9, 22]

$$\mathrm{BDe}(B) = p(B)\prod_{i=1}^{n}\prod_{\pi_i}\frac{\Gamma(m'(\pi_i))}{\Gamma(m'(\pi_i)+m(\pi_i))}\prod_{x_i}\frac{\Gamma(m'(x_i,\pi_i)+m(x_i,\pi_i))}{\Gamma(m'(x_i,\pi_i))}, \tag{4.5}$$

where $p(B)$ is the prior probability of the network structure B; the product over x_i runs over all instances of x_i (in the binary case these are 0 and 1); the product over π_i runs over all instances of the parents Π_i of X_i; $m(\pi_i)$ is the number of instances with the parents Π_i set to the particular values given by π_i; $m(x_i,\pi_i)$ is the number of instances with $X_i = x_i$ and $\Pi_i = \pi_i$; and $m'(\pi_i)$ and $m'(x_i,\pi_i)$ denote prior information about $m(\pi_i)$ and $m(x_i,\pi_i)$, respectively. Here, we consider K2 metric [9], which uses an uninformative prior that assigns $m'(x_i, \pi_i) = 1$ and $m'(\pi_i) = \sum_{x_i} m'(x_i, \pi_i)$.

According to our experience, Bayesian metrics tend to be too sensitive to noise in the data and often capture unnecessary dependencies. To avoid overly complex models, the space of network structures must usually be restricted by specifying a maximum order of interactions [22, 49]. On the other hand, minimum description length metrics favor simple models so that no unnecessary dependencies need to be considered. In fact, minimum description length metrics often result in overly simple models and require large populations to learn a model that captures all necessary dependencies.

Usually, a simple greedy search algorithm is used to construct a network that maximizes the scoring metric. The greedy algorithm starts with an empty network and in each iteration it applies a primitive graph operator to the current network that improves the network score the most. As primitive operators, edge additions, removals, and reversals are usually considered; nonetheless, according to our experience, edge removals and reversals do not significantly improve quality of the final model and that is why we only use edge additions. The network must remain acyclic and all operators leading to networks that violate this property are disallowed. The search is terminated whenever the current network cannot be improved anymore. In some cases, it is necessary to restrict the complexity of the network to contain dependencies of at most a specified order [22, 48]. In BOA the network from the previous generation can be used as a starting point for building the model in each generation as suggested by Etxeberria et al. [11]; this can significantly reduce the overall computational complexity of model building.

4.2.4 Sampling Bayesian Networks

Once the structure and parameters of a Bayesian network have been learned, BOA generates new candidate solutions by sampling the distribution encoded by the learned network [see (4.2)].

The sampling can be done using the probabilistic logic sampling of Bayesian networks [23], which proceeds in two steps. The first step computes an ancestral ordering of the nodes, where each node is preceded by its parents.

In the second step, the values of all variables of a new candidate solution are generated according to the computed ordering. Since the algorithm generates the variables according to the ancestral ordering, when the algorithm attempts to generate the value of each variable, the parents of the variable must have already been generated. Given the values of the parents of a variable, the distribution of the values of the variable is given by the corresponding conditional probabilities.

4.2.5 Scalability of BOA on Decomposable Problems

BOA can solve problems decomposable into subproblems of bounded order in a quadratic or subquadratic number of function evaluations [43, 55]. This can be shown by using facetwise theory, which originates in IlliGAL decomposition of genetic algorithm design [14–16]. In particular, the problem of estimating the number of evaluations until successful convergence is first split into the problem of determining:

(1) An adequate population size, N
(2) The number of generations, G, until the optimum is found.

For most standard settings, the overall number of evaluations is then equal to (or at least proportional to) $N \times G$.

The problem of determining an adequate population size can be decomposed into four facets [43, 50]:

(1) *Initial supply.* For each subproblem in an appropriate problem decomposition, the population size must be large enough to ensure that all instances of the particular subproblem have representatives in the initial population [15, 17, 24]. For an n-bit problem decomposable into subproblems of order k, the initial supply model leads to the population size that grows as $O(2^k(k + \log n))$ [15, 17].
(2) *Genetic drift.* In some decomposable problems, some subproblems have a stronger effect on solution quality than others. Then, solutions converge in several phases where in each phase only some subproblems provide distinguishable signal for selection. Nonetheless, even when a particular subproblem does not affect selection, some partial solutions corresponding to this subproblem may be lost due to the effects of stochastic operators of BOA (genetic drift) [2, 30, 62, 73], and the population size must be large enough to ensure that there is a sufficient initial supply for partial solutions that converge last. As a bounding case, problems with exponentially scaled subproblems can be considered, leading to the population-sizing bound of $O(n)$ [2, 30, 62, 73].
(3) *Decision making.* The quality of any partial solution of one of the subproblems is affected by the remaining solution and, therefore, the decision making between competing partial solutions is stochastic [24]. The population size must be large enough to ensure that the decision making propagates best partial solutions and eliminates the worst ones. To ensure that the expected number of incorrect bits is upper bounded by a constant, the decision making bound on the population size is $O(2^k\sqrt{n}\log n)$ [15, 21]. The above decision making bound decreases for problems where subproblems are not scaled uniformly because in such problems, the effective problem size becomes smaller.
(4) *Model building.* Without ensuring that the problem is properly decomposed, BOA cannot guarantee that the optimum will be found. The population size must be large enough to learn a probabilistic model that encodes a good decomposition of the problem [43, 55]. To ensure that an appropriate model is learned, the population size is bounded by $O(2^k n^{1.05})$ [43, 55]. Similarly as for decision making, model building becomes easier for nonuniformly scaled problems.

To estimate the number of generations until convergence, two bounding cases are considered:

(1) *Uniform scaling.* Here the influence on solution quality is approximately the same for all partial solutions and all subproblems converge in parallel (parallel convergence). In this case, the expected number of generations until convergence can be bounded by $O(\sqrt{n})$ [37, 72].
(2) *Exponential scaling.* Here the influence of subproblems is scaled so that at any generation, one subproblem overshadows the remaining ones and selection is driven by one or a few subproblems at a time (sequential convergence). In this case, the expected number of generations until convergence can be bounded by $O(n)$ [73].

Putting all facets of the population sizing and convergence theory together indicates that for decomposable problems of bounded difficulty the expected number of evaluations is from $O(n^{1.55})$ to $O(n^2)$, depending on the type of the problem. That means that BOA can solve decomposable problems of bounded difficulty in low-order polynomial time with respect to the number of evaluations until convergence to the global optimum.

Low-order polynomial solution to boundedly difficult decomposable problems is an important result because most other optimization algorithms fail to solve boundedly difficult decomposable problems scalably without prior information about adequate problem decomposition. For example, most standard genetic algorithms require exponential population sizes to solve some decomposable problems because standard variation operators, such as one-point and uniform crossover, are often misled even for decomposable problems of bounded difficulty [15, 71]. Additionally, it can be shown that hill climbers and other algorithms based on local search operators may require time proportional to $n^k \log n$ for many decomposable problems of bounded difficulty because of potential deception in signal from small perturbations of candidate solutions and the necessity of changing groups of k variables at a time [33]; of course, for large problems with moderate values of k, convergence in $\Theta(n^k \log n)$ evaluations is intractable.

4.3 Hierarchical BOA

A key feature of many complex systems that allows us to comprehend, analyze and build such systems is *hierarchy* [70]. By hierarchy, we mean a system composed of subsystems each of which is a hierarchy by itself until we reach some bottom level [70]. At each level of a hierarchy, interactions within each subsystem are of much higher magnitude than the interactions between the subsystems.

This section describes the hierarchical BOA (hBOA) [45, 47], which combines BOA with Simon's concept of hierarchical and nearly decomposable problems [70] to provide *scalable solutions for nearly decomposable and hierarchical problems*, including problems that cannot be decomposed on a single level.

4.3.1 Three Keys to Hierarchy Success

To solve difficult hierarchical problems, hBOA starts at the bottom level with subproblems of small order. At each level of optimization, hBOA uses promising solutions of the subproblems at the lower level as basic building blocks to juxtapose solutions at the current level. For each identified subproblem, several alternative promising solutions are preserved to serve as the basic building blocks at the next higher level.

Three important challenges must be considered for the design of robust and scalable solvers for difficult hierarchical problems [44, 45]:

(1) *Decomposition.* A competent hierarchical optimizer must be capable of decomposing the problem at each level properly by identifying most important interactions between the problem variables and modeling them appropriately.
(2) *Chunking.* A competent hierarchical optimizer must be capable of representing partial solutions at each level compactly to enable the algorithm to effectively process partial solutions of large order (this becomes most important for highest levels).
(3) *Diversity maintenance.* A competent hierarchical optimizer must be capable of effective diversity maintenance to preserve alternative partial solutions until it becomes clear which partial solutions may be eliminated.

To ensure *decomposition*, hBOA uses the same approach as BOA; it uses Bayesian networks built from the selected population of promising solutions. To ensure *chunking*, hBOA uses local structures to represent parameters of the learned Bayesian networks. Finally, to ensure *diversity maintenance*, hBOA uses restricted tournament replacement but other niching techniques can be used as well. Figure 4.3 shows the pseudocode of hBOA.

```
Hierarchical BOA (hBOA)
  t := 0;
  generate initial population P(0);
  while (not done) {
    select population of promising solutions S(t);
    build Bayesian network B(t) with local struct. for S(t);
    sample B(t) to generate offspring O(t);
    incorporate O(t) into P(t) using RTR yielding P(t+1);
    t := t+1;
  };
```

Fig. 4.3. Pseudocode of the hierarchical Bayesian optimization algorithm

The remainder of this section first describes *Bayesian networks with decision trees* used in hBOA for chunking. Next, *restricted tournament replacement* used for diversity maintenance in hBOA is described.

4.3.2 Bayesian Networks with Decision Trees for Chunking

To motivate the use of local structures in Bayesian Networks, let us assume that candidate solutions are represented by n-bit binary strings. There are 2^k independent probabilities $p(X_i|\Pi_i)$ for k parents, $|\Pi_i| = k$. Assuming that on higher levels of the hierarchy, hBOA must process partial solutions of moderate to large order, the number of independent probabilities will make the computation intractable if the parameters are represented as standard conditional probability tables. Clearly, the situation gets even worse for larger alphabets.

To make the learning of interactions tractable even for interactions of large order, local structures can be incorporated into BNs to enable a more efficient representation of conditional probabilities [12, 15]. This section describes *decision trees*, which are among the most powerful local structures.

In BNs with decision trees, the conditional probabilities $p(X_i|\Pi_i)$ for each variable X_i are encoded by a special decision tree T_i; for n variables, there are n decision trees. Each internal node of the decision tree T_i is labeled by a variable X_j where $j \neq i$. Children of a node labeled by X_j correspond to disjoint subsets of the potential values of X_j; for each value of X_j, there is one child corresponding to this value. Each traversal of a decision tree T_i for X_i thus corresponds to a constraint on the values of some other variables. Each leaf node of T_i then stores the probabilities of X_i given the constraint defined by the traversal of T_i that ends in this leaf.

For the binary alphabet, there are two children of any internal node (one child corresponds to a 0, whereas the other one corresponds to 1 and only one probability must be stored in each leaf because $p(X_i = 0|\Pi_i = \pi_i) + p(X_i = 1|\Pi_i = \pi_i) = 1$ for any instance π_i of Π_i.

To illustrate the use of decision trees in Bayesian networks, consider the example shown in Fig. 4.4, which considers a conditional probability table for X_1 where $\Pi_1 = \{X_2, X_3, X_4\}$ and all variables are binary. In this example, the full conditional probability table must contain eight values, whereas the decision tree can store the same information with only four parameters.

Decision trees allow a much more efficient representation of conditional probabilities, where in an extreme case, a dependency of order k may be represented by only $O(k)$ parameters, whereas for the full conditional probability table the number of parameters is at least $\Omega(2^k)$.

Analogically to standard Bayesian networks, the problem of learning Bayesian networks with decision trees can be split into two subproblems: (1) learn the structure and (2) learn the parameters.

To estimate the parameters of a Bayesian network with decision trees, hBOA uses the maximum likelihood estimate of the probabilities in the leaves

X_2	X_3	X_4	$p(X_1\|X_2, X_3, X_4)$
0	0	0	0.75
0	0	1	0.25
0	0	0	0.25
0	0	1	0.25
1	1	0	0.20
1	1	1	0.20
1	1	0	0.20
1	1	1	0.20

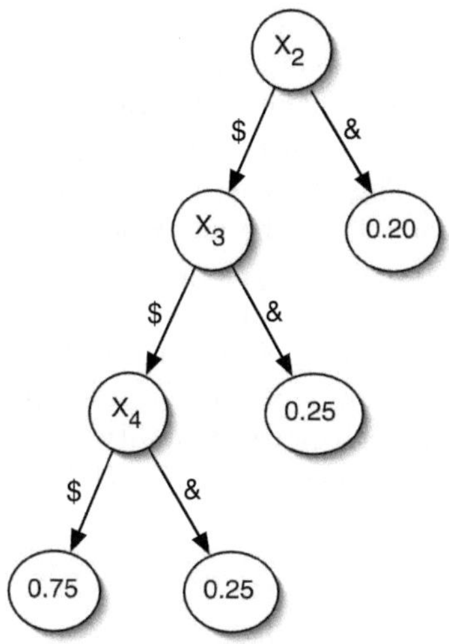

Fig. 4.4. A conditional probability table for $p(X_1|X_2, X_3, X_4)$ and a decision tree that reduces the number of parameters from 8 to 4

of all decision trees. Consider a decision tree T_i for X_i and a leaf in this decision tree that specifies a condition C (based on the traversal). Then, the maximum likelihood estimate of $p(X_i = x_i|C)$ where x_i is a potential value of X_i, is given by

$$p(X_i = x_i|C) = \frac{m(X_i = x_i, C)}{m(C)}, \tag{4.6}$$

where $m(X_i = x_i, C)$ denotes the number of instances with $X_i = x_i$ that satisfy C, and $m(C)$ denotes the number of instances that satisfy C.

To measure quality of competing network structures, scoring metrics for standard Bayesian networks can be adapted to Bayesian networks with decision trees [12, 15]. For example, the BDe score for a Bayesian network B with decision trees can be computed as [15]

$$\text{BDe}(B) = \prod_{i=1}^{n} \prod_{l \in L_i} \frac{\Gamma(m_i'(l))}{\Gamma(m_i(l) + m_i'(l))} \prod_{x_i} \frac{\Gamma(m_i(x_i, l) + m_i'(x_i, l))}{\Gamma(m_i'(x_i, l))}, \tag{4.7}$$

where L_i is the set of leaves in the decision tree T_i for X_i; $m_i(l)$ is the number of instances in D which end up the traversal through the tree T_i in the leaf l; $m_i(x_i, l)$ is the number of instances that have $X_i = x_i$ and end up the traversal of the tree T_i in the leaf l; $m_i'(l)$ represents the prior knowledge about the value of $m_i(i, l)$; and $m_i'(x_i, l)$ represents the prior knowledge about the value of $m_i(x_i, l)$. Again, an uninformative prior $m_i'(x_i, l) = 1$ is used in the K2 variant of the BDe metric for Bayesian networks with decision trees.

As mentioned earlier, Bayesian metrics tend to be more sensitive to the noise in data and, in practice, they often lead to overly complex models. To bias the learning of Bayesian networks with decision trees toward simple models, we adjust the prior probability of each network to favor simpler models. This is done by first computing the description length of the parameters required by the network. One frequency in the data set of size N can be encoded using $\log_2 N$ bits; however, only half of the bits suffice to encode the frequencies with sufficient accuracy [13]. Therefore, to encode all parameters, $0.5(\sum_i |L_i|) \log_2 N$ bits are needed, where $\sum_i |L_i|$ is the total number of leaves in all decision trees. To favor simpler networks to the more complex ones, we decrease the prior probability of each network exponentially fast with respect to the description length of this network's parameters [12]

$$p(B) = c2^{-0.5(\sum_i |L_i|) \log_2 N}, \tag{4.8}$$

where c is a normalization constant required for the prior probabilities of all possible network structures to sum to one. The normalization constant does not affect the result because we are interested in only relative quality of networks and not the absolute value of their marginal likelihood.

To learn the structure of a Bayesian network with decision trees, a simple greedy algorithm is used [15, 22]. The greedy algorithm starts with an empty network, which is represented by single-node decision trees. Each iteration splits one leaf of any decision tree that improves the score of the network most until no more improvement is possible. For more details on learning BNs with decision trees, see [12, 15, 43].

The sampling of a Bayesian network with decision trees can be done using probabilistic logic sampling [23] analogically to the case with standard Bayesian networks in BOA.

For more information on using local structures in BNs, please see [12, 15].

4.3.3 Restricted Tournaments for Diversity Maintenance

A number of niching techniques exist that could be applied to maintain useful diversity in hBOA. Here we use restricted tournament replacement (RTR) [20], which incorporates the newly sampled candidate solutions one by one, using the following three-step procedure for each new solution X:

1. Randomly select a subset W of w candidate solutions from the original population
2. Let Y be a solution from W that is most similar to X (based on genotypic distance)
3. Replace Y with X if X is better; otherwise, discard X

A robust rule of thumb is to set $w = \min\{n, N/20\}$, where n is the number of decision variables in the problem and N is the population size [43].

Although RTR does not ensure that the population will be divided among the niches evenly, it provides a fairly robust and powerful nicher, which appears to be well-suited for typical hBOA applications.

4.3.4 Scalability of hBOA on Hierarchical Problems

The convergence of hBOA on hierarchical problems proceeds sequentially from the bottom to the top level. At each level, the correct building blocks at this level must be discovered and their competitors must be eliminated. The number of evaluations required by hBOA to discover the correct building blocks at each level can be upper-bounded by the overall number of fitness evaluations required to solve the problem at the current level only.

Assuming that the subproblems at each level are of similar order and difficulty, BOA scalability theory indicates that at each level l the number of evaluations can be upper bounded by $O(n_l^{1.55})$, where n_l is the number of subproblems from the lower level, which serve as the basic building blocks at the current level.

Assuming that the number of levels changes with problem size according to a function $L(n)$, the resulting bound is $O(n^{1.55}L(n))$, where $L(n)$ denotes the number of levels in an n-bit problem.

The following section analyzes performance of BOA and hBOA by presenting a number of experimental results.

4.4 Experiments

This section presents experimental results for BOA and hBOA. First, test problems are described and their difficulty is discussed. Next, the experimental approach is discussed. Finally, the results are presented and discussed.

4.4.1 Test Problems

Several artificial classes of problems were used to test BOA and hBOA on the boundary of their design envelopes. The problems were designed so that if BOA and hBOA cannot solve nearly decomposable and hierarchical problems, their computational complexity will fail to scale up polynomially with problem size.

To test BOA, three test problems for fixed-length binary strings were considered. Two of these problems – dec-3 and trap-5 – are uniformly scaled decomposable problems of bounded difficulty for which it is necessary that BOA finds an adequate problem decomposition; if an adequate decomposition is not found, the algorithm is expected to scale up exponentially [71]. Consequently, solving dec-3 and trap-5 scalably indicates that BOA can learn an adequate problem decomposition and that it should be capable of solving other decomposable problems of bounded difficulty and anything easier [15]. The last test problem for BOA is expdec-3, which represents an exponentially scaled decomposable problem of bounded difficulty based on dec-3. A detailed description of the three test problems for BOA follows:

(1) *Dec-3: Concatenated deceptive of order 3.* In dec-3 [10], the input string is first partitioned into independent groups of 3 bits each. This partitioning is unknown to the algorithm, and it does not change during the run. A 3-bit deceptive function is applied to each group of 3 bits and the contributions of all deceptive functions are added together to form the fitness. Each 3-bit deceptive function is defined as follows:

$$\mathrm{dec}(u) = \begin{cases} 1 & \text{if } u = 3, \\ 0 & \text{if } u = 2, \\ 0.8 & \text{if } u = 1, \\ 0.9 & \text{if } u = 0, \end{cases} \tag{4.9}$$

where u is the number of ones in the input string of 3 bits. An n-bit dec-3 function has one global optimum in the string of all ones and $2^{n/3} - 1$ other local optima. To solve dec-3, it is necessary to consider interactions among the positions in each partition because when each bit is considered independently, the optimization is misled away from the optimum [5, 48, 71].

(2) *Trap-5: Concatenated trap of order 5.* Trap-5 can be defined analogically to dec-3, but instead of 3-bit groups, 5-bit groups are considered. The contribution of each group of 5 bits is computed as

$$\mathrm{trap}_5(u) = \begin{cases} 5 & \text{if } u = 5, \\ 4 - u & \text{otherwise}, \end{cases} \tag{4.10}$$

where u is the number of ones in the input string of 5 bits. An n-bit trap-5 has one global optimum in the string of all ones and $2^{n/5} - 1$ other local optima. Trap-5 also necessitates that all bits in each group are treated together, because statistics of lower order are misleading.

(3) *Expdec-3: Exponentially scaled concatenated deceptive of order 3.* Dec-3 and trap-5 are uniformly scaled decomposable problems where all partitions of the decomposition converge simultaneously. Expdec-3 modifies dec-3 to investigate the behavior of BOA on exponentially scaled problems, where the partitions converge sequentially (domino convergence) and selection focuses on only one or a few partitions at a time [73]. Specifically, expdec-3 scales the partitions so that the signal decreases exponentially in a specified sequence of partitions. To ensure that a particular partition in the sequence overshadows the contributions of all the subsequent partitions in the sequence, it is sufficient to multiply the ith partition by c^i, where c satisfies $0.1c^{\frac{n}{3}} > \frac{n}{3} - 1$. The inequality restricts c so that the smallest signal coming from ith partition is greater than

the sum of the maximum signals coming from the remaining partitions on positions 1 to $(i-1)$ in the sequence. Any constant c that satisfies the above inequality can be used without affecting performance of BOA. Like dec-3, expdec-3 has one global optimum in the string of all ones and $2^{n/3}-1$ local optima. Nonetheless, here the model does not have to consider all interactions, but only those that correspond to subproblems that are in the process of convergence at the particular iteration.

To test hBOA, two difficult hierarchical test problems – HIFF and hTrap – for fixed-length binary strings were considered. Both problems test whether hBOA is capable to ensure the three keys to hierarchy success to solve difficult hierarchical problems scalably. If hBOA fails to solve HIFF and hTrap scalably, it can be expected to fail on many other difficult hierarchical problems. On the other hand, if hBOA is able to identify appropriate decomposition at each level, represent the identified decomposition efficiently, and maintain useful diversity, hBOA can be expected to scale up with a low-order polynomial also for other hierarchically decomposable problems. A description of the two functions follows:

(1) *HIFF: Hierarchical if-and-only-if.* Hierarchical if-and-only-if (HIFF) [74] represents a problem that is intractable using single-level decomposition, but that can be efficiently solved using a competent hierarchical optimizer. In HIFF, the string length must be an integer power of 2, that is, $n = 2^l$ where l is the number of levels.
The lowest level consists of the bits in the input string. At each level, consequent pairs of bits contribute to the objective function using if-and-only-if (iff), which returns 1 for inputs 00 and 11, and returns 0 otherwise. The pairs of bits are then mapped to the next level; 00 is mapped to 0, 11 is mapped to 1, and everything else is mapped to the null symbol '-'. The mapping and evaluation of pairs of bits then continues from the second level until the top level, which consists of a single bit and is not evaluated. To make the overall contribution at each level of the same magnitude, the contributions of pairs of bits on ith level from the bottom are multiplied by 2^i.
Similarly as for test problems for BOA, the pairs of bits that contribute to the overall fitness and are mapped to the next level can be chosen arbitrarily but they must be fixed during the entire run. The further the coupled string positions are, the more difficult HIFF becomes for standard recombination operators that consider only interactions in tightly coded partitions.
HIFF has two global maxima, one in the string of all ones and one in the string of all zeros. HIFF does not have any other strict maxima. Nonetheless, HIFF contains large-order interactions that make it difficult to juxtapose either global optimum and cannot be effectively decomposed on a single level. For more details on HIFF, see [74].

(2) *hTrap: Hierarchical trap.* Like HIFF, hierarchical traps (hTraps) [42] cannot be tractably solved using single-level decomposition, but can be efficiently solved using a competent hierarchical optimizer. hTraps are created by combining trap functions of order 3 over multiple levels of difficulty. For hTraps, the string length should be an integer power of 3, that is, $n = 3^l$ where l is the number of levels.

In the variant of hTrap used in this work, on the lowest level, groups of 3 bits contribute to the overall fitness using a generalized 3-bit trap

$$\text{trap}_3(u) = \begin{cases} f_{\text{high}} & \text{if } u = 3, \\ f_{\text{low}} - u\frac{f_{\text{low}}}{2} & \text{otherwise}, \end{cases} \tag{4.11}$$

where $f_{\text{high}} = 1$ and $f_{\text{low}} = 1 + 0.1/l$. Note that for these values of f_{high} and f_{low}, the optimum of the trap is 000.

Each group of 3 bits corresponding to one of the traps is then mapped to a single symbol on the next level; a 000 is mapped to a 0, a 111 is mapped to a 1, and everything else is mapped to the null symbol '-'. The bits on the next level again contribute to the overall fitness using 3-bit traps defined above [see (4.11)], and the groups are mapped to an even higher level. This continues until the top level is evaluated that contains 3 bits total. However, on the top level, a trap with $f_{\text{high}} = 1$ and $f_{\text{low}} = 0.9$ is applied. As a result, the optimum of hTrap is in the string of all ones despite the superior performance of blocks of zeros on any level except for the top one. Any group of bits containing the null symbol does not contribute to the overall fitness.

To make the overall contribution at each level of the same magnitude, the contributions of traps on ith level from the bottom are multiplied by 3^i.

hTraps have many local optima, but only one global optimum in the string of all ones. Nonetheless, any single-level decomposition into subproblems of bounded order will lead away from the global optimum. That is why hTraps necessitate an optimizer that can build solutions hierarchically by juxtaposing good partial solutions over multiple levels of difficulty until the global optimum if found.

For more details on hTraps, see [42].

4.4.2 Description of Experiments

To study scalability, we consider a range of problem sizes for each test problem. The results are then used to investigate the growth of the number of function evaluations until successful convergence to the global optimum with respect to the problem size.

For each problem and problem size, bisection is ran to determine the minimum population size to ensure reliable convergence to the global optimum in 30 out of 30 independent runs [50]. Binary tournament selection is used to select promising solutions. In BOA, the population of new candidate solutions is half the size of the original population and the worst individuals in the original population get replaced by new solutions. In hBOA, the population of new solutions has the same size as the original population, but RTR (where $w = \min\{n, N/20\}$) is used to incorporate new solutions. BOA use Bayesian networks with full conditional probability tables to model and sample candidate solutions, whereas hBOA uses Bayesian networks with decision trees. To construct the model, BOA uses BIC metric [67], whereas hBOA uses the BDe metric for decision trees [15] modified by subtracting a complexity penalty term described in Sect. 4.3.2 [43, 51]. A greedy network construction algorithm is used in all cases.

Performance of BOA on trap-5 is compared to that of a simple genetic algorithm with uniform crossover. The simple genetic algorithm uses binary tournament selection to select promising solutions. The probability of crossover is $p_c = 0.6$. To focus on the effects of selectorecombinative search, we use no mutation.

A simple hill climber is also compared to BOA on trap-5. The hill climber starts with a random binary string. In each iteration, it applies bit flip mutation to the current binary string where each bit is flipped with a small probability p_m. If the modified solution outperforms the current one, the new solution replaces the old one. The run is terminated when the optimum is found. We use $p_m = 5/n$, which is the optimal flipping probability for trap-5 [30].

4.4.3 Results

Figure 4.5 shows the number of evaluations until BOA finds the optimum of dec-3 of $n = 60$ to $n = 240$ bits. The figure also compares the performance of BOA with the dimensional scalability model discussed earlier, which estimates the growth of the number of evaluations as $O(n^{1.55})$. Analogical results are shown in Fig. 4.6, which considers BOA for trap-5 ($n = 100$ to $n = 250$).

Figure 4.7 compares BOA performance on trap-5 with that of the simple genetic algorithm (sGA) with uniform crossover and the hill climber (HC) with bit-flip mutation.

Figure 4.8 shows the number of evaluations until BOA finds the optimum of expdec-3 of $n = 60$ to $n = 210$ bits. The figure also compares the performance of BOA with the dimensional scalability model, which estimates the growth of the number of evaluations for expdec-3 as $O(n^2)$.

Figure 4.9 shows the number of evaluations until hBOA finds the optimum of HIFF of $n = 16$ to $n = 512$ bits. Figure 4.10 shows the number of evaluations until hBOA finds the optimum of hTrap of $n = 27$ to $n = 243$ bits. In both cases, the results are compared to the theoretical model, which approximates the growth of the number of evaluations as $O(n^{1.55} \log n)$, because for HIFF and hTrap the number of levels grows proportionally to $\log n$.

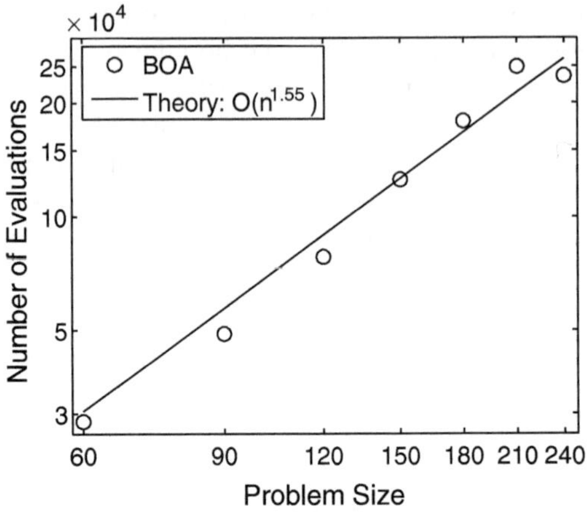

Fig. 4.5. Scalability of BOA on dec-3

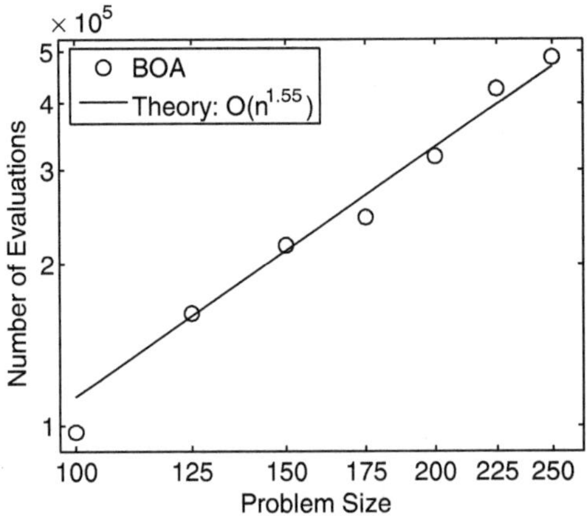

Fig. 4.6. Scalability of BOA on trap-5

4.4.4 Discussion

The results confirm that BOA and hBOA can solve test problems in a quadratic or subquadratic number of evaluations with respect to the number of decision variables, as is suggested by the scalability theory for BOA and hBOA.

For dec-3 and trap-5 (see Figs. 4.5 and 4.6), the number of evaluations required by BOA is expected to grow proportionally to $n^{1.55}$ or faster, and the results confirm this result. For expdec-3 (see Fig. 4.8), the number of evaluations is expected to grow proportionally to n^2, and the results again confirm this result. The comparison of BOA with the hill climbing and the simple genetic algorithm (see Fig. 4.7) confirms that BOA can solve decomposable problems with a significantly lower asymptotic complexity.

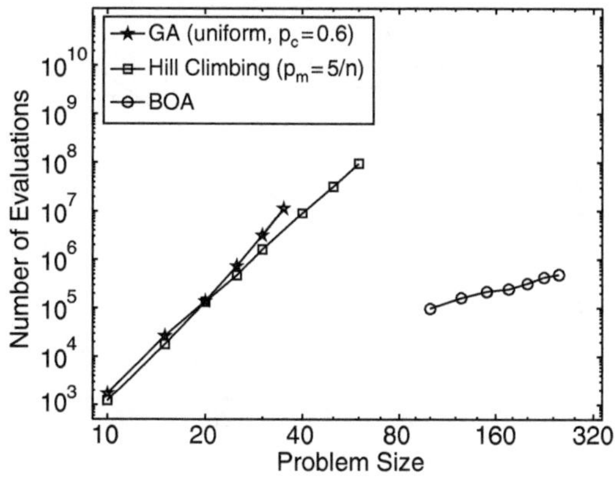

Fig. 4.7. BOA vs. HC and sGA on trap-5

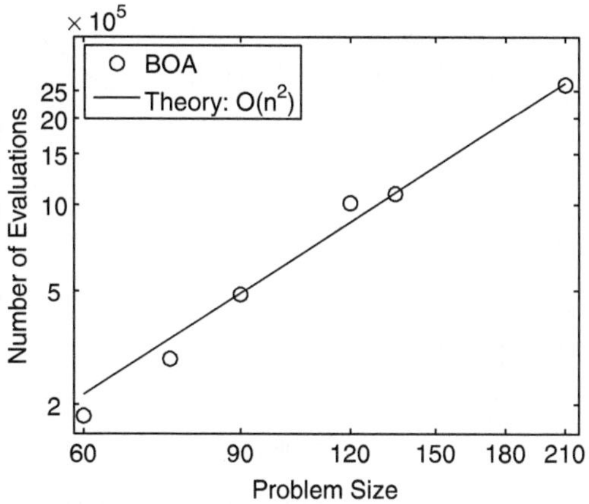

Fig. 4.8. Scalability of BOA on expdec-3

For HIFF and hTrap (see Figs. 4.9 and 4.10), the number of levels grows as a logarithm of problem size. The number of evaluations is thus expected to grow proportionally to $n^{1.55} \log n$, which is confirmed by the results.

4.5 Related Work

The most important recent developments in the research on BOA and hBOA can be classified as follows:

(1) *Efficiency enhancement.* A number of efficiency enhancement techniques were proposed for BOA and hBOA with the goal of speeding up these two algorithms for large problems. Endogenous fitness modeling can be

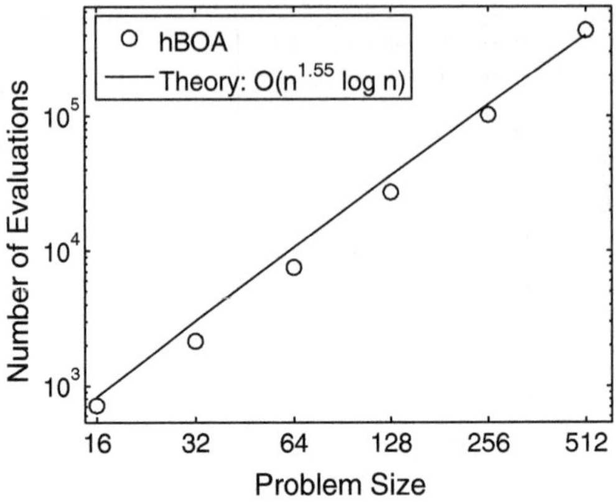

Fig. 4.9. Scalability of hBOA on HIFF

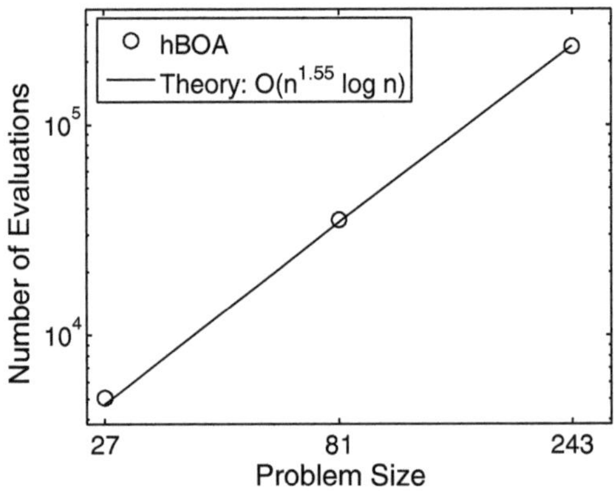

Fig. 4.10. Scalability of hBOA on hTrap

used to speed up fitness evaluation for problems where fitness evaluation is the bottleneck [54, 66]. Specialized local searchers can be incorporated to speed up BOA and hBOA [35, 46, 53]. Prior problem-specific knowledge of various forms can be used to increase efficiency [4, 35, 65, 69]. Sporadic model building can be used when model building is the bottleneck [57]. Parallel computers can be used to distribute both fitness evaluation [7] and model building [38, 40].

(2) *Parameter elimination.* BOA and hBOA were combined with the parameter-less genetic algorithm [19] to eliminate the need for setting an adequate population size manually [52]; consequently, BOA and hBOA do not require the user to set any parameters and still scale up with a low-order polynomial [52].

(3) *Multiobjective optimization.* Several multiobjective variants of BOA, hBOA and mixed BOA (mBOA) were proposed [26, 28, 56] and shown to be capable of solving difficult multiobjective decomposable problems.
(4) *Applications.* BOA and hBOA were successfully applied in a number of interesting application; here we mention a few of them. Santarelli et al. [64] applied hBOA to military antenna design; they showed that hBOA outperforms standard evolutionary algorithms and provides high quality solutions in the design of a complex, constrained feed network for an antenna array. Li and Aickelin [29] applied BOA to nurse scheduling, where the task was to assign nurses to shifts to satisfy a number of constraints; the results indicated that BOA outperformed human schedulers. Pelikan and Goldberg [46] applied hBOA to Ising spin glass in two and three dimensions and maximum satisfiability (MAXSAT), and showed that hBOA is capable of providing robust performance across a spectrum of difficult problem instances. Rothlauf et al. [63] applied BOA to telecommunication network design, and showed that BOA is able to provide high quality results even for poor problem encodings of telecommunication problems. Mühlenbein and Mahnig [35] and Schwarz and Ocenasek [68] applied BOA and LFDA to graph bipartitioning and showed that BOA can solve large problem instances competitively. Finally, Arst [1] applied BOA to groundwater remediation design; the results indicated that BOA can provide high-quality solutions 30–2,400% faster than the simple genetic algorithm and the extended compact genetic algorithm (ECGA) [18], which is another EDA that uses multivariate models.
(5) *Extension of BOA and hBOA to other representations.* Mixed BOA (mBOA) [39] was proposed as an extension of BOA with decision graphs; mBOA can learn and sample distributions over real-valued parameters. BOA was extended to the domain of genetic programming [31]. Finally, BOA was combined with adaptive mutation of evolution strategies and clustering to tackle real-valued problems [58].
(6) *Combination with XCS learning classifier system.* XCS [75] is an online accuracy-based classifier system that evolves a distributed problem solution represented by a population of rules. To combine local classifiers relevant to the current problem instance, XCS draws inspiration from simple genetic algorithms and uses standard crossover and mutation operators on pairs of relevant classifiers. Butz et al. [6] incorporated BOA into XCS to enable XCS to learn and process complex features effectively. The proposed method builds a Bayesian network with decision trees for the population of rules at certain intervals. To recombine local classifiers, model parameters are updated to match the current set of relevant classifiers and the model is sampled to generate new classifiers. The proposed hybrid was shown to significantly outperform standard XCS on a number of difficult hierarchical classification problems [6].

4.6 Future Work

There are three important areas of future work on BOA and hBOA:

- *Extension of BOA and hBOA to other representations.* Real-valued vectors, variable-length vectors, and program codes are among the most prominent representations to tackle. Several approaches to bridging BOA and hBOA with other representations were proposed [31, 32, 39, 58]; nonetheless, this line of research still represents one of the most important challenges in research on BOA, hBOA and other EDAs.
- *Efficiency enhancement of BOA and hBOA.* A number of efficiency enhancement techniques exist for BOA and hBOA [4, 35, 46, 53, 54, 57, 65, 66, 69], most of which are presented in other chapters of this book. Nonetheless, finding new ways for improving performance of BOA and hBOA represents an important challenge for future research in this area.
- *Applications.* New applications are another important component of future work in this area. BOA and hBOA are broadly capable optimizers and they should provide an important tool for practitioners across a broad range of disciplines.

4.7 How to Obtain BOA and hBOA

The source code of BOA and BOA with decision graphs can be found at `http://www-illigal.ge.uiuc.edu/sourcecd.html`. For information on obtaining hBOA, visit `http://www-hboa.ge.uiuc.edu/`.

4.8 Summary and Conclusions

This chapter presented, discussed and tested a competent hierarchical optimizer called the *hierarchical Bayesian optimization algorithm (hBOA)*, which solves nearly decomposable and hierarchical problems scalably and reliably with only little problem-specific knowledge. The chapter also presented the *Bayesian optimization algorithm (BOA)*, which is applicable to boundedly difficult problems decomposable on a single level.

Performance of BOA and hBOA can be further improved using a number of efficiency enhancement techniques that address both model building as well as objective-function evaluation. Prior knowledge exploitation represents one class of efficiency enhancement techniques and is particularly interesting in this setting because of the large potential of BOA and hBOA for incorporating such knowledge. BOA and hBOA were extended to tackle multiobjective optimization problems and discover a tradeoff between the objectives for decomposable multiobjective problems. BOA was hybridized with an XCS learning classifier system to provide XCS with a practical technique for discovering and combining complex features. Parameter-less variants of BOA and hBOA were proposed to eliminate the need for setting parameters manually. BOA and hBOA were successfully applied to a number of applications from spin glasses to nurse scheduling and military antenna design, confirming that BOA and hBOA are robust and scalable optimization algorithms applicable to a broad range of real-world problems.

Many complex real-world systems are nearly decomposable and hierarchical. hBOA combines concepts from genetic and evolutionary computation and machine learning to solve difficult nearly decomposable and hierarchical problems. That is why we expect hBOA and its modifications and extensions to significantly advance the area of optimization of difficult real-world problems and to provide practical solutions for difficult optimization problems across a broad range of disciplines of science, engineering and commerce.

Acknowledgments

This work was sponsored by the National Science Foundation under NSF CAREER grant ECS-0547013, the Air Force Office of Scientific Research, Air Force Materiel Command, USAF, under grant F49620-03-1-0129, the National Science Foundation under ITR grant DMR-99-76550 (at Materials Computation Center), and ITR grant DMR-0121695 (at CPSD), and the Department of Energy under grant DEFG02-91ER45439 (at Fredrick Seitz MRL). The US Government is authorized to reproduce and distribute reprints for government purposes notwithstanding any copyright notation thereon.

This work was also partially supported by the Research Award and the Research Board at the University of Missouri. Most experiments were done using the hBOA software developed by Martin Pelikan and David E. Goldberg at the University of Illinois at Urbana-Champaign.

References

[1] Arst, R. E. (2002). *Which are better, probabilistic model-building genetic algorithms (PMBGAs) or simple genetic algorithms (sGAs)? A comparison for an optimal groundwater remediation design problem.* PhD thesis, University of Illinois at Urbana-Champaign, Department of Civil Engineering, Urbana, IL

[2] Asoh, H., and Mühlenbein, H. (1994). On the mean convergence time of evolutionary algorithms without selection and mutation. *Parallel Problem Solving from Nature*, pp. 88–97

[3] Baluja, S. (1994). *Population-based incremental learning: A method for integrating genetic search based function optimization and competitive learning.* Tech. Rep. No. CMU-CS-94-163, Carnegie Mellon University, Pittsburgh, PA

[4] Baluja, S. (2002). Using a priori knowledge to create probabilistic models for optimization. *International Journal of Approximate Reasoning*, 31(3):193–220
[5] Bosman, P. A. N., and Thierens, D. (1999). Linkage information processing in distribution estimation algorithms. *Proceedings of the Genetic and Evolutionary Computation Conference (GECCO-99)*, I:60–67
[6] Butz, M. V., Pelikan, M., Llorà, X., and Goldberg, D. E. (2005). Extracted global structure makes local building block processing effective in XCS. *Proceedings of the Genetic and Evolutionary Computation Conference (GECCO-2005)*, 1:655–662
[7] Cantú-Paz, E. (2000). *Efficient and Accurate Parallel Genetic Algorithms.* Kluwer, Boston, MA
[15] Chickering, D. M., Heckerman, D., and Meek, C. (1997). A Bayesian approach to learning Bayesian networks with local structure. Technical Report MSR-TR-97-07, Microsoft Research, Redmond, WA
[9] Cooper, G. F., and Herskovits, E. H. (1992). A Bayesian method for the induction of probabilistic networks from data. *Machine Learning*, 9:309–347
[10] Deb, K., and Goldberg, D. E. (1994). Sufficient conditions for deceptive and easy binary functions. *Annals of Mathematics and Artificial Intelligence*, 10:385–408
[11] Etxeberria, R., and Larrañaga, P. (1999). Global optimization using Bayesian networks. In Rodriguez, A. A. O., Ortiz, M. R. S., and Hermida, R. S., (Eds.), *Second Symposium on Artificial Intelligence (CIMAF-99)*, pp. 332–339, Habana, Cuba. Institude of Cybernetics, Mathematics, and Physics and Ministry of Science, Technology and Environment
[12] Friedman, N., and Goldszmidt, M. (1999). Learning Bayesian networks with local structure. In Jordan, M. I., (Ed.), *Graphical models*, pp. 421–459. MIT, Cambridge, MA
[13] Friedman, N., and Yakhini, Z. (1996). On the sample complexity of learning Bayesian networks. *Proceedings of the Conference on Uncertainty in Artificial Intelligence (UAI-96)*, pp. 274–282
[14] Goldberg, D. E. (1994). First flights at genetic-algorithm Kitty Hawk. IlliGAL Report No. 94008, University of Illinois at Urbana-Champaign, Urbana, IL
[15] Goldberg, D. E. (2002). *The design of innovation: Lessons from and for competent genetic algorithms*, volume 7 of *Genetic Algorithms and Evolutionary Computation.* Kluwer, Boston, MA
[16] Goldberg, D. E., Deb, K., and Clark, J. H. (1992). Genetic algorithms, noise, and the sizing of populations. *Complex Systems*, 6:333–362
[17] Goldberg, D. E., Sastry, K., and Latoza, T. (2001). On the supply of building blocks. *Proceedings of the Genetic and Evolutionary Computation Conference (GECCO-2001)*, pp. 336–342
[18] Harik, G. (1999). Linkage learning via probabilistic modeling in the ECGA. IlliGAL Report No. 99010, University of Illinois at Urbana-Champaign, Illinois Genetic Algorithms Laboratory, Urbana, IL

[19] Harik, G., and Lobo, F. (1999). A parameter-less genetic algorithm. *Proceedings of the Genetic and Evolutionary Computation Conference (GECCO-99)*, I:258–265

[20] Harik, G. R. (1995). Finding multimodal solutions using restricted tournament selection. *Proceedings of the International Conference on Genetic Algorithms (ICGA-95)*, pp. 24–31

[21] Harik, G. R., Cantú-Paz, E., Goldberg, D. E., and Miller, B. L. (1997). The gambler's ruin problem, genetic algorithms, and the sizing of populations. *Proceedings of the International Conference on Evolutionary Computation (ICEC-97)*, pp. 7–12. Also IlliGAL Report No. 96004

[22] Heckerman, D., Geiger, D., and Chickering, D. M. (1994). Learning Bayesian networks: The combination of knowledge and statistical data. Technical Report MSR-TR-94-09, Microsoft Research, Redmond, WA

[23] Henrion, M. (1988). Propagating uncertainty in Bayesian networks by probabilistic logic sampling. In Lemmer, J. F., and Kanal, L. N., (Eds.), *Uncertainty in Artificial Intelligence*, pp. 149–163. Elsevier, Amsterdam London New York

[24] Holland, J. H. (1975). *Adaptation in natural and artificial systems.* University of Michigan, Ann Arbor, MI

[24] Howard, R. A., and Matheson, J. E. (1981). Influence diagrams. In Howard, R. A., and Matheson, J. E., (Eds.), *Readings on the principles and applications of decision analysis*, volume II, pp. 721–762. Strategic Decisions Group, Menlo Park, CA

[26] Khan, N. (2003). Bayesian optimization algorithms for multiobjective and hierarchically difficult problems. Master's thesis, University of Illinois at Urbana-Champaign, Urbana, IL

[27] Larrañaga, P., and Lozano, J. A., (Eds.) (2002). *Estimation of Distribution Algorithms: A New Tool for Evolutionary Computation.* Kluwer, Boston, MA

[28] Laumanns, M., and Ocenasek, J. (2002). Bayesian optimization algorithms for multi-objective optimization. *Parallel Problem Solving from Nature*, pp. 298–307

[29] Li, J., and Aickelin, U. (2003). A Bayesian optimization algorithm for the nurse scheduling problem. *Proceedings of the IEEE Congress on Evolutionary Computation 2003 (CEC-2003)*, pp. 2149–2156

[30] Lobo, F. G., Goldberg, D. E., and Pelikan, M. (2000). Time complexity of genetic algorithms on exponentially scaled problems. *Proceedings of the Genetic and Evolutionary Computation Conference (GECCO-2000)*, pp. 151–158. Also IlliGAL Report No. 2000016

[31] Looks, M., Goertzel, B., and Pennachin, C. (2005). Learning computer programs with the Bayesian optimization algorithm. *Proceedings of the Genetic and Evolutionary Computation Conference (GECCO-2005)*, 1:747–748

[32] Looks, M., Groetzel, B., and Pennachin, C. (2003). *Learning computer programs with the Bayesian optimization algorithm.* Technical report, Object Sciences Corporation and Novamente LLC

[33] Mühlenbein, H. (1992a). How genetic algorithms really work: I.Mutation and Hillclimbing. *Parallel Problem Solving from Nature*, pp. 15–25

[30] Mühlenbein, H. (1992b). How genetic algorithms really work: I.Mutation and Hillclimbing. In Männer, R., and Manderick, B., (Eds.), *Parallel Problem Solving from Nature*, pp. 15–25, Elsevier, Amsterdam

[35] Mühlenbein, H., and Mahnig, T. (2002). Evolutionary optimization and the estimation of search distributions with aplication to graph bipartitioning. *International Journal of Approximate Reasoning*, 31(3):157–192

[36] Mühlenbein, H., and Paass, G. (1996). From recombination of genes to the estimation of distributions I. Binary parameters. In Eiben, A., Bäck, T., Shoenauer, M., and Schwefel, H., (Eds.), *Parallel Problem Solving from Nature*, pp. 178–187, Springer, Berlin Heidelberg New York

[37] Mühlenbein, H., and Schlierkamp-Voosen, D. (1993). Predictive models for the breeder genetic algorithm: I. Continuous parameter optimization. *Evolutionary Computation*, 1(1):25–49

[38] Ocenasek, J., and Schwarz, J. (2000). The parallel Bayesian optimization algorithm. In *Proceedings of the European Symposium on Computational Inteligence*, pp. 61–67. Physica, Wurzburg (Wien)

[39] Ocenasek, J., and Schwarz, J. (2002). Estimation of distribution algorithm for mixed continuous-discrete optimization problems. In *2nd Euro-International Symposium on Computational Intelligence*, pp. 227–232

[40] Ocenasek, J., Schwarz, J., and Pelikan, M. (2003). Design of multithreaded estimation of distribution algorithms. *Proceedings of the Genetic and Evolutionary Computation Conference (GECCO-2003)*, pp. 1247–1258

[32] Pearl, J. (1988). *Probabilistic reasoning in intelligent systems: Networks of plausible inference.* Morgan Kaufmann, San Mateo, CA

[42] Pelikan, M. (2002). *Bayesian optimization algorithm: From single level to hierarchy.* PhD thesis, University of Illinois at Urbana-Champaign, Urbana, IL. Also IlliGAL Report No. 2002023

[43] Pelikan, M. (2005). *Hierarchical Bayesian optimization algorithm: Toward a new generation of evolutionary algorithms.* Springer, Berlin Heidelberg New York

[44] Pelikan, M., and Goldberg, D. E. (2000). Hierarchical problem solving by the Bayesian optimization algorithm. *Proceedings of the Genetic and Evolutionary Computation Conference (GECCO-2000)*, pp. 275–282

[45] Pelikan, M., and Goldberg, D. E. (2001). Escaping hierarchical traps with competent genetic algorithms. *Proceedings of the Genetic and Evolutionary Computation Conference (GECCO-2001)*, pp. 511–518

[46] Pelikan, M., and Goldberg, D. E. (2003a). Hierarchical BOA solves Ising spin glasses and MAXSAT. *Proceedings of the Genetic and Evolutionary Computation Conference (GECCO-2003)*, II:1275–1286

[47] Pelikan, M., and Goldberg, D. E. (2003b). A hierarchy machine: Learning to optimize from nature and humans. *Complexity*, 8(5):36–45

[48] Pelikan, M., Goldberg, D. E., and Cantú-Paz, E. (1999). BOA: The Bayesian optimization algorithm. *Proceedings of the Genetic and Evolutionary Computation Conference (GECCO-99)*, I:525–532
[49] Pelikan, M., Goldberg, D. E., and Cantú-Paz, E. (2000). Linkage problem, distribution estimation, and Bayesian networks. *Evolutionary Computation*, 8(3):311–341. Also IlliGAL Report No. 98013
[50] Pelikan, M., Goldberg, D. E., and Lobo, F. (2002a). A survey of optimization by building and using probabilistic models. *Computational Optimization and Applications*, 21(1):5–20
[51] Pelikan, M., Goldberg, D. E., and Sastry, K. (2001). Bayesian optimization algorithm, decision graphs, and Occam's razor. *Proceedings of the Genetic and Evolutionary Computation Conference (GECCO-2001)*, pp. 519–526. Also IlliGAL Report No. 2000020
[52] Pelikan, M., and Lin, T.-K. (2004). Parameter-less hierarchical BOA. *Proceedings of the Genetic and Evolutionary Computation Conference (GECCO-2004)*, 2:24–35
[53] Pelikan, M., Ocenasek, J., Trebst, S., Troyer, M., and Alet, F. (2004). Computational complexity and simulation of rare events of Ising spin glasses. *Proceedings of the Genetic and Evolutionary Computation Conference (GECCO-2004)*, 2:36–47
[54] Pelikan, M., and Sastry, K. (2004). Fitness inheritance in the Bayesian optimization algorithm. *Proceedings of the Genetic and Evolutionary Computation Conference (GECCO-2004)*, 2:48–59
[55] Pelikan, M., Sastry, K., and Goldberg, D. E. (2002b). Scalability of the Bayesian optimization algorithm. *International Journal of Approximate Reasoning*, 31(3):221–258
[56] Pelikan, M., Sastry, K., and Goldberg, D. E. (2005a). Multiobjective hBOA, clustering, and scalability. *Proceedings of the Genetic and Evolutionary Computation Conference (GECCO-2005)*, pp. 663–670
[57] Pelikan, M., Sastry, K., and Goldberg, D. E. (2005b). *Sporadic model building for efficiency enhancement of hBOA*. Technical report, University of Illinois at Urbana-Champaign, Illinois Genetic Algorithms Laboratory
[58] Pelikan, M., Goldberg, D. E., and Tsutsui, S. (2003). Getting the best of both worlds: Discrete and continuous genetic and evolutionary algorithms in concert. *Information Sciences*, 156 (3–4), 147–171
[59] Rissanen, J. J. (1978). Modelling by shortest data description. *Automatica*, 14:465–471
[60] Rissanen, J. J. (1989). *Stochastic complexity in statistical inquiry*. World Scientific, Singapore
[61] Rissanen, J. J. (1996). Fisher information and stochastic complexity. *IEEE Transactions on Information Theory*, 42(1):40–47
[62] Rothlauf, F. (2001). *Towards a theory of representations for genetic and evolutionary algorithms: Development of basic concepts and their*

application to binary and tree representations. PhD thesis, University of Bayreuth, Beyreuth, Germany
[63] Rothlauf, F., Goldberg, D. E., and Heinzl, A. (2000). *Bad codings and the utility of well-designed genetic algorithms.* IlliGAL Report No. 200007, University of Illinois at Urbana-Champaign, Illinois Genetic Algorithms Laboratory, Urbana, IL
[64] Santarelli, S., Goldberg, D. E., and Yu, T.-L. (2004). Optimization of a constrained feed network for an antenna array using simple and competent genetic algorithm techniques. *Proceedings of the Workshop Military and Security Application of Evolutionary Computation (MSAEC-2004)*
[65] Sastry, K. (2001). *Efficient atomic cluster optimization using a hybrid extended compact genetic algorithm with seeded population.* IlliGAL Report No. 2001018, University of Illinois at Urbana-Champaign, Illinois Genetic Algorithms Laboratory, Urbana, IL
[66] Sastry, K., Pelikan, M., and Goldberg, D. E. (2004). Efficiency enhancement of genetic algorithms via building-block-wise fitness estimation. *Proceedings of the IEEE Conference on Evolutionary Computation*, pp. 720–727
[67] Schwarz, G. (1978). Estimating the dimension of a model. *The Annals of Statistics*, 6:461–464
[68] Schwarz, J., and Ocenasek, J. (1999). Experimental study: Hypergraph partitioning based on the simple and advanced algorithms BMDA and BOA. In *Proceedings of the International Conference on Soft Computing*, pp. 124–130, Brno, Czech Republic. PC-DIR
[69] Schwarz, J., and Ocenasek, J. (2000). A problem-knowledge based evolutionary algorithm KBOA for hypergraph partitioning. In *Proceedings of the Fourth Joint Conference on Knowledge-Based Software Engineering*, pp. 51–58, IO, Brno, Czech Republic
[70] Simon, H. A. (1968). *The Sciences of the Artificial.* MIT, Cambridge, MA
[71] Thierens, D. (1995). *Analysis and design of genetic algorithms.* PhD thesis, Katholieke Universiteit Leuven, Leuven, Belgium
[72] Thierens, D., and Goldberg, D. (1994). Convergence models of genetic algorithm selection schemes. *Parallel Problem Solving from Nature*, pp. 116–121
[73] Thierens, D., Goldberg, D. E., and Pereira, A. G. (1998). Domino convergence, drift, and the temporal-salience structure of problems. *Proceedings of the International Conference on Evolutionary Computation (ICEC-98)*, pp. 535–540
[74] Watson, R. A., Hornby, G. S., and Pollack, J. B. (1998). Modeling building-block interdependency. *Parallel Problem Solving from Nature*, pp. 97–106
[75] Wilson, S. W. (1995). Classifier fitness based on accuracy. *Evolutionary Computation*, 3(2):149–175

5

Numerical Optimization with Real-Valued Estimation-of-Distribution Algorithms

Peter A.N. Bosman and Dirk Thierens

Summary. In this chapter we focus on the design of real-valued estimation-of-distribution algorithms (EDAs) for the task of numerical optimization. Here, both the problem variables as well as their encoding are real values. Concordantly, the type of probability distribution to be used for estimation and sampling in the EDA is continuous. In this chapter we indicate the main challenges in this area. Furthermore, we review the existing literature to indicate the current EDA practice for real-valued numerical optimization. Based on observations originating from this existing research and on existing work in the literature regarding dynamics of continuous EDAs, we draw some conclusions about the feasibility of existing EDA approaches. Also we provide an explanation for some observed deficiencies of continuous EDAs as well as possible improvements and future directions of research in this branch of EDAs.

5.1 Introduction

One of the main impulses that triggered the emergence of the EDA field has been the research into the dynamics of discrete GAs. EDAs provide an elegant way of overcoming some of the most important shortcomings of classical GAs. In general, for optimization to proceed efficiently, the induced search bias of the optimizer must fit the structure of the problem under study. From this point of view, the success of the EDA approach in the discrete domain is rather intuitive. Because probability distributions are used to explicitly guide the search in EDAs, the probability distribution itself is an explicit model for the inductive search bias. Estimating the probability distribution from data corresponds to tuning the model for the inductive search bias. Because a lot is known about how probability distributions can be estimated from data (Buntine, 1994; Lauritzen, 1996) the flexibility of the inductive search bias of EDAs is potentially large. In addition, the tuning of the inductive search bias in this fashion also has a rigorous foundation in the form of the well-established field of probability theory.

P.A.N. Bosman and D. Thierens: *Numerical Optimization with Real-Valued Estimation-of-Distribution Algorithms*, Studies in Computational Intelligence (SCI) **33**, 91–120 (2006)
www.springerlink.com

Estimating probability distributions, especially in the discrete case, is very closely related to the modeling of dependencies between random variables, specific settings for these random variables, or both. Such dependencies are clearly embodied in the use of factorized probability distributions (Edwards, 1995; Lauritzen, 1996; Friedman and Goldszmidt, 1996). As a result, the processing of these dependencies by the EDA in the discrete domain is also explicitly ensured under the assumption of a proper estimation of the (factorized) probability distribution. Dependencies in the discrete domain match exactly with the notion of linkage information or, synonymously, the structure of the problem. Hence, the competent estimation of probability distributions in EDAs allows these algorithms to adequately perform linkage learning and to ensure that such necessary conditions as proper building block mixing (Thierens and Goldberg, 1993) meets with the positive effects of the schema theorem (Holland, 1975) and the building block hypothesis (Goldberg, 1989; Goldberg, 2002a). Although clearly more computational effort is required to tune the inductive search bias by estimating the probability distribution, the payoff has been shown to be worth the computational investment (e.g., polynomial scale-up behavior instead of exponential scale-up behavior) (Pelikan et al., 1999; Etxeberria and Larrañaga, 1999; Pelikan and Goldberg, 2001; Pelikan and Goldberg, 2003).

The use of factorized probability distributions in an EDA works very well in discrete domain. In general, the EDA approach can be motivated perfectly from the requirement of fitting the inductive search bias to the structure of the optimization problem at hand. It is now interesting to investigate how the approach carries over to the continuous domain. The main important questions to answer are: (1) what does the model (i.e., probability distribution) for the inductive search bias look like in the continuous domain and (2) can this model be adapted properly to fit the structure of the problem?

The remainder of this chapter is organized as follows. First, in Sect. 5.2 we discuss real-valued numerical optimization problems and point out which sources of problem difficulty typically exist. Next, in Sect. 5.3 we consider the EDA approach in the continuous domain and show that in theory real-valued EDAs can be very efficient numerical optimizers. In Sect. 5.4 we then present a literature overview that describes the current state of the continuous subfield of EDAs. Subsequently, in Sect. 5.5 we reflect on the dynamics of the currently available real-valued EDAs in more depth. We also discuss whether and how improvements can be made over current approaches. Finally, we summarize and conclude this chapter in Sect. 5.6.

5.2 Problem Difficulty

The degree of difficulty of a numerical optimization problem is in a certain sense unbounded. Because the search space is by definition infinitely large, the number of substructures, i.e., the number of local optima and areas with various types of dependency, can also be infinite. The structure of a continuous problem can most generally be seen as the composition of its contour lines. The shape of these contour lines is unbounded and there are infinitely many of them. Hence, anything is possible. Fortunately, we are typically not interested in all possible problems. We assume that there is some form of "logical" structure, similar to assuming in the binary domain that the function is not a needle-in-a-haystack function. Hence, the problem structure we typically expect to be asked to tackle is a composition of simpler structures such as local optima and some form of dependency between the problem variables.

Still, even assuming "simpler" substructures are present that we can tailor our EDA to, it should be noted that in the continuous case the actual difficulty of the problem can still be arbitrary in the sense that the number of substructures that can be present in the problem is arbitrarily large, even if the problem has only one variable. This arbitrary problem difficulty in terms of presence of multiple substructures throughout the search space will play important role in evaluating the sensibility of the EDA approach as a whole as well as existing actual EDA instances in the continuous domain.

At the most general level the substructures that are typically to be expected in the continuous domain are similar to the ones in the discrete domain. The two basic substructures are multimodality and dependency. The main difference with the discrete case is that in the continuous case these structures can also be viewed upon from a different perspective, viz., one that entails slopes and peaks (i.e., hills or valleys). Since nothing is said about the number of slopes and peaks or their actual configuration with respect to orientation and location, this essentially allows for the construction of the same search spaces with local optima (i.e., multimodality) and dependencies.

Various sources of problem difficulty in continuous spaces are often explained most effectively by the design of an actual problem and its visualization. For the purpose of illustration of problem difficulty and because the discussion of the results obtained with various continuous EDAs is often related to existing benchmark problems, we now describe a set of five optimization problems. These problems represent a variety of difficulties in numerical optimization. The definitions of the numerical optimization problems are given in Table 5.1. For an intuitive impression of the characteristics of the problems, two-dimensional surface plots (with the exception of the readily-imaginable sphere function) are provided in Fig. 5.1.

Sphere

The sphere function is probably the most standard unimodal benchmark problem for numerical optimization. It involves the minimization of a single hyperparabola. The function is unimodal, has no dependencies and has smooth gradient properties. The sphere function is often used to study convergence. The minimum value for any dimensionality is 0 which is obtained if all y_i are set to a value of 0.

Table 5.1. Numerical optimization test problems

Name	Definition	Range
Sphere	Minimize $\sum_{i=0}^{l-1} y_i^2$	$y_i \in [-5, 5]$ $(0 \leq i < l)$
Griewank	Minimize $\frac{1}{4{,}000} \sum_{i=0}^{l-1} (y_i - 100)^2 - \prod_{i=0}^{l-1} \cos\left(\frac{y_i - 100}{\sqrt{i+1}}\right) + 1$	$y_i \in [-600, 600]$ $(0 \leq i < l)$
Michalewicz	Minimize $-\sum_{i=0}^{l-1} \sin(y_i) \sin^{20}\left(\frac{(i+1)y_i^2}{\pi}\right)$	$y_i \in [0, \pi]$ $(0 \leq i < l)$
Rosenbrock	Minimize $\sum_{i=0}^{l-2} 100(y_{i+1} - y_i^2)^2 + (1 - y_i)^2$	$y_i \in [-5.12, 5.12]$ $(0 \leq i < l)$
Summation cancellation	Maximize $100/(10^{-5} + \sum_{i=0}^{l-1} \lvert\gamma_i\rvert)$ where $\gamma_0 = y_0, \gamma_i = y_i + \gamma_{i-1}$	$y_i \in [-3, 3]$ $(0 \leq i < l)$

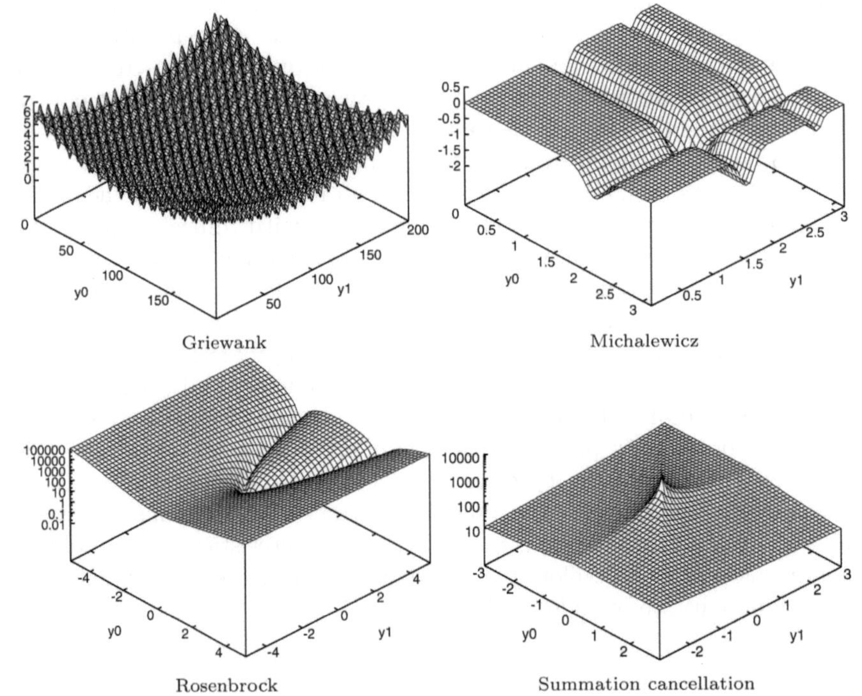

Fig. 5.1. Two-dimensional surface plots for four numerical optimization problems. The ranges for Griewank's function were zoomed to get a better indication of the many local optima. Rosenbrock's function and the summation cancellation function are shown on a logarithmic scale for a better impression of their problem features. Note that the summit of the peak for summation cancellation is actually 10^7, but the drawing resolution prohibits accurate visualization thereof

Griewank

Griewank's function is a function with many local optima. Basically, it is a parabola superimposed with a sine function to obtain many local optima. As a result, if large steps are taken in Griewank's function, the so observed coarse-grained gradient information will quickly lead to a region close to the minimum of the parabola. However, if only small steps are taken, the many local optima will prevent efficient optimization of this function, even when a random restart strategy is used. Furthermore, for a fixed precision, Griewank's function becomes easier to optimize as l increases if large steps are taken. The minimum value for Griewank's function for any dimensionality is 0, which is obtained if all y_i are set to a value of 100.

Michalewicz

Michalewicz's function is also a function with many local optima, albeit to a lesser degree than Griewank's function. An important difference is that Michalewicz's function has many long channels along which the minimum value throughout the channel is the same. The gradient information in such a channel therefore does not lead to the better local optima which are found at the intersections of the channels. Proper optimization of Michalewicz's function is therefore only possible if the channels of equal optimization value are explored or covered fully to find the intersections. The minimum value for Michalewicz's function depends on its dimensionality. A description of its solutions at which the minimum value is obtained for any dimensionality has not been reported in the literature.

Rosenbrock

Rosenbrock's function is highly nonlinear. It has a curved valley along which the quality of the solutions is much better than in its close neighborhood. Furthermore, this valley has a unique minimum of 0 itself for any dimensionality of Rosenbrock's function, which is obtained if all y_i are set to a value of 1. Rosenbrock's function has proved to be a real challenge for any numerical optimization algorithm. The gradient along the bottom of the nonlinear valley is very slight. Any gradient approach is therefore doomed to follow the long road along the bottom of the valley, unless a starting point is provided in the vicinity of the optimum. Furthermore, since the valley is nonlinear, simple gradient based approaches will oscillate from one side of the valley to the other, which does not result in efficient gradient exploitation. For an $\mathbb{IDEA}$, capturing the valley in a probabilistic model is difficult, even if all of the points within the valley are known. The reason for this is that the valley is nonlinear in the coding space.

Summation Cancellation

The summation cancellation problem was proposed by Baluja and Caruana (1995). This optimization problem has multivariate linear interactions between the problem variables. This should allow algorithms that are capable of modeling linear dependencies to outperform algorithms that are not capable of doing so. Furthermore, the degree of multivariate interaction is as large as possible since each γ_i in the problem definition is defined in terms of all y_j with $j < i$. Finally, the optimum is located at a very sharp peak, which implies that the optimization algorithm needs to have a large precision and needs to be able to prevent premature convergence in order to reach the global optimum. The minimum value for this function for any dimensionality is 10^7, which is obtained if all y_i are set to a value of 0.

5.3 Optimal EDAs

The definition of an optimal algorithm is subject to a certain viewpoint. In search and optimization, a common definition of optimality is that the algorithm is capable of finding an optimal solution (Russel and Norvig, 2003). A more refined definition of optimality in the context of a specific class of optimization problem is that for any problem instance there exists no algorithm that is able to find the optimal solution faster than the optimal algorithm. The EDA method is driven mainly by the estimation of a probability distribution. The estimated probability distribution is meant to approximate the true probability distribution. The true probability distribution is the one that describes perfectly the set of the selected solutions in the case of an infinitely large population size. If the method of selection is Boltzmann selection, it can be proved that the resulting EDA using the true probability distribution is optimal in the sense that it will indeed converge to the optimal solution (Mühlenbein and Höns, 2005). Hence, we can call an EDA optimal if the estimated probability distribution equals the true probability distribution.

Drawing samples from the Boltzmann distribution requires exponential computation time and hence, the optimal Boltzmann EDA is not a practical search algorithm. In practice, selection methods such as tournament selection and truncation selection are commonly used. Without loss of generality, assume that the goal of the EDA is to minimize fitness. Now, the optimal probability distribution associated with truncation selection is a distribution that is uniform over all solutions $\boldsymbol{y}$ that have a fitness $\mathfrak{F}(\boldsymbol{y})$ that is at most the value of the fitness of the worst selected solution. This essentially corresponds to pruning the search space to only those regions that are at least as interesting as the currently available worst selected solution. The probability distribution can be imagined as essentially maintaining exactly all solutions that have a fitness of at most the value of the fitness of the worst selected solution. Hence, sampling from this probability distribution entails nothing more than just returning a single solution from the (infinitely large) set of maintained solutions.

Now observe the series of the fitness values of the worst selected solution in subsequent generations. Using truncation selection this series is monotonously decreasing. Moreover, the series becomes strictly monotonously decreasing if in generation t all solutions with a fitness at least equal to the worst selected fitness in generation $t-1$ are pruned from the selected solutions. To prevent obtaining an empty selection set, this additional discarding should not be done for the solution(s) with the best fitness. Using this slight variation to the truncation selection scheme the EDA approach with the true probability distribution will clearly converge to the optimum with probability one because optimal solutions are never discarded from the probability distribution and the worst selected fitness values strictly decreases every generation.

The size of the search space that is still to be regarded, is encoded in the probability distribution. Using truncation selection and an infinitely large population, $100(1-\tau)\%$ of all solutions that make up this remaining search space are discarded. Hence, the size of the search space still to be explored is reduced exponentially as a function of the generations passed. In practice however, we do not have an infinite population size. Still the use of the optimal probability distribution is likely to be extremely efficient in terms of the number of evaluations actually required to find a solution of a certain quality. To illustrate the potential effectiveness of EDAs in continuous spaces, Fig. 5.2 shows the convergence behavior on three of all optimization problems for a dimensionality of $l = 5$ if the optimal probability distribution is known and used in the EDA. The optimal probability distribution is hard to formalize analytically, hence the method used uses rejection sampling. In other words, the method generates random solutions and only accepts those that have a fitness value smaller or equal to the worst selected solution. For Michalewicz' function and the summation cancellation function we plotted the difference with the optimal fitness value to be able to plot their convergence graphs as a minimization procedure that searches for the minimum value of 0. From the results in Fig. 5.2, we indeed find extremely efficient optimization behavior.

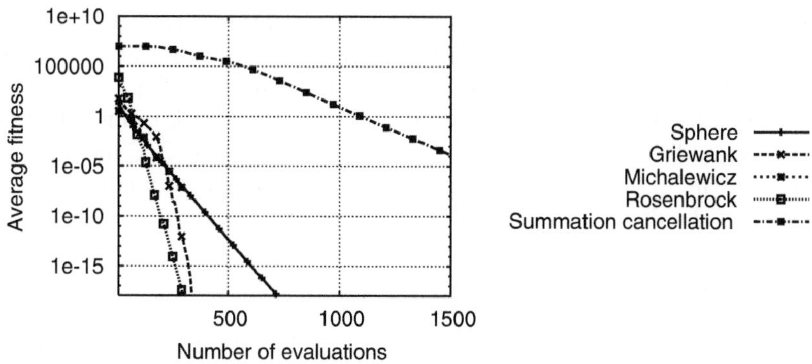

Fig. 5.2. Average convergence behavior (100 runs) of the EDA with true density on a selection of problems and a dimensionality of $l = 5$

The fitness improves exponentially fast with the number of evaluations required. Hence, only extremely few evaluations are actually needed to obtain close-to-optimal results, regardless of the actual optimization problem. Concluding, in theory, as a result of the solid background, EDAs for continuous domains work well also next to EDAs for discrete domains as long as the true probability distribution can be closely approximated.

In practice we in general do not have access to the optimal probability distribution. Moreover, in the general sense the optimal probability distribution is arbitrarily complex as the problem itself can be arbitrarily complicated (see Sect. 5.2). This makes that sampling from the true probability distribution can take up an arbitrary long time. Hence, to obtain an actual EDA to work with, we must approximate the probability distribution using practical techniques. In the continuous case, the most common approaches are the normal probability density function (pdf) or combinations thereof. It is not surprising that the first EDAs in continuous spaces were based exactly on these common parametric pdfs. The most important question is of course how the extremely efficient optimization behavior of the EDA using the true probability distribution changes in the continuous domain if we use approximated probability distributions instead.

5.4 An Overview of Existing EDAs

In this section we provide a literature survey of EDAs for continuous domains. In Sect. 5.4.1, we consider factorized probability distributions. In Sect. 5.4.2, we consider mixtures of factorized probability distributions. The majority of the literature concerning EDAs is based on these two types of probability distribution. We end with Sect. 5.4.3 where we consider other classes of probability distribution.

5.4.1 Factorized Probability Distributions

Similar to the discrete case, estimating a factorization, typically a Bayesian one as introduced in the previous chapter, makes sense since at the top level it assists in finding dependencies between the variables (i.e., the use of a joint distribution versus the use of separate marginal distributions). The same greedy arc-addition algorithm and scoring metrics such as BIC metric can be used as in the discrete case. Note that in the continuous domain the nature or shape of the dependency depends on the actual pdf that is factorized.

Normal pdf

The first real-valued EDAs used maximum-likelihood estimations of the normal pdf. The use of the normal pdf was a logical first choice since this pdf is widely used, relatively simple and unimodal.

Definition

The normal pdf $P^{\mathcal{N}}_{(\boldsymbol{\mu}_{\boldsymbol{i}},\boldsymbol{\Sigma}^{\boldsymbol{i}})}(Y_{\boldsymbol{i}})$ is parameterized by a vector $\boldsymbol{\mu}_{\boldsymbol{i}}$ of means and a symmetric covariance matrix $\boldsymbol{\Sigma}^{\boldsymbol{i}}$ and is defined by

$$P^{\mathcal{N}}_{(\boldsymbol{\mu}_{\boldsymbol{i}},\boldsymbol{\Sigma}^{\boldsymbol{i}})}(Y_{\boldsymbol{i}})(\boldsymbol{y}) = \frac{(2\pi)^{-|\boldsymbol{i}|/2}}{(\det \boldsymbol{\Sigma}^{\boldsymbol{i}})^{1/2}} \mathrm{e}^{-1/2(\boldsymbol{y}-\boldsymbol{\mu}_{\boldsymbol{i}})^{\mathrm{T}}(\boldsymbol{\Sigma}^{\boldsymbol{i}})^{-1}(\boldsymbol{y}-\boldsymbol{\mu}_{\boldsymbol{i}})}. \tag{5.1}$$

Parameter Estimation

A maximum likelihood estimation for the normal pdf is obtained from a vector $\boldsymbol{\mathcal{S}}$ of samples if the parameters are estimated by the sample average and the sample covariance matrix (Anderson, 1958; Tatsuoka, 1971):

$$\hat{\boldsymbol{\mu}}_{\boldsymbol{i}} = \frac{1}{|\boldsymbol{\mathcal{S}}|}\sum_{j=0}^{|\boldsymbol{\mathcal{S}}|-1}(\boldsymbol{\mathcal{S}}_j)_{\boldsymbol{i}}, \qquad \hat{\boldsymbol{\Sigma}}^{\boldsymbol{i}} = \frac{1}{|\boldsymbol{\mathcal{S}}|}\sum_{j=0}^{|\boldsymbol{\mathcal{S}}|-1}((\boldsymbol{\mathcal{S}}_j)_{\boldsymbol{i}} - \hat{\boldsymbol{\mu}}_{\boldsymbol{i}})((\boldsymbol{\mathcal{S}}_j)_{\boldsymbol{i}} - \hat{\boldsymbol{\mu}}_{\boldsymbol{i}})^{\mathrm{T}}. \tag{5.2}$$

To estimate the conditional pdfs $P^{\mathcal{N}}(Y_i|Y_{\boldsymbol{\pi}_i})$ required when constructing Bayesian factorizations, let $\boldsymbol{W}^{\boldsymbol{j}}$ be the inverse of the symmetric covariance matrix, that is $\boldsymbol{W}^{\boldsymbol{j}} = (\boldsymbol{\Sigma}^{\boldsymbol{j}})^{-1}$. Matrix $\boldsymbol{W}^{\boldsymbol{j}}$ is commonly called the *precision matrix*. It can be shown that for a maximum likelihood estimate of $P^{\mathcal{N}}(Y_i|Y_{\boldsymbol{\pi}_i})$ the maximum-likelihood estimations in (5.2) can be used (Bosman and Thierens, 2000b):

$$\hat{P}^{\mathcal{N}}(Y_i|Y_{\boldsymbol{\pi}_i})(y_{(i,\boldsymbol{\pi}_i)}) = \frac{1}{(\breve{\sigma}_i\sqrt{2\pi})}\mathrm{e}^{\frac{-(y_i-\breve{\mu}_i)^2}{2\breve{\sigma}_i^2}}, \tag{5.3}$$

$$\text{where} \begin{cases} \breve{\sigma}_i = \dfrac{1}{\sqrt{\hat{\boldsymbol{W}}^{(i,\boldsymbol{\pi}_i)}_{00}}} \\ \breve{\mu}_i = \dfrac{\hat{\mu}_i\hat{\boldsymbol{W}}^{(i,\boldsymbol{\pi}_i)}_{00} - \sum_{j=0}^{|\boldsymbol{\pi}_i|-1}(y_{(\boldsymbol{\pi}_i)_j} - \hat{\mu}_{(\boldsymbol{\pi}_i)_j})\hat{\boldsymbol{W}}^{(i,\boldsymbol{\pi}_i)}_{(j+1)0}}{\hat{\boldsymbol{W}}^{(i,\boldsymbol{\pi}_i)}_{00}}. \end{cases}$$

Properties

The number of parameters to be estimated equals $1/2|\boldsymbol{i}|^2 + 3/2|\boldsymbol{i}|$. Different from the discrete case, the number of parameters to be estimated therefore does not grow exponentially with $|\boldsymbol{i}|$ but quadratically.

The density contours of a normal factorized probability distribution are ellipsoids. Depending on the dependencies modeled by the factorization, these ellipsoids can be aligned along any axis. If there is no dependency between a set of random variables, the projected density contours in those dimensions are aligned along the main axes. In either case, a normal pdf is only capable of efficiently modeling *linear* dependencies.

EDAs

The first EDA for real-valued random variables was an adaptation of the original binary PBIL algorithm. The algorithm uses l normal pdfs, one for each of the l random variables (Rudlof and Köppen, 1996). To accommodate for these normal pdfs, the probability vector from the original PBIL algorithm is replaced with a vector that specifies for each variable the mean and variance of the associated normal pdf. The means are updated using a similar update rule as in the original binary PBIL. The variances are initially relatively large and are annealed down to a small value using a geometrically decaying schedule. New solutions are generated by drawing values from the normal pdfs for each variable separately.

A second adaptation of PBIL to the continuous case was introduced by Sebag and Ducoulombier (1998). Similar to the approach by Rudlof and Köppen (1996), they proposed to use a normal pdf for each variable. However, the variance is now updated using the same update rule as for the mean.

For real-valued random variables, Bayesian factorizations using normal pdfs were proposed simultaneously by Bosman and Thierens (2000b) within the probabilistic model-building EA framework and by Larrañaga et al. (2000) in a variant of MIMIC that uses normal pdfs, termed MIMIC_C, and in the Estimation of Gaussian Network Algorithm (EGNA). As a first approach, Bosman and Thierens (2000b) used an algorithm by Edmonds (1967) to find a Bayesian factorization of minimal entropy in which each variable has at most one parent. Also, the optimal dependency-tree algorithm used in COMIT and the greedy chain-learning algorithm used in MIMIC were used (Bosman and Thierens, 2000a; Bosman and Thierens, 2000b). In a later publication (Bosman and Thierens, 2001a), the BIC metric was proposed in combination with a greedy factorization-learning algorithm. In the work by Larrañaga et al. (2000), finding a Bayesian factorization starts with a complete factorization graph. A likelihood-ratio test is then performed for each arc to determine whether or not that arc should be excluded from the graph. A greedy factorization-learning algorithm based on the BIC metric that starts from the univariate factorization was also used.

The use of univariate factorizations for real-valued random variables was studied and compared against the use of Bayesian factorizations by various researchers (Bosman and Thierens, 2000a; Bosman and Thierens, 2000b; Larrañaga et al., 2000; Bosman and Thierens, 2001a). In these studies, the use of univariately factorized normal probability distributions was shown to be inferior to the use of multivariate factorized normal probability distributions for optimization problems that have linear interactions between the problem variables. Specifically, far better results were obtained on the summation cancellation function. Although on this particular problem the EDAs even outperformed evolution strategies (ES), improvement over ES was not observed on all problems described in Sect. 5.2. On both the Michalewicz and the Rosenbrock function the EDA approach was even strongly inferior to ES.

In general, the EDA approach was observed to have good optimization performance on problems with linear dependencies and even on problems with many local optima, in the likes of the Griewank function, of both lower and higher dimensionality. However, the EDAs cannot efficiently solve optimization problems with nonlinear interactions between their variables. The main reason is that the interactions that can be modeled using the normal pdf are just linear. Hence, the structure of the problem is not fit well by the probability distribution. As a result, points surrounding the optimum get lost during optimization and the EDA is left only with solutions far away from the optimum. However, by estimating the probability distribution of these points with maximum likelihood, the EDA suffers from the drawback that it disregards gradient information. Instead, it keeps sampling points in the same area where the current selected solutions are while selection continuously decreases the size of this area. The EDA can therefore converge even while on a slope towards the optimum. Hence, even if the problem is unimodal, the EDA using maximum-likelihood estimations of the normal pdf can fail in finding an optimal solution. This happens for instance on the Rosenbrock problem, even for very large population sizes as is illustrated in Fig. 5.3. The success rate of the EDA using the maximum-likelihood normal pdf was 0% for both

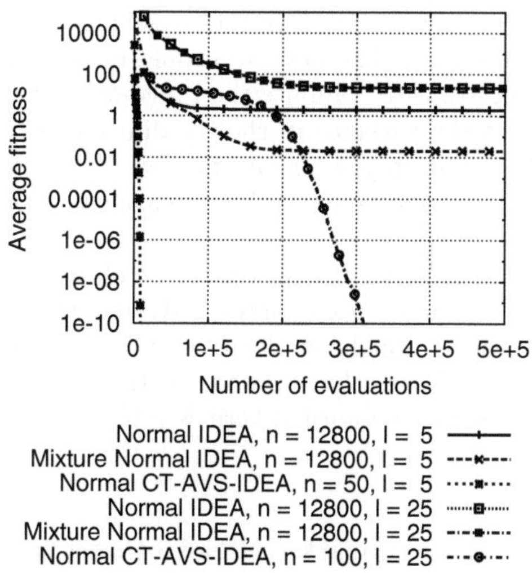

Fig. 5.3. Average convergence behavior (100 runs) of various continuous EDAs on the Rosenbrock function with a dimensionality of $l \in \{5, 25\}$. The results shown are the best results from all test settings: $n \in \{25, 50, 100, 200, 400, 800, 1,600, 3,200, 6,400, 12,800\}$, maximum number of evaluations $= 10^6$, number of clusters in mixture distribution $= 5$. Rates of success for $l = 5$: normal pdf 0%, mixture of normal pdfs 72%, normal pdf with variance scaling 100%. Rates of success for $l = 25$: normal pdf 0%, mixture of normal pdfs 0%, normal pdf with variance scaling 100%

dimensionalities. Conversely however, if the solutions do continue to surround the optimum, the use of the maximum-likelihood normal pdf is very efficient as is for instance the case on the sphere function where the initial variable ranges are centered around the optimum.

Normal Kernels pdf

Definition

The normal kernels pdf is obtained by centering one multivariate normal pdf at each point in the sample vector and by normalizing the sum of the densities:

$$P^{\mathcal{N}_K}_{(\boldsymbol{\Sigma}^i)}(Y_{\boldsymbol{i}})(\boldsymbol{y}) = \frac{1}{|\boldsymbol{S}|} \sum_{j=0}^{|\boldsymbol{S}|-1} P^{\mathcal{N}}_{((\boldsymbol{S}_j)_{\boldsymbol{i}}, \boldsymbol{\Sigma}^i)}(Y_{\boldsymbol{i}})(\boldsymbol{y}). \tag{5.4}$$

Parameter Estimation

Deciding how to choose the covariance matrix for each normal pdf is hard. A maximum-likelihood estimate is undesirable because in that estimate the variances are zero, corresponding to a density of infinity at the mean of each normal pdf. Therefore, the variances in the covariance matrix that is used for each normal pdf in the normal kernels pdf should be set in a different manner.

One way of doing so, is to compute the range of the samples in $\boldsymbol{S}$ in each dimension and to set the variance in the ith dimension to a value based on the range such that it decreases as the number of samples increases, e.g., $\alpha \cdot \text{range}_i / |\boldsymbol{S}|$. The value of α controls the smoothness of the resulting density estimation.

Although formulas exist for the univariate conditional form of the normal kernels pdf that is required to construct Bayesian factorizations (Bosman, 2003), both their computational complexity and the lack of sensibility of using maximum likelihood estimates have prevented the use of such factorizations in continuous EDAs using the normal kernels pdf.

Properties

The main advantage of the normal kernels pdf is that it has a natural tendency to fit the structure of the sample vector and is thereby capable of expressing complicated nonlinear dependencies. A related disadvantage however is that the quality of the density estimation heavily depends on the value of α. Intuitively, a larger α results in a smoother fit, but it is hard to predict beforehand what a good value for α would be. The normal kernels pdf has a tendency to *overfit* a sample collection. Without proper model selection and model fitting, the normal kernels pdf is hard to handle although it may be relatively fast in its use. The other possibility is to set the variances, or even covariances, for the normal kernels pdf adaptively. If the adaptation is done for each normal kernel separately, the resulting approach is equivalent to the use of evolution strategies (Bäck and Schwefel, 1993). Concluding, the normal kernels pdf certainly has interesting properties and potential to be used in IDEAs, but it is likely to be hard to handle when perceived as a manner of describing the structure of the data in a sample set.

EDAs

The normal kernels pdf was initially tried in an EDA by Bosman and Thierens (2000a) (see also (Bosman and Thierens, 2000b; Bosman, 2003)). The range-based approach based on the α parameter as mentioned above was used to set the variances of the normal kernels. Fine-tuning α was found to be highly problem-dependent. Not only is it difficult to set α to a useful value beforehand, a proper value for α is apt to change during the run of the EDA. Hence, good results are hard to obtain with this use of the normal kernels pdf.

Normal Mixture pdf

Definition

If we take w normal pdfs instead of only a single one or as many as $|\boldsymbol{S}|$, we have a trade-off between the cluster-insensitive normal pdf and the cluster-oversensitive normal kernels pdf. The normal mixture pdf for random variables $Y_{\boldsymbol{i}}$ is parameterized by w triples that each consist of a mixture coefficient $\beta_{\boldsymbol{i}}^j$ a vector of $|\boldsymbol{i}|$ means and a symmetric covariance matrix of dimension $|\boldsymbol{i}| \times |\boldsymbol{i}|$:

$$P^{\mathcal{N}_M}_{((\beta_{\boldsymbol{i}}^0,\boldsymbol{\mu}_{\boldsymbol{i}}^0,\boldsymbol{\Sigma}^{0,\boldsymbol{i}}),\ldots,(\beta_{\boldsymbol{i}}^{w-1},\boldsymbol{\mu}_{\boldsymbol{i}}^{w-1},\boldsymbol{\Sigma}^{w-1,\boldsymbol{i}}),)}(Y_{\boldsymbol{i}})(\boldsymbol{y}) = \sum_{j=0}^{w-1} \beta_{\boldsymbol{i}}^j P^{\mathcal{N}}_{(\boldsymbol{\mu}_{\boldsymbol{i}}^j,\boldsymbol{\Sigma}^{j,\boldsymbol{i}})}(Y_{\boldsymbol{i}})(\boldsymbol{y}). \quad (5.5)$$

Parameter Estimation

A maximum likelihood estimation cannot be obtained analytically for $w > 1$. Therefore, as an alternative approach, the EM algorithm (Dempster et al., 1977) can be used. The EM algorithm is a general iterative approach to computing a maximum-likelihood estimate. For the normal-mixture pdf, an EM-algorithm first initializes each mixture coefficient, all means and all covariance matrices to certain values and then updates all of these values iteratively until the algorithm converges or until a maximum number of iterations has been reached. We refrain from presenting a derivation of the update equations and refer the interested reader to the literature (Dempster et al., 1977; Bilmes, 1997).

An alternative manner to estimating a normal mixture pdf is to use clustering (Hartigan, 1975). First, the samples are clustered using a clustering method and subsequently a normal pdf is estimated in each cluster. The drawback of this approach is that from a probabilistic viewpoint, the resulting probability distribution estimation is almost certainly not a maximum likelihood estimate. However, the use of clustering does allow for the modeling of nonlinear dependencies between the variables.

Properties

The main advantage of the normal mixture pdf is that it provides a trade-off between the normal pdf and the normal kernels pdf. The normal mixture pdf is able to model nonlinear dependencies a lot more accurate than when a single normal pdf is used (see Fig. 5.4 for an example). Although the normal mixture pdf does not have a flexibility as great as the normal kernels pdf has, the normal mixture pdf is much easier to handle. The reason for this is that the only parameter to set is the number of clusters to construct, which is a lot more transparent than the value of α to set in the case of the normal kernels probability distribution.

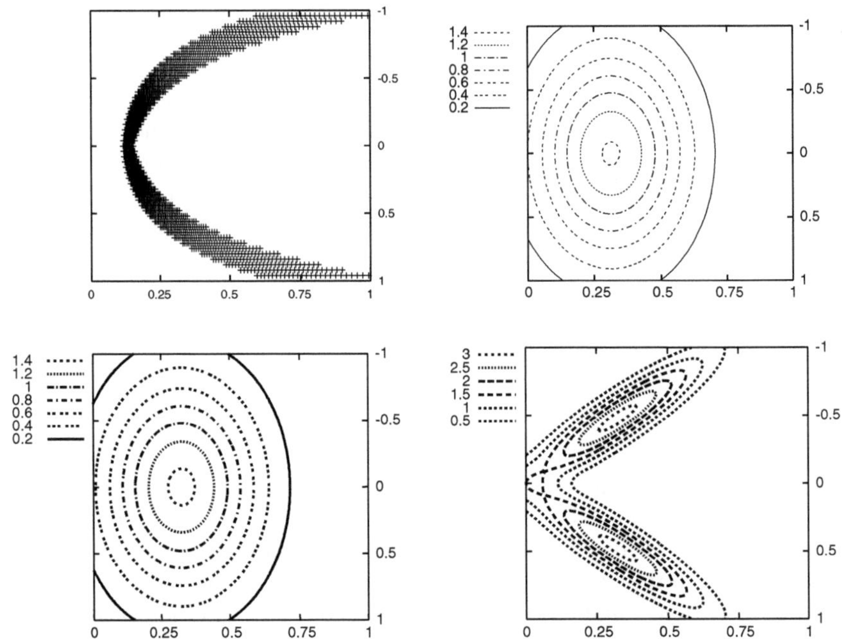

Fig. 5.4. *Top left*: A sample vector that contains nonlinear dependencies. *Top right*: density contours of a product of two one-dimensional normal pdfs (maximum-likelihood estimate). *Bottom left*: density contours of a multivariate joint two-dimensional normal pdf (maximum-likelihood estimate). *Bottom right*: density contours of two multivariate joint two-dimensional normal pdfs that have been fit (maximum-likelihood estimate) after the sample vector was clustered

A maximum likelihood approach is available to estimate the normal mixture pdf parameters. However, especially as the number of variables and the number of mixture components increases, the result using the EM algorithm becomes unreliable. Similarly, the usefulness of clustering also decreases as the number of variables increases. Hence, it makes sense to factorize the probability distribution so that the probability distribution over all random variables is a composition of normal mixture pdf estimates for smaller sets of variables. Furthermore, the EM algorithm is a time-consuming approach. The clustering approach on the other hand can be quite fast, depending on the clustering algorithm that is used.

EDAs

Although the required expressions for computing Bayesian factorizations of the normal mixture pdf have been derived (Bosman, 2003), the resulting formulas are rather involved and not efficient for practical use. Concordantly, no EDA has been reported in the literature as yet that uses such a factorization.

Gallagher et al. (1999) proposed an EDA that uses the adaptive mixture pdf by Priebe (1994) for each variable separately (i.e., using the univariately factorized probability distribution). Although their approach outperformed other algorithms, the set of test problems consisted only of two relatively simple two-dimensional problems.

Bosman (2003) experimented with an EDA that uses the EM-algorithm to estimate a normal mixture pdf. Due to the complicated nature of Bayesian factorizations when combined with the normal mixture pdf, only the univariate factorization and the unfactorized probability distribution were tested. In low-dimensional spaces the added flexibility of the normal mixture pdf compared to the single normal distribution resulted in a clear improvement when solving the Rosenbrock function using an unfactorized pdf. The growth of the complexity of the nonlinear dependency in the Rosenbrock function, however, quickly decreases the advantage of the higher flexibility offered by the use of the normal mixture pdf. Although good results were also obtained on other problems in low-dimensional spaces, the additional computational overhead was found to be considerable.

Although the conditional pdf required for a Bayesian factorization is rather complex, the approach for constructing a marginal-product factorization as is done in the ECGA (see the chapter on ECGA) and the resulting factorized normal mixture pdf are rather straightforward. The resulting probability distribution describes for mutually exclusive subsets of variables a normal mixture pdf than can be constructed in one of the aforementioned manners. Ahn et al. (2004) developed an EDA that uses exactly this approach where each normal mixture pdf is estimated using the clustering approach. Factorizing the probability distribution in this manner can allow for a more efficient modeling of independent subproblems. Although improvements were obtained over earlier real-valued EDA approaches, the resulting EDA was still not found to be efficient at solving the Rosenbrock function, especially as the dimensionality of the problem is increased.

Histogram pdf

Definition

Although conceptually the histogram pdf is arguably the most simple way to represent a continuous real-valued probability distribution, we refrain from giving a precise mathematical definition as it is unnecessarily complicated. The histogram pdf splits up the range of each variable into several parts. A probability is associated with each hyperbox (commonly called a *bin*) that represents the probability of a sample to lie anywhere inside the combination of ranges associated with that bin. Note that the number of splits per variable or the relative size of all subranges do not need to be the same although in the most traditional sense, such a fixed-width version of the histogram pdf is the most commonly used.

Parameter Estimation

A maximum-likelihood estimation is obtained in the same manner as for the integer (or binary) case, i.e., by computing the proportion of samples that fall into a certain bin.

Properties

A more detailed estimate of the true underlying probability distribution can be obtained if more bins are used. However, as the number of bins increases, the efficiency of modeling dependencies with the histogram pdf rapidly decreases as many bins will be empty. Moreover, since often the number of bins grows exponentially when expressing the joint probability of multiple random variables (for instance when using the fixed-width approach), histograms are actually quite inefficient in expressing dependencies.

Furthermore, as the number of bins is increased, the danger of over-fitting and lack of generalization increases as well. If more bins are used, the number of empty bins will increase. Drawing more samples from the estimated probability distribution will thus not produce any more samples in these areas, even though these areas might very well contain the global optimum.

EDAs

The algorithm by Servet et al. (1997) is an adaptation of the original PBIL algorithm. In the algorithm, a range is stored for each variable. For each variable then, a histogram pdf with two bins is maintained, where the first bin corresponds with the first half of the domain and the second bin corresponds with the second half. The probability vector from the original PBIL algorithm now specifies for each variable the probability with which a new value for that variable is generated in the second half of the domain currently stored for that variable. A domain is resized to contain exactly one of the two halves of that domain if the histogram starts to converge to that half of that domain.

Bosman and Thierens (2000a) first investigated the use of an EDA that is based on a univariately factorized histogram distribution. In their work they considered the fixed-width variant. Tsutsui et al. (2001) considered both the fixed-width as well as the fixed-height histogram variants in their work but also only in combination with the univariate factorization. The conclusions of both investigations are that the resulting EDAs are quite well capable of solving problems that do not exhibit dependencies between their variables. If dependencies do occur, the EDAs are not capable of exploiting these dependencies efficiently, which severely limits its practical use.

To overcome the exponential growth of the total number of bins as the joint probability distribution of multiple variables is taken into account, a specialized repeated-splitting technique as used in classification and regression trees can be used (Breiman et al., 1984). The available data is split in a single dimension. To this end all possible axis-parallel splits are investigated and the one that is the most beneficial in splitting up the data is selected. The process is then recursively repeated in the two subsets until no split significantly improves the quality of the overall probability distribution anymore. The resulting bins are not identically sized and are specifically tailored to fit the data, thereby reducing the number of empty bins to 0. This approach was used in an EDA by Pošík (2004). Although the EDA is capable of modeling unfactorized joint probabilities, the approach was found not to be able to minimize Rosenbrock's function. The main reason for this is likely the inability to generalize the probability distribution outside of the current range of selected solutions. Moreover, the approach was found not to be very robust against multimodality.

5.4.2 Mixture-of-Factorizations Probability Distributions

A mixture-probability distribution is a weighted sum of probability distributions, each of which is a function of all random variables. The weights in the sum are all positive and sum up to one to ensure that the summation is itself a probability distribution. Although we have already encountered mixture probability distributions when we considered factorized probability distributions in Sect. 5.4.1 (i.e., the normal mixture pdf), in this section we consider the mixture as the main structure of the probability distribution. In Sect. 5.4.1 the main structure was the factorization. The general form of the mixture probability distribution over all random variables $\boldsymbol{Y}$ is:

$$P(\boldsymbol{Y}) = \sum_{i=0}^{k-1} \alpha^i P^i(\boldsymbol{Y}), \tag{5.6}$$

where k is the number of mixture components, $\alpha^i \geq 0$ holds for all $i \in \{0, 1, \ldots, k-1\}$ and $\sum_{i=0}^{k-1} \alpha^i = 1$.

The mixture probability distribution can be seen as a logical extension of the factorization approach. It can be viewed upon as a way of using multiple factorizations by taking each probability distribution in the mixture to be a factorization. The use of mixture probability distributions intuitively appears to be an especially useful approach in the presence of multiple local optima as each component in the mixture can be used to focus on a single local optimum and thereby distribute the factorization search bias of the EDA over multiple regions of interest. To a lesser extent one of the virtues of a mixture probability distribution is that by using a combination of (simpler) probability distributions, more involved probability distributions can be constructed that for instance model nonlinear dependencies better. However, this is already the case when using a factorization of a mixture pdf as already demonstrated in Fig. 5.4 and hence is less of a virtue contributed by the mixture probability distribution on its own.

By Means of Clustering

Using clustering to subdivide the set of solutions for which to estimate a probability distribution followed by the subsequent estimation of probability distributions for each cluster separately was already discussed when we reviewed the normal mixture pdf in Sect. 5.4.1. Before its use by Ahn et al. (2004) in the construction of the factorized normal mixture probability distribution, the use of clustering to construct mixture probability distributions in EDAs was initially investigated in the binary domain by Pelikan and Goldberg (2000) using the k-means clustering algorithm (Hartigan, 1975). This approach was concurrently investigated in the context of real-valued EDAs by Bosman and Thierens (2001a) (see also (Bosman, 2003)). In their work, a different, slightly faster clustering algorithm was used, called the leader algorithm (Hartigan, 1975). The probability distributions in the mixture are factorized normal pdfs. Hence, the resulting probability distribution is actually a normal mixture probability distribution. The results generally indicate an increase in optimization performance especially on the Rosenbrock function (see Fig. 5.3 for an illustration). However, since no means has been proposed to detect the number of clusters that is actually required, the approach requires significantly more evaluations on problems where a mixture probability distribution is not required. Unfortunately, the improvement of using the mixture of factorized normal pdfs over the use of a single normal pdf on the Rosenbrock function decreases rapidly as the dimensionality of the problem increases. The reason for this is that the number of clusters required to model the nonlinear dependencies between the problem variables properly, increases also with an increase in the problem dimensionality. As this in turn requires a larger population size to ensure enough solutions in each cluster to estimate a factorized normal pdf, the increase in search efficiency in terms of the number of required evaluations to solve the problem, decreases.

The difference with the factorized normal mixture pdf used by Ahn et al. (2004) is that because in their probability distribution the top-structure is a factorization, if the number of variables between which nonlinear dependencies exist is limited and the groups of such nonlinearly dependent variables are additively decomposable in the problem to be optimized, the factorization can match this additive decomposability structure whereas the mixture pdf can match the (smaller) nonlinear dependencies. This additive decomposability and the smaller nonlinear dependency can clearly not be exploited effectively by the top-level mixture probability distribution. Indeed, on an additively decomposable variant of the Rosenbrock function, the approach by Ahn et al. (2004) outperforms the approach by Bosman and Thierens (2001a). However, it should be clear that the incentive behind the mixture probability distribution is inherently different, namely to distribute the search bias over the search space and thus to be able to cope with multiple regions of interest in the search space simultaneously. The mixture probability distribution approach is hence more general as it also allows for instance to use the factorized normal mixture pdf in the EDA by Ahn et al. (2004) in each cluster, effectively combining the positive features of the EDA by Ahn et al. (2004) with the incentive behind the approach by Bosman and Thierens (2001a).

5.4.3 Other Probability Distributions

Although the use of the parametric pdfs in combination with factorizations and the use of clustering techniques to obtain mixture probability distributions as discussed in earlier sections are representative of the most common techniques, other, often more involved, techniques exist.

Top-Level Discretization and Bottom-Level Continuous Modeling

A different use of the normal kernels pdf was proposed by Ocenasek and Schwarz (2002). In their EDA, a discrete decision tree (Friedman and Goldszmidt, 1996) is built for each variable by discretizing all other real-valued random variables using axis-parallel splits. The variable that is found to influence the current variable of interest the most is used to split up the data range. The procedure is recursively repeated after splitting the data. The rationale behind this approach is that once no more splitting is found to be beneficial, the leaf nodes in the decision tree correspond to subranges in the data set that can be properly estimated using univariately factorized pdfs. The pdf used by Ocenasek and Schwarz (2002) is the normal kernels pdf with $\alpha = 1$. Since now the variances are specifically tailored to the subsets constructed by the decision trees, this particular use of the normal kernels pdf is much more robust. Although the resulting EDA was found to obtain much better results on the Rosenbrock function, the overall performance of the algorithm was not found to be significantly better than previously developed EDAs. Moreover, the optimum of the Rosenbrock could still not be reached satisfactorily (Kern et al., 2004).

Probability Distributions Based on Latent Variables

A few alternative approaches currently used in real-valued EDAs are based on the use of latent, or hidden, variables. These techniques attempt to model the underlying data source by projecting the data onto another domain while attempting to retain the most important features. Often, the dimensionality of the data is then reduced.

An example of such techniques is the well-known principal component analysis (PCA) (Jolliffe, 1986). In PCA, $l' < l$ vectors are chosen such that the variance in those vectors is the largest when projecting the one-dimensional data onto these vectors. The latent variables are the newly introduced variables that are used to model the data. Another approach in which latent variables are used, is the Helmholtz machine. A Helmholtz machine is closely related to neural networks and consists of a layer of input variables representing the l dimensions of the data and provides for multiple layers of latent variables. Connections between these layers allow for the learning of a model that describes the data, as well as the generation of new data.

Bishop (1999) indicated how PCA can be used to estimate probability distributions and how to generate new samples from the estimated probability distributions. Using normal pdfs, the PCA-based estimated probability distribution over the selected solutions, is an one-dimensional normal probability distribution. This approach has been used for real-valued optimization (Shin and Zhang, 2001; Cho and Zhang, 2001; Shin et al., 2001). The authors also used Helmholtz machines in combination with normal pdfs. The results obtained are comparable to those obtained with factorized probability distributions, but the number of latent variables is fixed beforehand, whereas the approaches using factorized probability distributions are able to learn the structure of the probabilistic model from data.

In the approach by Cho and Zhang (2002) a mixture of factor analyzers is used. Standard factor analysis is a latent variable model that is based on a linear mapping between the random variables and the latent variables, resulting in a normal distribution that is modeled over the original random variables. An EM-algorithm is used to find parameter estimates for the latent variables in each mixture component as well as the mixture coefficients themselves. The number of mixture components is fixed beforehand as are again the number of latent variables. The results for numerical optimization indicate better performance for the mixture over the single factor analysis and other nonmixture real-valued probabilistic model-building EAs, but the structures of the latent-variable models were composed by hand.

The recent most EDA based on latent variable models was introduced by Cho and Zhang (2004). The probability distribution used in their EDA is based on the variational Bayesian independent component analyzers mixture model by Choudrey and Roberts (2003). Although the formal equations involved are out of scope for this chapter, we note that this recent way of estimating probability distributions overcomes many of the problematic issues that traditional models such as mixture of normal distributions estimated with EM algorithms have. In principle, the model is capable of modeling any density function as long as enough components are used in the mixture. Traditionally, there exists a problem with computing the required number of components. In the model by Choudrey and Roberts (2003) this problem is solved using Bayesian inference. The use of this extremely powerful model in an EDA resulted in an improvement over earlier proposed real-valued EDAs. However, although the approach obtained much better results on the Rosenbrock function, it was still not able to converge to the optimum for the 10-dimensional case using a population size of 6,400. Although the model is in theory capable of capturing the structure of the search space, the number of solutions that is required to actually construct this probability distribution well enough is extremely large.

5.5 Analysis and Redesign of Real-Valued EDAs

5.5.1 Analysis

Following the traditional motivation from discrete spaces and the lessons learned from discrete GA research (Goldberg, 2002b), one assumes that initially enough information is supplied to the EDA, i.e., all the building blocks are initially in the population. This is typically ensured by making the population large enough. All that the EDA is now required to do is to detect the building blocks and mix them properly. From the previous chapters we already know that by using maximum-likelihood estimates for discrete spaces in combination with factorization selection techniques, EDAs can be built that very efficiently meet with this requirement and consequently achieve efficient optimization. By using a proper factorization, the important substructures of the optimization problem at hand are identified and specific combinations of bits can be assigned high probabilities of replication when generating new solutions. Hence, the important substructures never get lost during optimization and are combined effectively to reach the optimal solution.

All real-valued EDAs for numerical optimization that we discussed so far employ a probability distribution estimation technique that is equally bent on describing the set of selected solutions as well as possible. Whenever possible, maximum-likelihood estimates are used. An approximation of the maximum-likelihood estimate is used otherwise. One could validly argue that this approach thus follows a direct translation of the EDA approach in the discrete domain into the continuous domain. Alternatively, one can argue that this approach conforms completely to the design of the optimal EDA that uses the true probability distribution as discussed in Sect. 5.3. Still, real-valued EDAs so far have made for a far smaller success story than have EDAs for the discrete domain, even though the range of actual EDA approaches is wider in the continuous domain. For instance, no EDA discussed so far in this chapter is able to optimize the Rosenbrock function efficiently even though this problem is unimodal and smooth. This has recently led researches to take a closer look at what is happening with EDAs in the continuous domain.

The apparent failure of real-valued EDAs to solve the Rosenbrock function was first noticed by Bosman and Thierens (2001b) (see also (Bosman, 2003)). The explanation given for this failure was that because no assumptions are made on the source of the data from which to estimate the probability distribution, real-valued EDAs disregard gradient information. In other words, even part of an interesting region is discovered and the selected solutions are located on a slope towards the optimum, the EDA can converge *on this slope* since it only effectively searches inside the area covered by the selected solutions. Through maximum-likelihood estimates there is no means of generalizing the estimated density outside this area. The premature convergence on the Rosenbrock function was discussed further more recently by Bosman and Grahl (2005) as well as by Yuan and Gallagher (2005). Premature convergence was also observed by Kern et al. (2004) on the even simpler tilted-plane function where the optimum is located at a boundary of the search space. For a real-valued EDA using the normal pdf with maximum-likelihood estimates it was proved by Grahl et al. (2005a) (see also (Grahl et al., 2005b)) that the mean of this EDA can indeed only move a limited distance through the search space. This distance is dependent on the selection intensity and the initial variance (i.e., spread of the initial population). Unless the optimum lies within this bounded distance, the variance will go to 0 too quickly causing the EDA to converge prematurely.

One could now argue that this is an unfair analysis as it violates the assumption that initially enough data is available, i.e., that the optimum inside the region covered by the initial population. If this is not the case, one should consider the analogy of solving the one-max problem with a binary EDA when for some variables not a single 1 is present in the population. It is however important to note that in the continuous domain that such essential information (i.e., the structure of the problem to be optimized, typically the building blocks in the discrete case) easily gets lost during optimization. In general, the reason for loss of essential information is that the problem structure is not matched by the search bias of the optimizer. For instance, if a binary EDA is not capable of detecting the individual deceptive trap functions, the additively decomposable trap function cannot be optimized efficiently because the building blocks will have a too small probability of surviving variation.

In the continuous domain however, capturing the essential information in the probability distribution is much more difficult (Bosman and Grahl, 2005). Problem structure in the continuous domain is exemplified by the contour lines of the function to be optimized. Hence, the true distribution of the selected individuals can be of virtually any shape. This means that we need both the universal approximation property of our probability distribution as well as an infinite sample size to ensure that the problem structure is effectively exploited by the EDA. However, such universal approximation is computationally intractable. In practice, a continuous EDA will have to rely on tractable probability distributions such as the normal pdf. This means that the contours of the probability distribution may no longer match with the contour lines of the function to be optimized. Consequently, the notion of dependency as observed in the estimated probability distribution does not have to match with the actual dependency of the variables according to the optimization problem at hand. Hence, linkage information in the continuous domain cannot by definition be extracted from the estimated probability distribution. What is even more important to realize is that by using relatively simple pdfs the property of having (local) optima inside the area covered by the selected solutions may easily be lost during optimization. This will put the EDA back in the position from where it can provably not find the optimum using maximum-likelihood estimates, even when the selected solutions are located on a slope.

5.5.2 Redesign

To overcome the problem identified above, three approaches can generally speaking be taken. We will discuss these three approaches in turn.

More Involved Probability Distributions

The capacity of the class of probability distribution used in the EDA can be increased. As pointed out in Sect. 5.4.3, the approach by Cho and Zhang (2004) is probably the best example of this approach. But even though the model used here is extremely flexible, the population size that is generally required to get the estimated probability distribution right and not lose the encapsulation of optima during optimization is enormous. Moreover, the computational costs of estimating such involved probability distributions is considerable.

Explicit Gradient Exploitation

To allow the EDA to search outside of its range the explicit use of gradient information can be enforced by hybridizing the EDA with local search operators. Although such a hybrid approach can indeed be effective (Bosman and Thierens, 2001b), such hybridization is possible for all EAs. Hence, we seek a more specific solution to the problem at hand. Moreover, if desired, the resulting improved EDA can then still be hybridized.

Variance Scaling

The third approach is strikingly simple in its nature, but it is specifically designed for use in real-valued EDAs and has already been shown to be a very effective principle (Ocenasek et al., 2004; Bosman and Grahl, 2005; Yuan and Gallagher, 2005). The basic idea is to set the variance of the probability distribution during the search not (only) according to the maximum-likelihood approach, but (also) according to other criteria such as the success rate.

In the approach by Ocenasek et al. (2004) the decision-trees combined with the normal kernels pdf is used (Ocenasek and Schwarz, 2002). In addition however, a scaling factor is maintained. When drawing new solutions from the estimated probability distribution, the variance of a normal kernel is scaled by multiplication with this factor. The size of the scaling factor depends on the success rate of the restricted tournament selection operator (i.e., the number of times an offspring solution is better). An adaptive scheme is used that changes the scaling factor to ensure that this success rate stays in-line with the 1/5 success rule for evolution strategies (ES) (Bäck and Schwefel, 1993).

Yuan and Gallagher (2005) showed that by scaling the variance of the estimated normal probability distribution by a factor of 1.5, an EDA based on this variance-scaled normal probability distribution is capable of finding the optimum of the Rosenbrock function. Although in itself this result is striking, a fixed scaling factor will in general not be optimal. In the approach by Bosman and Grahl (2005) a simple, but effective, adaptive-variance-scaling scheme is proposed for use in an EDA that uses the factorized normal pdf. Similar to the approach by Ocenasek et al. (2004) a scaling factor is maintained and upon drawing new solutions from the probability distribution, the covariance matrix is multiplied by this scaling factor. Updating this scaling factor is done differently however. If the best fitness value improves in one generation, the current size of the variance allows for progress. Hence, a further enlargement of the variance may allow for further improvement in the next generation and the scaling factor is increased. If on the other hand the best fitness does not improve, the range of exploration may be too large to be effective and the scaling factor is (slowly) decreased. In addition to variance scaling Bosman and Grahl (2005) also show how a test can be constructed specifically for the normal distribution that allows to distinguish between situations where variance scaling is required and where variance scaling is not required. The test is based on (ranked) correlation between density and fitness.

The results of the above mentioned EDAs that employ the scaling of the variance are significantly better than those of the earlier EDAs that only employ maximum-likelihood estimates. The algorithms are for instance capable of finding the optimum of the Rosenbrock function efficiently. The results of the algorithm by Bosman and Grahl (2005) on the Rosenbrock function are shown in Fig. 5.3. The key to their success is that the EDAs without variance scaling are very efficient if the structure of the problem is not too complicated, such as is the case for the sphere function. If the structure gets more involved, there is hardly any means to generalize the probability distribution over the entire search space. With the variance-scaling approach the EDA is equipped with the possibility to shift its focus by means of variance adaptation and to go and find local structures that it can efficiently exploit without the additional variance scaling.

The results obtained with variance-scaling bring the EDAs in strong competition with one of the leaders in (evolutionary) continuous optimization, the CMA-ES (Hansen and Ostermeier, 2001; Hansen et al., 2003; Kern et al., 2004). Strikingly, the EDA approach using the normal pdf is at the top level similar to the CMA-ES. Both algorithms compute values for the parameters of the normal pdf and subsequently draw new solutions from this normal pdf. Hence, the CMA-ES can validly be argued to actually be an EDA. The difference between the approaches lies in how the parameters are set. In the pioneering EDA approaches, maximum-likelihood estimates are used on the basis of the set of selected solutions. It has been observed however that this more often does not lead to efficient optimization. The CMA-ES on the other hand has been found to be a very efficient optimizer. The CMA-ES actually uses the same estimate for the mean (i.e., the center of mass (also called sample average) of the selected solutions). The differences lies in the way the covariance matrix is built. Strikingly, it is exactly in the changing of the covariance matrix (by scaling) that the EDA approach has recently been found to have room for improvement. In the CMA-ES, the covariance matrix is built using information about how the mean of the population has shifted the previous generation. This information is combined with the (by multiplication slowly fading) memory or cumulation of shifts in earlier generations to obtain what is called the *evolution path*. In this approach there is no need to explicitly check for success in generating new better solutions as the shift towards better regions automatically influences the evolution path, which is subsequently used to determine the covariance matrix.

As a result of using the mean shift, the CMA-ES is clearly capable of exploiting gradient properties of a problem, which is exactly what was hypothesized to be lacking in maximum-likelihood EDAs (Bosman and Thierens, 2001b; Bosman, 2003). A final striking correspondence between the CMA-ES and the EDAs that use variance-scaling is that the covariance matrix in the CMA-ES is actually factorized into the multiplication of a covariance matrix and what is called a global step-size. The global step-size is also computed on the basis of the evolution path. This factorization is identical to the variance-scaling approach in EDAs, the only difference being in terminology and the way in which the scaling factor (or global step-size) is computed.

It is interesting to see that although the background and motivation for the (real-valued) EDA approach and the CMA-ES approach are different, the developments in these areas appear to be converging onto a similar approach. Future research will have to point out which of the approaches is the overall best way to go and whether the benefits of both approaches can be integrated to arrive at even better evolutionary optimizers.

5.6 Summary and Conclusions

In this chapter we have discussed the design of real-valued EDAs for numerical optimization. We have indicated what types of problem difficulty we typically expect and what the optimal EDA for numerical optimization looks like. Subsequently we provided an overview of existing EDA approaches and indicated that already there exists a vast array of approaches as many different probability distributions have been studied.

Although the capacity of the class of probability distribution used in the EDA has grown significantly, real-valued EDAs that attempt to estimate the probability distribution as well as possible (i.e., typically using maximum-likelihood estimates) from the set of selected solutions seem not to work well on problems such as the Rosenbrock problem. Although this problem is unimodal and has nice smooth gradient properties, EDAs tend to converge prematurely while still on the slope towards the optimum.

The main problem is that in continuous problems, the structure of a problem is characterized by the contours of the function to be optimized. These contours can take any shape. Hence fitting the problem structure in the continuous case requires the intractable universal approximation property and huge amounts of data, rending the EDA approach inefficient. It is therefore much more convenient to rely on the use of simpler probability distributions. However, by doing so, important regions of the search space may get lost during optimization because they are assigned extremely low densities due to the mismatch between the density contours of the estimated probability distribution and the density contours of the fitness function. To still be able to use simpler probability distributions alternative means of overcoming problems of premature convergence should be found. To this end it is much more convenient to view problem structure as an arrangement of slopes and peaks in the search space. These simpler substructures are much easier to take into account and to build inductive search biases for. Actually, the current range of EDAs is already capable of searching such simpler substructures efficiently. Hence, what should in addition be provided is the means to shift between these substructures and ultimately focus on the most interesting one. To this end variance-scaling approaches have recently been proposed and have been shown to lead to much improved EDAs.

This significant improvement signals that there is a new open road to explore for the design of real-valued EDAs for numerical optimization; and this one has a fast lane.

References

Ahn, C. W., Ramakrishna, R. S. and Goldberg, D. E. (2004). Real-coded Bayesian optimization algorithm: Bringing the strength of BOA into the continuous world, *in* K. Deb et al. (Eds.), *Proceedings of the Genetic and Evolutionary Computation Conference – GECCO–2004*, Springer, Berlin Heidelberg New York, pp. 840–851

Anderson, T. W. (1958). *An Introduction to Multivariate Statistical Analysis*, Wiley, New York

Bäck, T. and Schwefel, H.-P. (1993). An overview of evolutionary algorithms for parameter optimization, *Evolutionary Computation* **1**(1): 1–23

Baluja, S. and Caruana, R. (1995). Removing the genetics from the standard genetic algorithm, *in* A. Prieditis and S. Russell (Eds.), *Proceedings of the Twelfth International Conference on Machine Learning*, Morgan Kauffman, Madison, WI, pp. 38–46

Bilmes, J. (1997). *A gentle tutorial of the EM algorithm and its application to parameter estimation for Gaussian mixture and hidden markov models,* Technical Report ICSI TR-97-021, Department of Electrical Engineering, University of Berkeley, Berkeley, CA

Bishop, C. M. (1999). Latent variable models, *in* M. I. Jordan (Ed.), *Learning in Graphical Models*, MIT, Cambridge, MA

Bosman, P. A. N. (2003). *Design and Application of Iterated Density-Estimation Evolutionary Algorithms*, PhD thesis, Utrecht University, Utrecht, The Netherlands

Bosman, P. A. N. and Grahl, J. (2005). Matching inductive search bias and problem structure in continuous estimation of distribution algorithms, Technical report, Mannheim Business School, Department of Logistics, http://www.bwl.uni-mannheim.de/Minner/hp/_files/forschung/reports/tr-2005-03.pdf

Bosman, P. A. N. and Thierens, D. (2000a). Continuous iterated density estimation evolutionary algorithms within the IDEA framework, *in* M. Pelikan et al. (Eds.), *Proceedings of the Optimization by Building and Using Probabilistic Models OBUPM Workshop at the Genetic and Evolutionary Computation Conference – GECCO – 2000*, Morgan Kaufmann, San Francisco, CA, pp. 197–200

Bosman, P. A. N. and Thierens, D. (2000b). Expanding from discrete to continuous estimation of distribution algorithms: The IDEA, *in* M. Schoenauer et al. (Eds.), *Parallel Problem Solving from Nature – PPSN VI*, Springer, Berlin Heidelberg New York, pp. 767–776

Bosman, P. A. N. and Thierens, D. (2001a). Advancing continuous IDEAs with mixture distributions and factorization selection metrics, *in* M. Pelikan and K. Sastry (Eds.), *Proceedings of the Optimization by Building and Using Probabilistic Models OBUPM Workshop at the Genetic and Evolutionary Computation Conference – GECCO–2001*, Morgan Kaufmann, San Francisco, CA, pp. 208–212

Bosman, P. A. N. and Thierens, D. (2001b). Exploiting gradient information in continuous iterated density estimation evolutionary algorithms, *in* B. Kröse et al. (Eds.), *Proceedings of the Thirteenth Belgium–Netherlands Artificial Intelligence Conference BNAIC-2001*, pp. 69–76

Breiman, L., Friedman, J. H., Olshen, R. A. and Stone, C. J. (1984). *Classification and Regression Trees*, Wadsworth, Pacific Grove, CA

Buntine, W. (1994). Operations for learning with graphical models, *Journal of Artificial Intelligence Research* **2**: 159–225

Cho, D.-Y. and Zhang, B.-T. (2001). Continuous estimation of distribution algorithms with probabilistic principal component analysis, *Proceedings of the 2001 Congress on Evolutionary Computation – CEC – 2001*, IEEE, Piscataway, NJ, pp. 521–526

Cho, D.-Y. and Zhang, B.-T. (2002). Evolutionary optimization by distribution estimation with mixtures of factor analyzers, *Proceedings of the 2002 Congress on Evolutionary Computation – CEC – 2002*, IEEE, Piscataway, NJ, pp. 1396–1401

Cho, D.-Y. and Zhang, B.-T. (2004). Evolutionary continuous optimization by distribution estimation with variational Bayesian independent component analyzers mixture model, *in* X. Yao et al. (Eds.), *Parallel Problem Solving from Nature – PPSN VIII*, Springer, Berlin Heidelberg New York, pp. 212–221

Choudrey, R. A. and Roberts, S. J. (2003). Variational mixture of Bayesian independent component analyzers, *Neural Computation* **15**(1): 213–252

Dempster, A. P., Laird, N. M. and Rubin, D. B. (1977). Maximum likelihood from incomplete data via the em algorithm, *Journal of the Royal Statistic Society Series B* **39**: 1–38

Edmonds, J. (1967). Optimum branchings, *Journal of Research of the National Bureau of Standards* **71B**: 233–240. Reprinted in Math. of the Decision Sciences, *American Mathematical Society Lecture notes in Applied Mathematics*, 11: 335–345, 1968

Edwards, D. (1995). *Introduction to Graphical Modelling*, Springer, Berlin Heidelberg New York

Etxeberria, R. and Larrañaga, P. (1999). Global optimization using Bayesian networks, *in* A. A. O. Rodriguez et al. (Eds.), *Proceedings of the Second Symposium on Artificial Intelligence CIMAF–1999*, Institute of Cybernetics, Mathematics and Physics, pp. 332–339

Friedman, N. and Goldszmidt, M. (1996). Learning Bayesian networks with local structure, *in* E. Horvits and F. Jensen (Eds.), *Proceedings of the Twelfth Conference on Uncertainty in Artificial Intelligence UAI–1996*, Morgan Kaufmann, San Francisco, CA, pp. 252–262

Gallagher, M., Frean, M. and Downs, T. (1999). Real–valued evolutionary optimization using a flexible probability density estimator, *in* W. Banzhaf et al. (Eds.), *Proceedings of the Genetic and Evolutionary Computation Conference – GECCO–1999*, Morgan Kaufmann, San Francisco, CA, pp. 840–846

Goldberg, D. E. (1989). *Genetic Algorithms in Search, Optimization and Machine Learing*, Addison-Wesley, Reading, MA

Goldberg, D. E. (2002a). *The Design of Innovation: Lessons from and for Competent Genetic Algorithms*, Vol. 7 of *Series on Genetic Algorithms and Evolutionary Computation*, Kluwer, Dordrecht

Goldberg, D. E. (2002b). *The Design of Innovation: Lessons from and for Competent Genetic Algorithms*, Vol. 7 of *Series on Genetic Algorithms and Evolutionary Computation*, Kluwer, Dordrecht

Grahl, J., Minner, S. and Rothlauf, F. (2005a). An analysis of iterated density estimation and sampling in the UMDAc algorithm, *Late-Breaking Papers of the Genetic and Evolutionary Computation Conference – GECCO – 2005*

Grahl, J., Minner, S. and Rothlauf, F. (2005b). Behaviour of UMDAc with truncation selection on monotonous functions, *Proceedings of the 2005 Congress on Evolutionary Computation – CEC – 2005*, IEEE, Piscataway, NJ

Hansen, N. and Ostermeier, A. (2001). Completely derandomized self-adaptation in evolution strategies, *Evolutionary Computation* **9**(2): 159–195

Hansen, N., Müller, S. D. and Koumoutsakos, P. (2003). Reducing the time complexity of the derandomized evolution strategy with covariance matrix adaptation (CMA–ES), *Evolutionary Computation* **11**(1): 1–18

Hartigan, J. (1975). *Clustering Algorithms*, Wiley, New York

Holland, J. H. (1975). *Adaptation in Natural and Artifical Systems*, University of Michigan Press, Ann Arbor, MI

Jolliffe, I. T. (1986). *Principal Component Analysis*, Springer, Berlin Heidelberg New York

Kern, S., Müller, S. D., Hansen, N., Büche, D., Ocenasek, J. and Koumoutsakos, P. (2004). Learning probability distributions in continuous evolutionary algorithms – a comparative review, *Natural Computing* **3**(1): 77–112

Larrañaga, P., Etxeberria, R., Lozano, J. A. and Peña, J. M. (2000). Optimization in continuous domains by learning and simulation of Gaussian networks, *in* M. Pelikan et al. (Eds.), *Proceedings of the Optimization by Building and Using Probabilistic Models OBUPM Workshop at the Genetic and Evolutionary Computation Conference – GECCO-2000*, Morgan Kaufmann, San Francisco, CA, pp. 201–204

Lauritzen, S. L. (1996). *Graphical Models*, Clarendon, Oxford

Mühlenbein, H. and Höns, R. (2005). The estimation of distributions and the minimum relative entropy principle, *Evolutionary Computation* **13**(1): 1–27

Ocenasek, J. and Schwarz, J. (2002). Estimation of distribution algorithm for mixed continuous-discrete optimization problems, *2nd Euro-International Symposium on Computational Intelligence*, pp. 227–232

Ocenasek, J., Kern, S., Hansen, N., Müller, S. and Koumoutsakos, P. (2004). A mixed Bayesian optimization algorithm with variance adaptation, *in* X. Yao et al. (Eds.), *Parallel Problem Solving from Nature – PPSN VIII*, Springer, Berlin Heidelberg New York, pp. 352–361

Pelikan, M. and Goldberg, D. E. (2000). Genetic algorithms, clustering, and the breaking of symmetry, *in* M. Schoenauer et al. (Eds.), *Parallel Problem Solving from Nature – PPSN VI*, Springer, Berlin Heidelberg New York, pp. 385–394

Pelikan, M. and Goldberg, D. E. (2001). Escaping hierarchical traps with competent genetic algorithms, *in* L. Spector et al. (Eds.), *Proceedings of the GECCO–2001 Genetic and Evolutionary Computation Conference*, Morgan Kaufmann, San Francisco, CA, pp. 511–518

Pelikan, M. and Goldberg, D. E. (2003). Hierarchical BOA solves Ising spin glasses and MAXSAT, *in* E. Cantú-Paz et al. (Eds.), *Proceedings of the GECCO-2003 Genetic and Evolutionary Computation Conference*, Springer, Berlin Heidelberg New York, pp. 1271–1282

Pelikan, M., Goldberg, D. E. and Cantú-Paz, E. (1999). BOA: The Bayesian optimization algorithm, *in* W. Banzhaf et al. (Eds.), *Proceedings of the GECCO-1999 Genetic and Evolutionary Computation Conference*, Morgan Kauffman, San Francisco, CA, pp. 525–532

Pošík, P. (2004). Distribution tree-building real–valued evolutionary algorithm, *in* X. Yao et al. (Eds.), *Parallel Problem Solving from Nature – PPSN VIII*, Springer, Berlin Heidelberg New York, pp. 372–381

Priebe, C. E. (1994). Adaptive mixtures, *Journal of the American Statistical Association* **89**(427): 796–806

Rudlof, S. and Köppen, M. (1996). Stochastic hill climbing with learning by vectors of normal distributions, *in* T. Furuhashi (Ed.), *Proceedings of the First Online Workshop on Soft Computing (WSC1)*, Nagoya University, Nagoya, Japan, pp. 60–70

Russel, S. and Norvig, P. (2003). *Artificial Intelligence: A Modern Approach*, Prentice-Hall, Englewood Cliffs, NJ

Sebag, M. and Ducoulombier, A. (1998). Extending population–based incremental learning to continuous search spaces, *in* A. E. Eiben et al. (Eds.), *Parallel Problem Solving from Nature – PPSN V*, Springer, Berlin Heidelberg New York, pp. 418–427

Servet, I., Trave-Massuyes, L. and Stern, D. (1997). Telephone network traffic overloading diagnosis and evolutionary computation technique, *in* J. K. Hao et al. (Eds.), *Proceedings of Artificial Evolution '97*, Springer, Berlin Heidelberg New York, pp. 137–144

Shin, S.-Y. and Zhang, B.-T. (2001). Bayesian evolutionary algorithms for continuous function optimization, *Proceedings of the 2001 Congress on Evolutionary Computation – CEC – 2001*, IEEE, Piscataway, NJ, pp. 508–515

Shin, S.-Y., Cho, D.-Y., and Zhang, B.-T. (2001). Function optimization with latent variable models, *in* A. Ochoa et al. (Eds.), *Proceedings of the Third International Symposium on Adaptive Systems ISAS–2001 – Evolutionary Computation and Probabilistic Graphical Models*, Institute of Cybernetics, Mathematics and Physics, pp. 145–152

Tatsuoka, M. M. (1971). *Multivariate Analysis: Techniques for Educational and Psychological Research*, Wiley, New York

Thierens, D. and Goldberg, D. (1993). Mixing in genetic algorithms, *in* S. Forrest (ed.), *Proceedings of the fifth conference on Genetic Algorithms*, Morgan Kaufmann, San Mateo, CA, pp. 38–45

Tsutsui, S., Pelikan, M. and Goldberg, D. E. (2001). Evolutionary algorithm using marginal histogram in continuous domain, *in* M. Pelikan and K. Sastry (Eds.), *Proceedings of the Optimization by Building and Using Probabilistic Models OBUPM Workshop at the Genetic and Evolutionary Computation Conference – GECCO – 2001*, Morgan Kaufmann, San Francisco, CA, pp. 230–233

Yuan, B. and Gallagher, M. (2005). On the importance of diversity maintenance in estimation of distribution algorithms, *in* H.-G. Beyer et al. (Eds.), *Proceedings of the Genetic and Evolutionary Computation Conference – GECCO – 2005*, ACM, New York, USA, pp. 719–726

6

A Survey of Probabilistic Model Building Genetic Programming

Yin Shan, Robert I. McKay, Daryl Essam, and Hussein A. Abbass

Summary. There has been a surge of research interest in estimation of distribution algorithms (EDA). Several reviews on current work in conventional EDA are available. Although most work has focused on one-dimensional representations that resembles the chromosomes of genetic algorithms (GA), an interesting stream of EDA using more complex tree representations has recently received some attention. To date, there has been no general review of this area in the current literature. This chapter aims to provide a critical and comprehensive review of EDA with tree representation and closely related fields.

6.1 Introduction

6.1.1 Estimation of Distribution Algorithms

Evolutionary computation (EC) [69], motivated by evolution in the real world, has become one of the most widely used Machine Learning techniques, because of its effectiveness and versatility. It maintains a population of solutions, which evolves subject to selection and genetic operators (such as recombination and mutation). Each individual in the population receives a measure of its fitness, which is used to guide selection. EC covers a wide range of techniques, but we will frame our discussion primarily in terms of genetic algorithms (GA) [14] and genetic programming (GP) [17, 33].

Lately, there has been a surge of research interest in estimation of distribution algorithms (EDA) [37, 42], which is partially motivated by EC. EDA is also known as probabilistic model-building genetic algorithms (PMBGA) [49] or iterated density estimation evolutionary algorithms (IDEA) [13]. EDAs explicitly encode the knowledge accumulated in the course of the search in well-structured models, typically probabilistic models, and thus it becomes possible to explicitly exploit that knowledge to adaptively improve the efficiency of the search. More specifically, these models are inductively learnt from good individuals (training examples), and are sampled to create the new individuals of the next generation. A population is not usually maintained between generations, and genetic operators are omitted from EDAs, either partially or

Y. Shan et al.: *A Survey of Probabilistic Model Building Genetic Programming*, Studies in Computational Intelligence (SCI) **33**, 121–160 (2006)
www.springerlink.com

completely. Instead, EDA is designed to capture the interactions among genes, which represent the internal structure of problem solutions, and in this way it estimates the distribution of good solutions directly, rather than employing genetic operators. The major differences between various EDA methods lie in the different formalism of models, depending on the types of knowledge, i.e. the interactions among genes, intended to be represented.

There are several reasons for the increasing interest in EDA. The first reason is the theoretical attraction. The highly complex and dynamic consequences of genetic operators are extremely hard to understand and predict, thus hindering efforts to further improve the performance of EC systems. Replacing genetic operators and populations with a simple yet powerful model makes it simpler to understand system behaviour. In some simple cases, EDA is a quite accurate approximation of conventional EC [27, 42]. Secondly, in terms of practical usefulness, superior performance of EDAs relative to conventional EC has been reported in a number of publications [9, 10, 48, 50, 59].

6.1.2 Basic Algorithm of EDA

EDA uses a probabilistic model to estimate the distribution of promising solutions, and to further guide the exploration of the search space. By iteratively building and sampling from the probabilistic model, the distribution of good solutions is (hopefully) approximated.

The formal description of EDA can be found in [13]. For the completeness of this survey, we restate it as follows. Briefly, assume $\boldsymbol{Z}$ is the vector of variables we are interested in. $D^H(\boldsymbol{Z})$ is the probability distribution of individuals whose fitnesses are greater than or equal to some threshold H (without loss of generality, we assume we are dealing with a maximisation problem). Now if we know $D^{H_{\text{opt}}}(\boldsymbol{Z})$ for the optimal fitness H_{opt}, we can find a solution by simply drawing a sample from this distribution. However, usually we do not know this distribution. Due to the lack of prior information on this distribution, we start from a uniform distribution. In the commonest form of the algorithm, we generate a population $\mathcal{P}$ with n individuals and then select a set of good individuals $\mathcal{G}$ from $\mathcal{P}$. Since $\mathcal{G}$ contains only selected individuals, it represents the search space that is worth further investigation. We now estimate a probabilistic model $\mathcal{M}(\zeta, \theta)$ from $\mathcal{G}$. ζ is the structure of the probabilistic model $\mathcal{M}$ while θ is the associated parameter vector. With this model $\mathcal{M}$, we can obtain an approximation $\bar{D}^H_{\mathcal{M}}(\boldsymbol{Z})$ of the true distribution $D^H(\boldsymbol{Z})$. To further explore the search space, we sample distribution $\bar{D}^H_{\mathcal{M}}(\boldsymbol{Z})$, and the new samples are then re-integrated into population $\mathcal{P}$ by some replacement mechanism. This starts a new iteration.

In this process, due to the lack of knowledge of the true distribution, a usually well-studied probabilistic model $\mathcal{M}(\zeta, \theta)$ to approximate the distribution has to be introduced. Therefore, a vital part in the EDA algorithm is the appropriate choice of the model with respect to the true model. This is the reason that current EDA research is largely devoted to finding appropriate

1. Generate a population $\mathcal{P}$ randomly.
2. Select a set of fitter individuals $\mathcal{G}$ from $\mathcal{P}$.
3. Estimate a probabilistic model $\mathcal{M}$ over $\mathcal{G}$.
4. Sample from the probabilistic model $\mathcal{M}$ to obtain a set of new individuals $\mathcal{G}'$.
5. Incorporate $\mathcal{G}'$ into population $\mathcal{P}$.
6. If the termination condition is not satisfied, go to 2.

Fig. 6.1. High level algorithm of EDA

models and is hence one of the natural ways to differentiate EDA methods, i.e. with respect to their probabilistic models. The particular choice of this introduced probabilistic model is strongly related to specific assumptions about building blocks.

All the EDA methods share a similar high level algorithm presented in Fig. 6.1. EDA starts with a randomly generated population $\mathcal{P}$. A probabilistic model $\mathcal{M}$ is learnt from a set of individuals $\mathcal{G}$ selected from this population. A new set of individuals $\mathcal{G}'$ is generated by sampling from the learnt model $\mathcal{M}$. The new population is formed by incorporating $\mathcal{G}'$ into the original population $\mathcal{P}$. The next iteration starts again from this new population.

Given this high level algorithm, there are many variants. For example, each iteration may create a new population, or may simply replace part of the old population with newly generated individuals; the system may learn a model from scratch or update the previous model; and so on.

6.1.3 Extending EDA to GP style Tree Representation

Although EDA is becoming one of the most active fields in EC, the solution representation in most EDA is a GA style linear representation (one-dimensional array, known as a chromosome in the GA literature). The more complex tree representations, resembling GP individuals, have received only limited exploration [14, 53, 63, 64, 71, 82], perhaps as a result of their intrinsic complexity. This is unfortunate, because tree representations provide a natural and expressive way of representing solutions for many problems. Although there are only a limited number of publications in this area, it appears that authors are sometimes not fully aware of the work of others. This chapter aims to provide a comprehensive and critical survey of these works, as a resource for future workers. For simplicity, we refer to the idea of applying EDA approaches to estimate the distribution of GP-style tree form solutions as EDA–GP, while referring to the EDA dealing with GA style linear strings as conventional EDA. The term EDA–GP reflects the connection between EDA and GP, but is not intended to imply that EDA–GP employs genetic search. There are several GP works, for example [44, 45, 72], which do not strictly employ tree-representation. Due to the significant differences in the representation from the tree structure, these works are not covered in the chapter, and to minimise confusion, the term GP in this chapter generally refers to GP with a tree representation.

Extending conventional EDA to tree-representations is non-trivial, because of the complexity of tree structures. Tree representations provide an alternative way of representing solutions, which are important to a large class of problems, but at the same time introduce important complications, including larger search spaces and more complex interactions among tree nodes. A number of different models have been introduced to represent the important interactions. To summarise, there are two very distinct classes of models, prototype tree models and grammar models. This survey's organisation is based on this classification.

6.1.4 Structure of the Chapter

Although EDA–GP is in its infancy, the underlying ideas of EDA–GP are not entirely new. At first, this type of learning was applied as a supplement to conventional GP systems, as a mechanism for learning modularity. Later, this research became more and more independent of conventional GP, and methods which systematically learn mathematical models were proposed.

The remainder of this chapter is organised as follows. The next section is a brief introduction to GP. The lessons from GP research in searching for good probabilistic models for EDA–GP are presented in Sect. 6.3. The two main streams of EDA–GP work will be presented in the two sections which follow. The first stream, presented in Sect. 6.4, stems from probabilistic incremental program evolution (PIPE) [63], and covers other approaches based on the probabilistic model proposed in PIPE. The second stream, presented in Sect. 6.5 covers grammar-based representations. In Sect. 6.6, an alternative perspective for viewing EDA–GP, i.e. the perspective of probabilistic graph models, is presented. Ant Colony Optimisation (ACO) is a closely related field to EDA–GP. The relevant work in this field is reviewed in Sect. 6.7. The last section provides the conclusion and addresses future research issues.

6.2 Genetic Programming and Its Tree Representation

GP [18, 33, 65] is an evolutionary computation approach (EC). GP can be most readily understood by comparison with GA [14, 28]. The basic algorithm of GP can be found in Fig. 6.2. This algorithm is very similar to that of GA.

1. Generate a population $\mathcal{P}$ randomly.
2. Select a set of fitter individuals $\mathcal{G}$ from population $\mathcal{P}$.
3 Apply genetic operators on the set of selected individuals $\mathcal{G}$ to obtain a set of children $\mathcal{G}'$.
4. Incorporate the children $\mathcal{G}'$ into population $\mathcal{P}$.
5. If the termination condition is not satisfied, go to 2.

Fig. 6.2. Basic Algorithm of GP

Essentially, an initial population is randomly generated. Then, to each population in turn, genetic operators are imposed on stochastically selected individuals from one population to obtain its successor population.

Rather than evolving a linear string as GA does, Genetic Programming evolves computer programs, which are usually tree structures. The fitness of each individual is a function of the effectiveness of the program. The genetic operators, such as crossover, are refined to act upon sub-trees, so as to ensure that the child is syntactically correct.

Given the basic algorithm in Fig. 6.2, some important details of this algorithm will be highlighted in the following subsections. Step 1 is concerned with the initial population. It is presented in Sect. 6.2.1. In step 2, most of the common selection methods in GA may be applied. There are usually three genetic operators involved in step 3, which are presented in Sect. 6.2.2. In step 4, two possible approaches can be used, either the generational approach or the steady-state approach, presented in Sect. 6.2.3. There is some research in GP which incorporates some form of learning into conventional GP. These studies are related to this review and therefore are briefly mentioned in Sect. 6.2.4. A number of important GP issues which are relevant to this survey are discussed in Sect. 6.2.5.

6.2.1 Generation of Initial Population

Given two predefined sets – the terminal and non-terminal (function) sets – the generation of the initial population is a process of iteratively creating individuals by stochastically choosing symbols from these two sets and assembling them into a valid tree structure. There are many variants, primarily aimed at biasing the shapes of the trees produced, but the details are beyond the scope of this chapter. The terminal and non-terminal sets are the alphabet of the programs to be made. The terminal set consists of the variables and constants of the programs, while the non-terminal set consists of the functions of the program. For example, the symbolic regression problems are a class of benchmark problems in GP [33]. They require GP to search for a computer program (or in this case, more accurately, a mathematical function) to approximate the given input and output. For these problems, the terminal set could be $\{x, y, z\}$ and the non-terminal set could be $\{+, -, \times, /, \sin, \cos, \log, \mathrm{e}^\wedge\}$.

6.2.2 Genetic Operators

There are three basic genetic operators in GP and GA: reproduction, crossover and mutation.

Reproduction

Reproduction is straightforward. It simply copies the individual and places it into the new population.

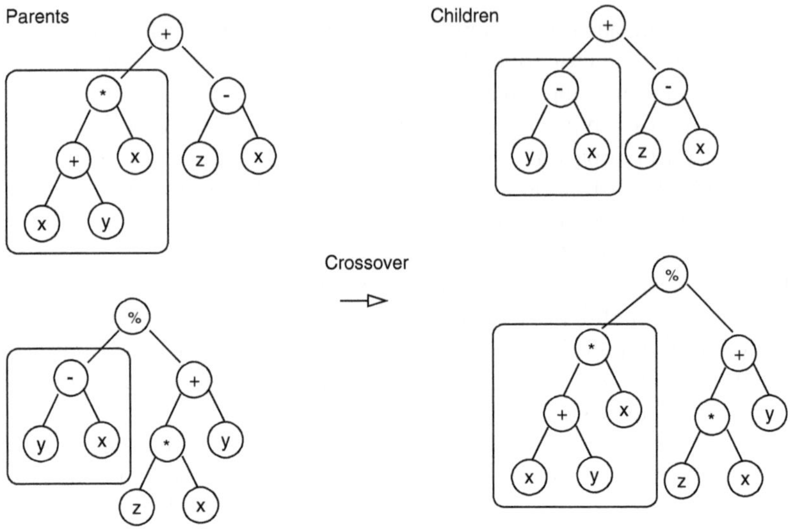

Fig. 6.3. Crossover in GP

Crossover

Crossover combines the genetic material of two parents by swapping certain parts from both parents. Given two parents which are obtained by some selection method, the crossover performs two steps:

- Select randomly a sub-tree in each parent.
- Swap the selected sub-trees between parents.

This is illustrated in Fig. 6.3. The sub-trees selected in parents are highlighted in the rectangle box on the left hand side. On the right hand side, those two sub-trees have been swapped to generate children.

Mutation

Mutation acts on only one individual. It introduces a certain amount of randomness, to encourage exploration. Given one parent obtained by some selection method, mutation performs three steps:

- Randomly select a sub-tree in the parent.
- Remove the selected sub-tree.
- Randomly generate a new sub-tree to replace the removed sub-tree.

This is illustrated in Fig. 6.4. To obtain the child, the selected sub-tree highlighted in the rectangular box on the left hand side is replaced by the newly generated sub-tree on the right hand side.

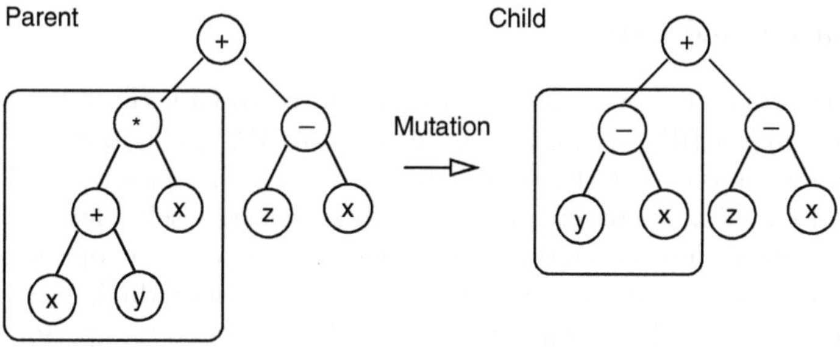

Fig. 6.4. Mutation in GP

6.2.3 Incorporation of New Individuals

There are two alternatives for implementing step 5 of the algorithm in Fig. 6.2, the generational approach and the steady-state approach. In each iteration, the generational approach discards the entire old population $\mathcal{P}$ and replaces it with a newly created set of individuals $\mathcal{G}'$. Each iteration is a generation. In contrast, for the steady-state approach, once the individuals $\mathcal{G}'$ are created, they are incorporated back into the population $\mathcal{P}$ directly, i.e. the old population $\mathcal{P}$ is maintained and some of its individuals are replaced by the new individuals according to some rules. Therefore, there are no clearly identifiable generations.

6.2.4 Learning in Genetic Programming

In conventional GP research, the knowledge encoded in the population, and the value of using it, have long been a focus of study. Consequently, numerous methods have been proposed to utilise this knowledge, for example by adaptive genetic operators, and learning modularity. Studies in this direction provides some insight into how GP works and as a result have had an important influence on EDA–GP research.

Modularity and Building Blocks

Modularity and building blocks are related to the processes of hierarchical problem solving and decomposition. Building blocks (BBs) are defined as sub-trees which appear more frequently in good individuals. If building blocks can be correctly identified and used, the performance of GP may be significantly improved.

This line of research includes automatically defined functions (ADF) [33], genetic library builder (GLiB) [4] and adaptive representation (AR) [57]. The basic idea is that during the search, good sub-trees are identified, either heuristically or by means of evolution, and are then explicitly encapsulated in some form as one entity, so that they can be re-used later on.

Permutation and Crossover

Permutation and crossover are also closely related to BBs. Since the discovery and utilisation of BBs are important aspects of GP, it is important to adapt genetic operators so that they can help to preserve and promote BBs.

In [3], two self-adaptive crossover operators, selective self-adaptive crossover and self-adaptive multi-crossover, were proposed. These new operators adaptively determine where crossover will occur in an individual. Experimental results demonstrate that both of these self-adaptive operators perform as well or better than standard GP crossover.

Recombinative guidance for GP is proposed in [29]. In this approach, all the performance values for all the sub-trees of a GP tree are calculated. These values are then used to decide which sub-tree will be chosen to apply GP operations on. Although GP with recombinative guidance performs well on some problems, the definition of a sub-tree value is problem dependent.

6.2.5 Some Issues in GP

Although the basic ideas of GA and GP are similar, they have unique characteristics because of the representation differences. We focus on some of the characteristics of GP that are relevant to this survey. In order to highlight these characteristics, we need to sometimes make a comparison with classic GA, on which conventional EDA is based. Classic GA in this discussion is roughly the GA system with fixed length, semantics attached to each locus, and no genotype to phenotype mapping.

Semantics

When evolving a linear structure, as a classic GA does, it is usually assumed that the semantics are attached to each locus (each position of the GA linear string is called a locus). For example, when using a classic GA to solve a travelling salesman problem, one common encoding method has each locus representing one step, in which one specific city is visited. In other words, the first locus is interpreted as the city ID that is to be visited in the first step, the second locus is the second city to be visited, etc.

However, it is very hard, if not impossible, to build an analogy for this in GP. GP tries to assemble a set of symbols which have meaning *on their own*. Thus the meaning does not change, no matter where the symbol is. Consequently, the effect of the node has to be understood in its surrounding context, not by the absolute position of the symbol.

For example, in the symbolic regression problem, each symbol has its own meaning; for example, symbol $\times$ means multiplication. This meaning does not need to be interpreted according to its position. No matter whether $\times$ sits on either the root node or a node at some other depth level, $\times$ has the same meaning.

Building Blocks

The Schema Theorem and its related building block hypothesis [14, 28] provide a mathematical perspective to analyse GA and interpret its behaviour. These ideas have also been extended to GP research.

In GA, a schema is defined as any string composed of 0's, 1's and *'s. Each schema represents the set of bit strings containing the indicated 0's and 1's, with each "*" interpreted as a "don't care". For example schema 0*1 represents the set of bit strings composed of 001 and 011 exclusively. As we can see, in this definition, a schema implicitly takes position as a reference. This is related to the semantics discussed earlier. In theoretical GP research, such as [33, 46, 79], because the semantics are attached to the symbols, the schemas are non-rooted and thus the position reference is removed. In these studies, a schema is defined as a sub-tree of some kind, and the absolute position of the sub-tree is not considered. The quality of a non-rooted schema is hence determined by its own structure, not just by where it is. Some more recent studies [34, 56] have introduced rooted schema. However, it seems likely that this approach is motivated primarily by the mathematical tractability of schema modelling.

This subtle change in the definition of GP schema makes GP schema research more relevant than just taking the original GA definition into GP. Taking an example from symbolic regression problems, the symbol $\times$ returns completely different values depending on its surrounding context – two operands in this case. This suggests that the effect of the symbol closely depends on the surrounding context.

Bloat and Introns

In GP, individuals tend to bloat, i.e. to grow rapidly in size over the course of a run while the fitnesses do not improve. This has been well recognised in GP research [2, 8]. Bloat often involves introns. When evolving solutions in GP, it is common that some parts of an individual do not contribute to its fitness, and thus can be removed when evaluating fitness. These parts of the individual are called introns. There is no consensus on the cause of introns or their impact on GP performance, but their existence is amply confirmed. In contrast, this phenomenon does not exist in classic GA, which usually employs a fixed length linear representation

6.3 Lessons from GP Research in Searching for Models

Because the basic idea of EDA is to approximate the true distribution of solutions using a model $\mathcal{M}$, it is vital to choose an appropriate model $\mathcal{M}$. Consequently, conventional EDA research, which often employs a GA-style linear string to encode individuals, heavily focuses on finding such a suitable probabilistic model $\mathcal{M}$. The GA literature supports the belief that there are dependencies between genes (also known as linkages). In the EDA literature, many different kinds of probabilistic models have been proposed to represent linkage.

Estimation of the distribution of tree form solutions is far more complex than simply applying conventional EDA to a tree representation. One of the major difficulties lies in finding an appropriate probabilistic model. We may take lessons from GP research in this respect. Based on current GP understanding, we are convinced that a good model for EDA–GP should have the properties listed later.

6.3.1 Internal Hierarchical Structure

The tree representation has an internal structure – an intrinsic hierarchical structure. The model for EDA–GP needs to be able to represent this structure. Thus when sampling the model, i.e. when generating individuals from the model, a valid tree structure should be guaranteed. In this context, a valid tree is a tree that can be evaluated and complies with the given constraints, such as legitimate terminal and non-terminal symbol sets, depth limit and typing.

6.3.2 Locality of Dependence

In conventional EDA, the model employed does not usually assume specific dependencies between adjacent loci - a reasonable assumption for GA representations. However, in a tree structure, dependence exhibits strong locality, and this is the primary dependence which we should consider in tree representation. For example, dependencies (relationships) between parent and child nodes are expected to be stronger than among other nodes.

Another perspective for viewing this locality of dependence comes from semantics. When evolving a linear structure, as conventional EDA does, usually we assume the semantics are attached to the loci. For example, when using GA to solve the travelling salesman problem, one common encoding method uses each locus to represent a single step in which one specific city is visited. By contrast, in a tree representation, the meaning of each node is clearly defined by the symbol attached, and the effect of the node has to be interpreted in its surrounding context. For example, in the symbolic regression problems, the symbol $\times$ returns completely different values depending on the two operands. The model chosen for evolving the tree structure has to be able to represent this strong local dependence, as well as dependence on a larger scale.

6.3.3 Position Independence

Position independence is actually very closely related to locality of dependence. It emphasises that in GP tree representations, the absolute position of a symbol does not play a large role. Because of locality of dependence, a substructure may occur in various locations in different GP trees, but still have an identical or similar contribution to the overall fitness. This belief is well reflected in various GP schema theories [33, 46, 79], as discussed in Sect. 6.2.5.

The prevalence of introns [2] in GP representations provides a second reason for emphasising the importance of positional independence in EDA–GP models. When evolving solutions in GP, it is common that some parts of an individual, called introns, do not contribute to the fitness, and thus can be removed when evaluating fitness. If EDA–GP models are position-dependent, introns which move the location of a particular building block complicate the learning process, since each location has to be learnt separately; in a position-independent model, occurrences of a building block in different locations can reinforce each other.

An example may help to clarify the importance of position independence. Suppose we have a symbolic regression problem which requires GP to approximate the following function (i.e. the input and output of the function is given to GP and GP is required to find a function which has similar behaviour):

$$f(x) = x^3 + x^2 + x.$$

The individual in Fig. 6.5 is a possible solution. In this solution, the sub-trees in the boxes A and B, which produce x^2 is apparently a building block for the problem. However, the fact that this building block appears in both box A and box B suggests that position does not need to be a hard constraint in the probabilistic model, as is the case with EDA–GP. The sub-tree in box C does not contribute to the overall fitness of this individual. Thus, it is an intron and can be removed. Adding or removing box C would not change the fitness but it will change the position of the sub-tree in box B. The sub-tree in box B is always a building block for this problem no matter how many introns are inserted in front of it. This suggests that introns may change the position of the building block and thus is another cause of position independence. This example shows that it may be desirable for probabilistic models of EDA–GP to have a flexible positional reference, as it may assist the discovery and promotion of building blocks.

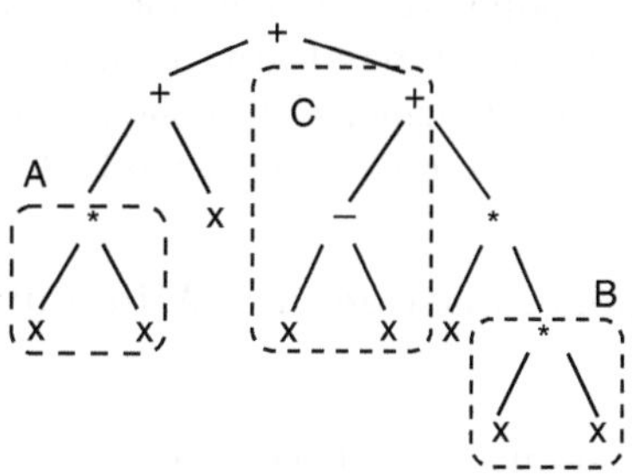

Fig. 6.5. Position independence and modularity in a tree-shaped individual

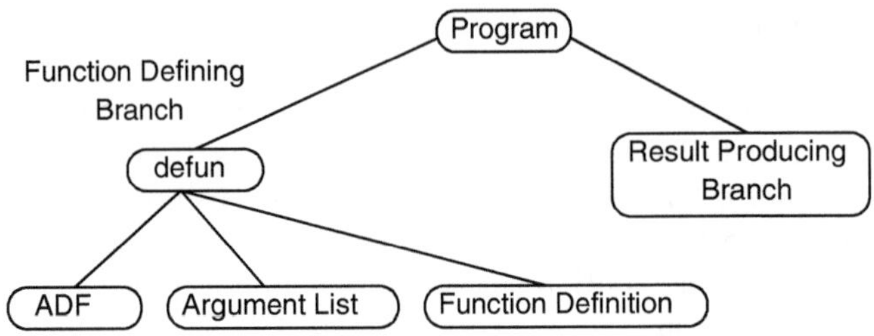

Fig. 6.6. Automatically defined functions of GP

6.3.4 Modularity

In a tree representation, it is common that building blocks, which are relatively independent sub-solutions, may need to be shared among tree branches and individuals. Therefore, one building block may occur multiple times, either in one individual or across the population.

This has been validated by numerous studies, such as ADF [33], GLiB [4] and AR [57]. In these studies, the "useful" sub-trees are identified, either by means of evolution or by some other heuristics, and are then encapsulated as intermediate building blocks so that they can be used multiple times within both one individual and also shared across the population. By identifying and reusing the useful structures, these studies have reported improvement over conventional GP.

An example of an ADF can be found in Fig. 6.6. When using ADFs, a GP tree has two parts, the result producing branch and a function defining branch. The function defining branch, on the left hand side in Fig. 6.6, defines a sub-tree as a function so that it can be called multiple times in the result producing branch. The result producing branch is a normal GP tree which can return the value of the entire program, but in addition to using the predefined functions, it can also refer to the functions defined by the function defining branches.

Let us again look at the example used earlier. The same sub-tree in Fig. 6.5 occurs twice, in the boxes A and B. It is desirable that these common structures are regarded as one single building block which can be shared within the individual and across the population. Furthermore, this makes learning more efficient, because it does not need to re-learn the common structure in different positions. This example clearly demonstrates the importance of modularity as a desirable characteristic of the probabilistic model for EDA–GP.

6.3.5 Complexity, Generalisation, and Minimum Encoding Inference

GP individuals usually do not have fixed complexity. This is one of the important properties of GP. In this, it differs radically from GA, which uses a fixed length string to encode individuals. GP is aimed at problems where no prior knowledge is available regarding the complexity of their solution. Hence, theoretically, there is no complexity constraint on individual GP trees.

Even if we do impose some limits, such as maximum depth or maximum number of nodes, the individual complexity still varies significantly from individual to individual, from generation to generation. A fixed-complexity model, resembling some models in conventional EDA in having a fixed size in terms of number of variables, may have difficulty in reflecting this variation. Note that variable complexity does not necessarily imply greater complexity. In a fixed complexity model, the initial model must be sufficiently complex to model solutions. Otherwise search will fail. Hence, there is a temptation to start with a high complexity model, leading to a high search cost. By contrast, a variable complexity system can start with a low complexity model, relying on the learning method to increase its complexity as required.

Another perspective of looking at complexity is generalisation. Solutions to problems are usually required to be not only correct, but robust. In particular, most GP problems involve some form of learning in the presence of noise, hence the well-known relationship between generalisation and complexity means that, assuming other things being equal, simpler models are preferred. This implies a necessity to trade off model complexity against model accuracy. While a wide range of approaches to this trade-off have been used in GP, in the more theoretically-based EDA approaches, these trade-offs have usually been based on minimum encoding inference, more commonly referred to as the minimum message length (MML) principle [74–76] or minimum description length (MDL) principle [54]. In this chapter, the term MML is employed to refer to both. The basic idea of MML is that a preferred model should give the minimum cost for describing the given data. In more detail, it should minimise the sum of the description length of the model, and the description length of the error of the model, given the data. The first part represents the complexity of the model and the second part is the cost of representing the data with respect to the model. It is very intuitive that the preferred model should be simple (short description length of the model) and accurate (short description length of error). Further, although a too-complex model may have better accuracy, it is very probable that it will not generalise well, i.e. it will not work well on unseen data. Conversely, a too-simple model may have very bad accuracy, and thus should also be penalised. MML is an important information-theoretic inference method, which has both intuitive appeal, and a solid theoretical support.

In order to effectively extend conventional EDA to tree representation, we may take lessons from GP research, which also employs the same representation. From the perspective of current GP research, we have identified the above five properties to be important for probabilistic models of EDA–GP.

6.4 PIPE Model

The first stream of EDA–GP work will be reviewed in this section. The earliest EDA–GP work was PIPE [63]. PIPE was motivated by the corresponding work in conventional EDA and GP. A number of studies based on the probabilistic model of PIPE have followed, including ECGP [64] and estimation of distribution programming (EDP) [82].

Interestingly, PIPE and its related works can fit into the same framework as conventional EDA. This is possible because the prototype tree of PIPE is a model assuming independence between random variables, while EDP considers pairwise dependences only, and ECGP extends this to multivariate dependence.

6.4.1 Probabilistic Incremental Program Evolution

PIPE [63] uses the probabilistic prototype tree (PPT) to represent the probability distribution of tree form programs. Its basic algorithm is consistent with the EDA illustrated in Fig. 6.1, and the learning method to update the PPT resembles the probability learning methods of PBIL [6].

PIPE iteratively generates successive populations of tree form programs according to an adaptive probability distribution over all possible programs of a given maximum depth, represented as a PPT. For example, in Fig. 6.7 which is adopted from [63], the left hand side is a PPT in which each node is a probability vector (more abstractly, each node is a random variable), indicating the probability of occurrence of different symbols. The right hand side is one of the possible GP trees, sampled from the PPT.

More specifically, the basic procedure of PIPE is as follows. Firstly, probabilities in the prototype tree are initialised uniformly. Starting from the root, we keep visiting nodes on the prototype tree until a valid tree, acting as an individual, is generated. When visiting each node, we choose a symbol according to the probabilities in the probability table of that node. In this way, a set of individuals (i.e. a population) is generated. Good individuals are then

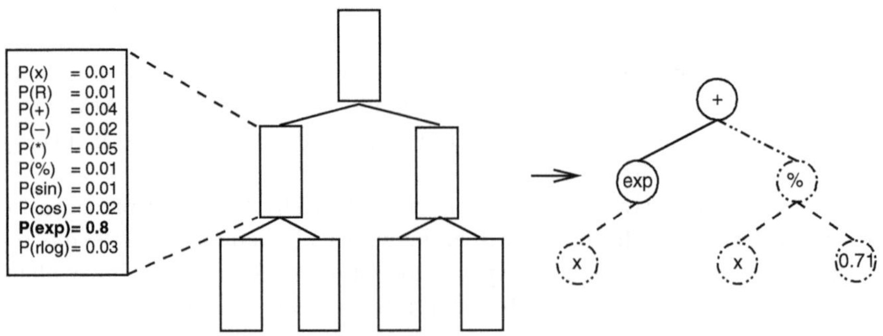

Fig. 6.7. Probabilistic incremental program evolution (adopted from [6])

selected from the population, and all of the probabilities of the entries in the probability table, which were used to generate those selected individuals, are increased. In other words, the probabilities of generating those good individuals are increased. Therefore, in each iteration, the probability distribution is updated using the best programs. Thus the structures of promising individuals are learnt and encoded in the PPT. This is the learning process for one step of a prototype tree. A new population is then generated from this updated prototype tree and a new iteration starts.

PIPE is significant as the first research in this area. It can be regarded as a tree-based extension of the linear string-based PBIL [6]. In PIPE, each node of the PPT is treated as an independent random variable, in that its probability is learnt independently from the other nodes. Therefore, what PIPE tries to learn is the probability of particular functions in particular loci. PIPE also implicitly assumes that the building blocks are position dependent. That is, in the PPT, the useful sub-trees/building blocks are attached to specific positions and cannot be moved to other positions.

6.4.2 Extended Compact Genetic Programming

ECGP [64] is a direct extension of ECGA [26] to the tree representation which is based on the PIPE prototype tree.

In ECGA, marginal product models (MPMs) are used to model the interaction among genes, represented as random variables, given a population of GA individuals. MPMs are represented as measures of marginal distributions on partitions of random variables. For example, in ECGA, an MPM for a four-bit problem might be

$$[1,3][2][4].$$

This represents that the first and third genes have intensive interaction, and the second and fourth genes are independent. That MPM would consist of the following marginal probabilities. $\{p(x_1 = 0, x_3 = 0)$, $p(x_1 = 0, x_3 = 1)$, $p(x_1 = 1, x3 = 0)$, $p(x_1 = 1, x_3 = 1)$, $p(x_2 = 0)$, $p(x_2 = 1)$, $p(x_4 = 0)$, $p(x_4 = 1)\}$, where x_i is the value of the ith gene.

This idea has been extended to the GP tree representation in ECGP. ECGP is based on the PIPE prototype tree, and thus each node in the prototype tree is a random variable. ECGP decomposes or partitions the prototype tree into sub-trees, and the MPM factorises the joint probability of all nodes of the prototype tree, to a product of marginal distributions on a partition of its sub-trees.

A greedy search heuristic is used to find an optimal MPM mode under the framework of minimum encoding inference (see Sect. 6.3.5).

ECGP can represent the probability distribution for more than one node at a time. Thus, it extends PIPE in that the interactions among multiple nodes are considered.

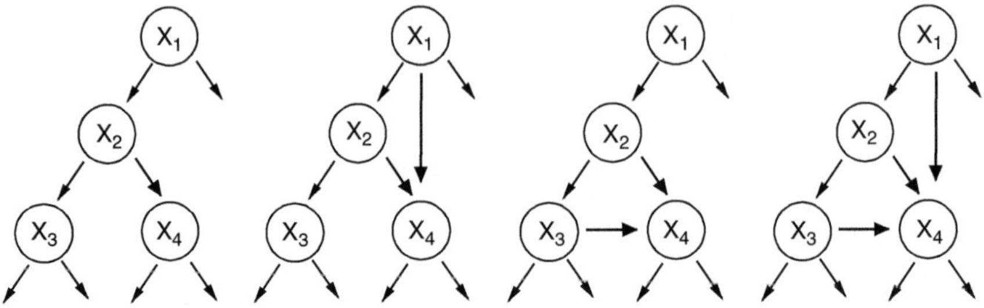

Fig. 6.8. Probability distribution model in EDP

6.4.3 Estimation of Distribution Programming

EDP is another extension of PIPE. Instead of treating each node as an independent random variable, EDP tries to model the conditional dependency among adjacent nodes in the PIPE prototype tree.

It is argued in [82] that strong dependence should exist between each particular node and its parent, grandparent and sibling nodes. (This is consistent with the argument of locality of dependence made in Sect. 6.3.2.) Some possible combinations of these dependences are illustrated in Fig. 6.8. The basic structure is again a PIPE prototype tree. For example, if node X_4 is under examination, then the thick lines indicate the important dependences. The right-most model is the most comprehensive model, which captures all the dependences believed to be important, while the left-most one is the most simplified model, in which only the dependence between one node and its immediate parent node is considered. Because of the computational overhead, among these possible models, the left-most model is implemented in [82]. In this work, the conditional probability between a child node and its parent are estimated, based on the selected individuals. Thus, EDP extends PIPE in that the probability distribution between two nodes is considered.

6.4.4 Summary

A visualised comparison of PIPE, ECGP and EDP can be found in Fig. 6.9. Each grey circle stands for a node of a PIPE prototype tree. The dependences considered in each model are illustrated by the bounded regions.

The left most figure in Fig. 6.9 corresponds to PIPE. Each node is treated as an independent variable. The figure in the middle is ECGP. The prototype tree is divided into sub-trees and there is no dependence considered between different sub-trees. The right most figure is EDP, which models the conditional probability of the node given its parent.

In summary, all these works are based on the prototype tree of PIPE, in which the probability tables of a prototype tree are organised in tree form. Therefore, these PIPE-based works can handle well the GP-style tree structure. In the original PIPE, each node is an independent random variable and

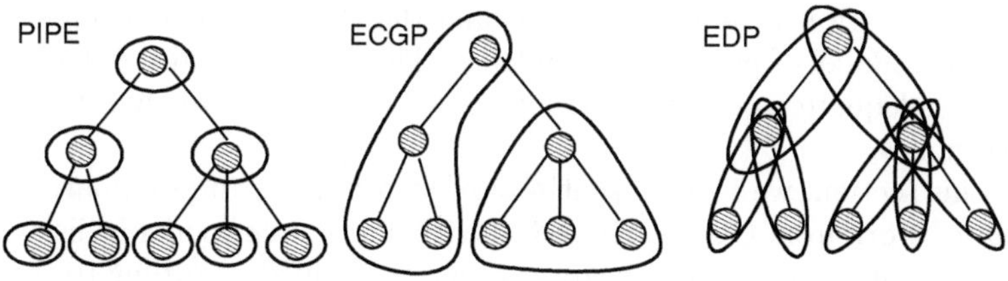

Fig. 6.9. Comparison of PIPE prototype tree-based probabilistic models of EDA–GP

thus its probability does not depend on any neighbouring nodes. The extensions, made in EDP and ECGP, make it possible to consider interactions among nodes. Further, the PIPE prototype tree does not have a problem in handling individuals with varying complexity. PIPE and related methods can also be very computationally efficient, due to the simplicity of using the prototype tree as their probabilistic model. However, because of the position reference embedded in the prototype tree, we cannot see any obvious way for PIPE-based methods to efficiently recognise building blocks with no fixed position.

One perspective on EDA–GP is to view it as a search through the space of all distributions on tree structures, gradually narrowing towards a distribution concentrated on solutions to the problem in question. Distributions over tree structures are a probabilistic generalisation of sets of tree structures – in other words, of languages. But if we view EDA–GP as a search through the space of probabilistic languages, the natural question arises, can we model this as a search through the space of some family of probabilistic grammars? And if so, can we use a family of grammars as the base models for representing the distributions? This leads naturally to the grammar-based methods for EDA–GP.

6.5 Grammar Model

Grammar model-based EDA–GP has a close connection with grammar guided genetic programming (GGGP). In GGGP, the grammar, as a formal model, effectively imposes a constraint on the search space, but the main search mechanism is still conventional genetic search. Grammar model-based EDA–GP takes grammars as probabilistic models, just like any other probabilistic model used in EDA research. Grammars are well-studied formalisms, originally proposed to model the internal hierarchical structure of languages, either natural languages or formal languages. They are particularly suitable for modelling GP-style tree structures because GP-style tree structures are just another kind of hierarchical structure.

Grammar model-based EDA–GP work can also be fitted into the same framework as conventional EDA. In a grammar model, each rule describes the dependence between the LHS symbols and the RHS symbols. Therefore, it is primarily a model of pairwise dependence. Through the chain of dependency, it is also adequate to describe structures which have more than two closely related nodes.

In this section, we first present a brief introduction to grammar, followed by related work from GGGP. EDA–GP with a grammar model is discussed in the last two subsections. These subsections are divided according to the types of grammar learning methods used.

6.5.1 Stochastic Context-free Grammar

There are a variety of grammar models in the field of natural language processing (NLP). Among them, context-free grammars (CFG) and stochastic context-free grammars (SCFG) [39] are most relevant to this chapter. The CFG is the commonest grammar formalism used in GGGP. SCFG can be viewed as a straightforward extension of the CFG, i.e. a CFG with an extra probability component.

Minimum encoding inference methods, such as minimum message length (MML) [74, 75] and minimum description length (MDL) [54], are widely used for inferring grammar models, in particular, inferring SCFGs [16, 70].

Formally, a SCFG M consists of:

- A set of non-terminal symbols N.
- A set of terminal symbols (or alphabet) Σ.
- A start non-terminal $S \in N$.
- A set of productions or rules R. The productions are of the form

  $$X \rightarrow \lambda,$$

  where $X \in N$ and λ is in the union of all powers of the elements of $N \cup \Sigma$, i.e. $\lambda \in (N \cup \Sigma)^*$. X is called the left-hand side (LHS) of the production, while λ is the right-hand side (RHS).
- Production probabilities $p(r)$ for all $r \in R$. For any given LHS symbol X, the sum of the probabilities of rules which have LHS X must be 1, i.e.

  $$\sum_{r \text{ has LHS } X} p(r) = 1.$$

The naming convention is as follows. The non-terminal symbols are distinguished by starting with a capital letter. Terminal symbols are defined by lower-case letters. A string that may be composed of both terminal and non-terminal symbols is represented by lower-case Greek letters.

Figure 6.10 is an example of a SCFG for a symbolic regression problem. This grammar can generate all the mathematical functions which involve operators $+, -, \times, /, \sin, \cos, \log, e^\wedge$. The symbol "|" represents disjunction which means the particular LHS may be rewritten with any of the RHSs connected

S	$\rightarrow$ Exp		(0)
Exp	$\rightarrow$ Exp Op Exp	|	(1)
	$\rightarrow$ Pre Exp	|	(2)
	$\rightarrow$ x		(3)
Op	$\rightarrow +$	|	(4)
	$\rightarrow -$	|	(5)
	$\rightarrow \times$	|	(6)
	$\rightarrow /$		(7)
Pre	$\rightarrow$ sin	|	(8)
	$\rightarrow$ cos	|	(9)
	$\rightarrow$ eˆ	|	(10)
	$\rightarrow$ ln		(11)

Fig. 6.10. A commonly used SCFG for symbolic regression problems. The probabilities attached to the rules follow a uniform distribution, and are therefore omitted

by |. For example, LHS symbol Exp may be rewritten with either rule 1, 2 or 3. Similarly, Op may be written as $+, -, \times$ or $/$ under rules, 4, 5, 6 and 7, respectively. For this example, we assume that the probabilities follow a uniform distribution, and therefore the probabilities attached to the rules are omitted.

Derivation

For any strings γ and δ in $(N \cup \Sigma)^*$, if string $\gamma S \delta$ can be transformed to string $\gamma \alpha \delta$ by rule $S \rightarrow \alpha$, we say that $\gamma S \delta$ *directly derives* $\gamma \alpha \delta$ in grammar M, or $\gamma \alpha \delta$ is directly derived from $\gamma S \delta$. This is denoted as follows.

$$\gamma S \delta \overset{S \rightarrow \alpha}{\Rightarrow} \gamma \alpha \delta.$$

If there exists a sequence of direct derivations $\alpha_0 \Rightarrow \alpha_1, \alpha_1 \Rightarrow \alpha_2, \ldots,$ $\alpha_{n-1} \Rightarrow \alpha_n$, where $\alpha_0 = \alpha$, $\alpha_n = \beta$, $\alpha_i \in (N \cup \Sigma)^*$, and $n \geq 0$, which transforms string α to string β, we say α *derives* β, or β is derived from α, denoted as follows.

$$\alpha \overset{*}{\Rightarrow} \beta$$

Such a sequence is called a derivation. Thus a derivation corresponds to a sequence of applying productions to generate a string. A derivation can be represented in a parse tree format, called a *derivation tree* or *parse tree*. For example, Fig. 6.11 is a derivation tree of string $x - (x + x)$.

In SCFGs, the probability of a derivation (or a derivation tree) is the product of the probabilities of all the production rules involved. Formally, the probability of a derivation $\alpha_0 \Rightarrow \alpha_1, \alpha_1 \Rightarrow \alpha_2, \ldots, \alpha_{n-1} \Rightarrow \alpha_n$, where each step of derivation is a direct derivation, is

$$p(\alpha_0 \overset{*}{\Rightarrow} \alpha_n) = \prod_{i=0}^{i=n-1} p(X_i = \lambda_i), \tag{6.1}$$

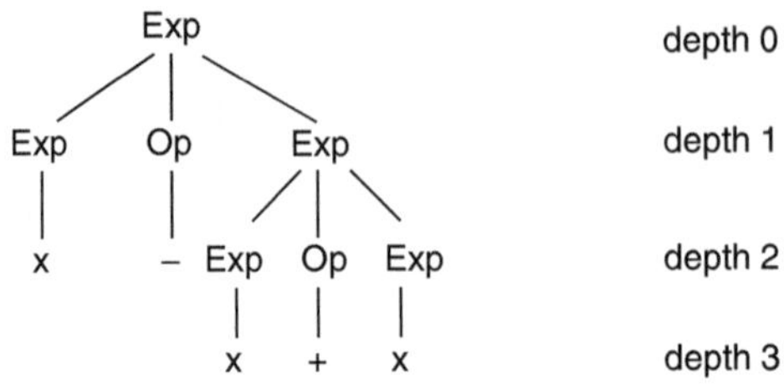

Fig. 6.11. Derivation tree of string $x - (x + x)$

where production rule $X_i = \lambda_i$ is used to derive α_{i+1} from α_i and $p(X_i = \lambda_i)$ is the probability of using the rule $X_i = \lambda_i$.

CFG theory and GP theory use the term "parse tree" inconsistently with each other; to avoid confusion, this chapter will use the term "derivation tree" (which is not otherwise used in GP) rather than "parse tree". Note: although a derivation corresponds to only one string, a string without bracketing to represent its internal structure, may (in general) be derived in a number of different ways from a given grammar. In NLP, this phenomenon, known as ambiguity, causes severe difficulty when inferring grammars from sentence examples. However, in GGGP, and in the research presented in this chapter, this is generally not an issue, because in both fields of research, the derivations of the individuals/examples from the given grammar are preserved, and thus correct, unique derivations are known.

In this chapter, sampling an SCFG grammar means deriving a set of strings from the given SCFG grammar. When deriving a string, an LHS may have more than one RHS. As previously defined, an SCFG has a probability component attached to each rule (more accurately to each RHS). The RHS is chosen based on this probability.

For example, a common SCFG grammar for the symbolic regression problem is illustrated in Fig. 6.10. An individual, which may be derived or sampled from this grammar, is illustrated in Fig. 6.11. The individual in Fig. 6.11 is derived by applying the following sequence of rules. Starting from the starting symbol S, rule 0 is chosen to rewrite S and obtain

$$\text{Exp Op Exp}.$$

After probabilistically choosing rules 3, 5 and 1 to respectively rewrite these three symbols, the following string is obtained:

$$x - (\text{Exp Op Exp}).$$

Symbols "x" and "$-$" are terminals because there is no rule which can rewrite them. Therefore, only the last three symbols are written with rules 3, 4 and 3, respectively. The eventual result is:

$$x - (x + x).$$

As defined in (6.1), the probability of the earlier derivation can be calculated as the product of the probabilities of all the rules involved. The rules involved in this derivation were used in the following order

$$0\ 3\ 5\ 1\ 3\ 4\ 3.$$

The probability of rule 0 is 1 because there is no other alternative. Given the assumption of uniform distribution, because "Exp" can be written with rule 1, 2 or 3, the probability of rule 3 is 1/3. Similarly we may work out the probability of other rules. Thus, the probability of this derivation is

$$1 \times \frac{1}{3} \times \frac{1}{4} \times \frac{1}{3} \times \frac{1}{3} \times \frac{1}{4} \times \frac{1}{3} \approx 0.00077.$$

As mentioned earlier, some symbols can be rewritten with more than one rule. For example, the symbol "Exp" can be rewritten using either rule 1, 2 or 3. In conventional GGGP, a uniform distribution of the probabilities of rules is assumed. Therefore, the choice of which rule to apply is made uniformly randomly. However, it does not have to be so. In some work, such as [77], weights, which roughly correspond to the probabilities attached to rules, are assigned to rules, so that this choice can be made to stochastically favour some rules over others.

Learning Grammars with Minimum Encoding Inference

Since grammars originated in the field of NLP, and it is an important means to model language, grammar inference (learning) has been extensively studied in NLP [62]. In EDA–GP research, we are particularly interested in the inference of SCFGs.

An SCFG can be understood as a CFG with probabilities attached. Therefore, it has two parts: the structure (the CFG part) and the probability part. To learn an SCFG model, we have to learn these two parts. The inference of the structure (CFG) [5, 60, 61] and of the probabilities given CFG [35, 36, 51] were initially studied separately. However, recent works address the inference of the full SCFG simultaneously, including both the structure and the probabilities of SCFG [15, 24, 32, 47, 70]. The methods used to infer the SCFG are usually based on minimum encoding inference (see Sect. 6.3.5).

Inferring an SCFG is a difficult problem. Given the MML principle, most of the current methods in SCFG inference are based on greedy search. In greedy search, the SCFG is altered using some variation operators. If, after a given number of variations, a better SCFG is found, it is accepted as the basis for the next iteration. Otherwise, the search usually stops. During the search, the MML principle is used to compare competing models.

From this perspective, grammar-model-based EDA may be viewed as a form of iterated grammar inference method, which tries to iteratively find better grammar models favouring high fitness individuals (i.e. grammar models which have an increased probability of generating high fitness individuals), by means of search methods, usually under an MML framework.

6.5.2 Conventional GGGP with Grammar Learning

GGGP [23, 77, 81] is a GP approach relying on a grammar constraint. All kinds of grammars can be used to describe the constraint but this thesis will mainly focus on CFG. Similar to Strongly-typed GP [41], grammars in GGGP provide a systematic way to handle typing.

GGGP usually has a predefined fixed grammar to impose the constraint. For example, the grammar in Fig. 6.10 is a commonly used grammar in GGGP for symbolic regression problem. The initial population is generated by sampling this grammar. The tree structure in Fig. 6.11 is a typical individual of GGGP. Once the initial population is obtained, the genetic operators, such as crossover and mutation, are applied. The use of a grammar requires that the genetic operators must respect the grammar constraint, i.e. after imposing a genetic operation, the child must still be grammatically correct. In other words, the child must still be consistent, after genetic operations, with the given grammar. To ensure this consistency, the crossover and mutation operators are modified. For example, only sub-trees which have the same symbol in their roots can be swapped in crossover. The selection procedure is identical to conventional GP. After selection, the next iteration starts.

Due to the introduction of a grammar, GGGP can constrain the search space so that only grammatically correct individuals can be generated. Thus, it can be used to reduce the search space or to introduce some background knowledge.

The recent grammar-based EDA research was prefigured by two studies which incorporated learning into the GGGP process. In conventional GGGP, it had been noticed early on that a grammar can not only be used to impose an initial bias, but can also be revised during search to further bias the search according to updated available experience. There are two studies in this field.

Whigham's Work

In [78], grammar refinement was introduced into conventional GGGP. More specifically, it functions as a conventional GGGP system but the grammar is refined during the search, and new individuals generated from the refined grammar are fed back into the population.

The refinement of the grammar has two components. The first updates the probabilities of rules. Merit, derived from probability, is attached to each rule to reflect the frequency of the rule use. This merit is updated according to its use in superior individuals. Then, when generating new individuals, the merit has a similar function to the probability in SCFG. The second component adds new production rules. New productions are learnt from the superior individuals, and they are chosen in a way to have minimum impact on the original grammar.

The population in the new generation not only has individuals obtained by applying genetic operators, but also individuals generated from the updated grammar.

As can be seen, this work is essentially a conventional GGGP, because its main search mechanism is conventional genetic search, and the incorporation of individuals generated from the refined grammar is only an aid to further bias the search. The grammar learning is also ad hoc. However, the importance of this work is that it first showed that grammars may be a good model for incrementally guiding the search.

Tanev's Work

A similar approach is proposed in [71]. Tanev incorporated learning Stochastic Context-sensitive Grammars (SCSG) into conventional GGGP.

This work is in the context of a dynamic environment. At the end of a run, a grammar is learnt from the best-of-run individuals. It is then moved to the new environment. In the new environment, part of the new population is generated from the learnt grammar, and the mutation operator must also respect the learnt grammar.

The grammar used here is an SCSG, which in this circumstance can be loosely understood as an SCFG with a context constraint. With the extra constraint, whether a rule is admissible is not only decided by matching the LHS, but also by matching the context. In [71], the probabilities and context are learnt in order to favour some rules, and to restrict the places where a rule can be applied.

More specifically, SCSG grammar learning happens at two levels. One is the probability learning. The other is the context learning. A fixed amount of context is added to the grammar to make the grammar more specific, so that some areas of the search space can be more intensively investigated.

This research is an enhancement of GGGP with the aid of a grammar model. The grammar learning occurs only once, and the main search mechanism is still genetic search. Through a comparison with conventional GGGP, this work empirically verifies that grammars can be used to efficiently bias search to promising areas and thus obtain superior performance.

6.5.3 Learning Parameters of a Grammar Model

As Whigham's work showed that grammar learning could form a useful adjunct to the genetic operators of genetic programming, it is not surprising that other researchers proceeded to investigate whether the genetic operators were necessary at all. Most grammar-based EDA–GP works have concentrated on some form of SCFGs. We know that an SCFG model can be understood as a normal CFG with a probability component. Therefore, the inference of an SCFG model usually consists of inferring these two parts, namely the structure of the SCFG (which is essentially a CFG) and its associated probability component. Accordingly, we identify two streams of EDA–GP with grammar models. One stream learns the probability only. The other learns both structure and probability. The former will be discussed in this section and the latter will be left to the Sect. 6.5.4.

We note in passing, that theoretically, there is no intrinsic difference between grammar structure learning and probability learning. The grammar structure is actually only a form of probability table, where absence of a particular rule implies that the probability attached to the rule is zero. Therefore, theoretically we may specify a (possibly infinite) grammar which includes all possible rules, and where probability learning will assign probability 0 to the unnecessary or incorrect rules. Distinguishing grammar structure learning and probability learning is largely a matter of implementation convenience, efficiency and comprehensibility.

Stochastic Grammar-Based GP

Stochastic grammar-based GP (SG-GP) [53] is an interesting but often-overlooked work proposed by Ratle et al. To the best of our knowledge, it is the earliest attempt to introduce grammar models into EDA–GP.

In essence, SG-GP learns the probability of an SCFG while keeping the structure of the grammar fixed. The basic algorithm of SG-GP is consistent with the EDA algorithm described in Sect. 6.1.2, i.e. it is an iteration of model learning and sampling.

SG-GP starts with a CFG and weights attached to each rule (the probability can be obtained by normalising the weights). Initially, because we have no prior knowledge, all of the weights are set to equal values (corresponding to a uniform distribution). Generating individuals from the grammar is similar to generating individuals from an SCFG, as discussed previously. Also similarly, at each generation, the probabilities are updated. To do this, the weight of those rules which contribute to superior individuals are increased, while the weights of those rules involved in generating inferior individuals are decreased. More precisely, assuming rule r_i is used in a superior individual and r_j in an inferior individual, their weights w_i and w_j are updated as follows:

$$\begin{aligned} w_i &\leftarrow w_i(1+\mathrm{lr}) \\ w_j &\leftarrow \frac{w_j}{1+\mathrm{lr}}, \end{aligned} \tag{6.2}$$

where lr is a predefined learning rate.

There are two variants of SG-GP proposed in [53], namely scalar and vectorial SG-GP. What we have discussed is scalar SG-GP, while vectorial is a straightforward extension of scalar SG-GP. Note that in scalar SG-GP, tree depth does not play a role. Each rule can be used to generate any depth of the tree as long as its LHS matches the non-terminal. However, this causes serious problems when the number of rules is small, which is usually the case. For example, the grammar in Fig. 6.10 has only 11 rules. No matter how we update the probability attached to each rule, it is unlikely that this grammar would be able to hold enough information to progressively bias the search. In other words, the model is too simple to be able to adequately describe the solution space. For example, suppose one particular rule is only beneficial if it is applied at depth d, but not at depth d', where $d \neq d'$. In scalar SG-GP there is no way to record this information.

To alleviate this problem, vectorial SG-GP was proposed. In vectorial SG-GP, a weight vector is attached to each rule. Each element of the vector represents the weight of the rule at a particular depth. Therefore, the depth information is used to effectively increase the total number of rules, i.e. the model complexity is increased.

However, in both scalar and vector SG-GP the overall structure of the grammar is fixed and it does not change with the progress of the search. Clearly, in this method, because of the fixed grammar structure, i.e. the number of rules is fixed, the complexity of the grammar model does not change. Therefore, the search may either stop very quickly, especially when the number of rules is very small, which is usually the case, or the search may converge very slowly if too many redundant rules are involved in the probability learning.

6.5.4 Learning Structure and Parameter of Grammar Model

PEEL

Program evolution with explicit learning (PEEL) [68] uses a specialised stochastic parametric Lindenmayer system (L-system) [52]. The reason for this specialised SCFG is a desire to balance between expressiveness of the grammar model and learning efficiency. This L-system is equivalent to a standard SCFG with two extra conditions on the LHS.

The two conditions introduced are *depth* and *location*. So the modified form of a production rule is

$$X(d, l) \to \lambda \ \ (p).$$

The first parameter, depth d is an integer, indicating that when generating individuals, this rule may be applied only at level d (in this respect, PEEL closely resembles vectorial SG-GP). The second parameter is a location l, indicating the relative location where this rule may be applied, p is the attached probability. In this approach, matching the LHS does not mean merely matching the LHS symbol X, but also matching these two conditions on the left hand side.

The location information is encoded as the relative coordinate in the tree, somewhat similar to the approach of [20]. The position of each node in a rooted tree can be uniquely identified by specifying the path from the root to this specific node.

The learning in PEEL involves two iterative stages: probability learning and grammar structure refinement. The basic idea of probability learning is to increase the probabilities of production rules which are used to generate good individuals. If the probability learning alone is not sufficient, grammar refinement starts to add more rules by splitting existing rules, so that the complexity of the grammar model progressively increases. As the grammar model becomes more complex, its expressiveness increases, and thus more complicated interaction among tree nodes may be represented and learnt.

The initial grammar will usually be minimal, representing the search space at a very coarse level. Then, when necessary, through structure learning, rules are added to focus the search to a finer level. New rules are added by splitting existing rules, thus adding more context information. Once more context information has been added, one production can be split into several.

For example, we have the following grammar rule:

$$X(d=1, l=\#) \rightarrow \lambda(p=0.33).$$

This rule could be applied to rewrite a predecessor X at depth 1. $l = \#$ implies that no location information will be considered. Therefore, at depth 1, as long as predecessor X is matched, the rule would be applied with probability $p = 0.33$ no matter where the predecessor X is. Suppose, however, that the location does have a strong impact on the fitness of an individual, and thus should be learnt. Hence, it is necessary for the system to be able to increase the expressiveness of the grammar through refinement. For example, if we split the above rule into three:

$$\begin{aligned} X(d=1, l=0) &\rightarrow \lambda(p_0), \\ X(d=1, l=1) &\rightarrow \lambda(p_1), \\ X(d=1, l=2) &\rightarrow \lambda(p_2), \end{aligned}$$

then the location information can be represented. For example, at depth 1 ($d = 1$) an X at location 2 will be rewritten as λ with probability p_2 while at location 0, it will be rewritten with probability p_0.

Briefly, PEEL is an EDA–GP method which learns a specialised form of SCFG grammar. Both the probability component and the structure of the grammar is learnt in PEEL. The structure is learnt by incrementally splitting the existing rules so that more location information can be added.

Bosman's Work

Bosman's work [14] intentionally extends EDA to EDA–GP, and infers a full grammar (both grammar structure and probabilities).

In [14], the derivation trees are not preserved – an unusual practise in GGGP – so ambiguity occurs when re-parsing the individual. Therefore, a large amount of effort is put into correctly re-parsing the derived sentence (individual), and a highly complex method is used.

The basic idea of [14] is as follows. It starts with a minimum SCFG. In each iteration, the grammar is learnt and sampled to obtain the next population. The grammar structure learning method is rule expansion (with some constraints to ensure correct parsing). The learning method is a greedy search similar to most EDA work, and MDL is used to choose between competing models. In MDL, the model complexity term is measured by the number of symbols in the grammar. When estimating the probabilities of rules, depth is introduced as an extra dimension, similar to SG-GP.

Through the expansion of grammar rules, more production rules are added and their probabilities are estimated accordingly. Thus the grammar becomes more specific, enabling the system to progressively bias search.

GMPE

Grammar model-based program evolution (GMPE) [67], is based on grammar learning methods from the field of NLP. NLP provides a strong theoretical basis, and avoids re-inventing the wheel. GMPE tries to learn a standard SCFG model, which is then used to direct its search. MML is used to guide the search for an SCFG model. It appears that most types of building blocks previously considered in GP research can be explicitly represented and preserved in GMPE.

The grammar learning method of GMPE is mainly inspired by [70]. This approach uses a specific-to-general search method, where the inference of the SCFG model is regarded as searching for the optimal model among the possible model solutions. The search operator is a modified merge operator – the Initial-Grammar-Consistent merge operator (IGC merge) a variant of the standard merge operator [70], modified to enforce consistency between the learnt grammar and the initial grammar.

This merge operator takes two rules and unifies their LHSs to one symbol (a necessary check of consistency is omitted for simplicity of explanation).

$$
\begin{aligned}
&X_1 \rightarrow \lambda_1,\\
&X_2 \rightarrow \lambda_2,\\
&\Downarrow \text{merge}(X_1, X_2) = Z,\\
&Z \rightarrow \lambda_1,\\
&Z \rightarrow \lambda_2.
\end{aligned}
$$

After the merge, all the occurrences of X_1 and X_2 in the grammar are replaced by Z. As can be seen, before merge, X_1 can only be rewritten with λ_1, while after merge, X_1 and X_2 are not distinguishable, so that the places where X_1 appears can be rewritten with either λ_1 or λ_2. The same applies to X_2. It is not hard to see that the merge operator usually generalises the grammar. Before the merge, the two rules had different LHSs; after the merge, the two rules share LHSs, so the merged grammar can cover more strings.

The basic algorithm is as follows. A model which only covers the training samples is initially constructed, i.e. this grammar can generate only the selected individuals of the current population. Search operators, consisting of IGC merges, are then imposed on the model to vary it, usually to generalise it. Since it is not possible to enumerate all the possible sequences of merges, a hill-climbing method is adopted. The metric derived from MML determines which search operations to accept, and also when to stop the search. Thus, IGC merge is repeatedly applied to the grammar to generalise it until the grammar does not improve any more with respect to the MML metric.

GMPE is a highly flexible method. It employs standard SCFG models and a well-known learning method. Theoretically, various forms of the building blocks studied in the GP literature can be represented. However, GMPE is very computationally expensive because the MML metric for grammar learning has to be frequently calculated to compare competing grammars, and the number of competing models is enormous.

6.5.5 Summary

A number of typical grammar model-based EDA–GP approaches have been surveyed in this section. An informal impression of the relationships among these works may be gained from Fig. 6.12. It gives a conceptual overview of some of the main EDA–GP approaches: PIPE [63], EDP [82], vector SG-GP [53], ECGP [64], Bosman's work [14], PEEL [68] and GMPE [67].

Different approaches are organised according to the degree of constraint imposed on their handling of dependence, and their consequent computational cost. The constraint has two-dimensions, the type of interactions among nodes (i.e. the complexity of the interaction relationships that can be represented), and position dependence (i.e. whether the dependence between nodes is positional). These two properties are summarised in brackets under the name of each method. The lower part of the figure lists approaches based on grammar models, while the upper part lists systems based on the PIPE prototype tree.

Among all these approaches, the earliest – PIPE – is the most rigid in representing building blocks. In PIPE, probabilities sit at particular positions of the prototype tree and thus cannot be moved around, and the probabilities in each node are independent. GMPE, at the other extreme, is the most flexible

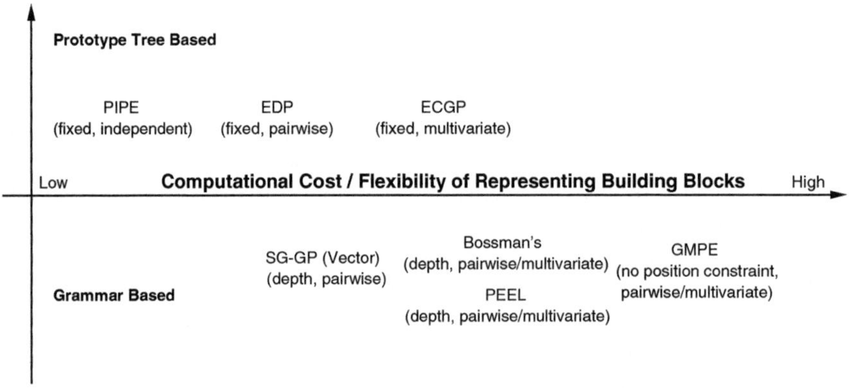

Fig. 6.12. Relations among EDA–GP methods

in representing building blocks, in that its probability model does not have any positional reference, and the probabilities on the grammar productions can represent pairwise or even multivariate interaction among nodes. Some methods in the middle of the spectrum, such as SG-GP, take only depth, instead of full position, as a reference.

From the perspective of computation cost, the methods on the left, such as PIPE, have the least computational overhead while those on the right, such as GMPE although very flexible in representing building blocks, are the most expensive to compute. There is no best algorithm as such. Whether the algorithm is suitable for a particular problem depends on the type of the building blocks needed, and the computing resources available.

6.6 Inference of Probabilistic Graph Model

Besides the traditional perspective of model dependence or the formalism of the model, there is an alternative perspective for viewing EDA. That is the perspective of probabilistic graph models. This view is applicable to both conventional EDA and EDA–GP. In this section, due to space constraints, we only discuss EDA–GP from this perspective. For more discussion of conventional EDA please refer to [66]. Because most models used in EDA can be regarded as a kind of probabilistic graph model, the EDA approaches can be conceptualised into three categories, based on the learning methods for the probabilistic graph model:

1. Inferring only the probability of the probabilistic graph model (assuming a fixed structure) [1, 53, 63, 82].
2. Inferring only the structure of the probabilistic graph model (assuming a crisp binary value of probability - not, to our knowledge, studied yet).
3. Inferring both the structure and probability of the probabilistic graph model [14, 64, 67, 68].

This perspective provides a unified view to understand EDA–GP regardless of the specific form of model. Interestingly, we are not aware of any work of category 2 in EDA–GP, though such approaches do exist in conventional EDA (LEM). However, we cannot see any intrinsic difficulty in applying this kind of method to EDA–GP. For example, inferring a crisp CFG model for EDA–GP is a way of achieving this.

6.7 Ant Colony Optimisation and Automatic Program Synthesis

EDA may also be viewed as a model-based search (MBS) [83]. In MBS algorithms, candidate solutions are generated using a parameterised probabilistic model, which is updated using the previously selected solutions in such a way that the search will concentrate in the regions containing high-quality solutions.

ACO [11] is another well-studied field of MBS. The similarity between ACO and EDA methods which only consider univariate probability distribution, such as PBIL [6], can be easily seen, because of the similarity in the probabilistic model and the probability updating methods. However, it is harder to make a direct analogy between EDA using more complicated dependency structures, and ACO, at the level of their probabilistic models and their mechanisms of updating probability, although they are almost identical at the abstract level. For a more detailed description of the MBS classification, and a discussion of ACO and EDA from the MBS perspective, please refer to [83].

In this section, given the major emphasis of this survey, a number of closely related works the use ACO to synthesise programs will be reviewed. With a few exceptions, most ACO-based automatic synthesis methods simply try to translate a conventional program synthesis problem by converting the program space to a representation, such as a graph, amenable to search by ACO. In the following subsections, as with the review of EDA–GP in the previous section, these works are grouped into grammar-based and non-grammar-based categories.

6.7.1 Grammar-Based Works

Ant-TAG [1] uses a grammar and ACO to synthesise programs. In this work, tree-adjunct grammars (TAGs) [30], instead of CFGs, are used. The individuals are assembled by combining elementary trees, which can be loosely understood as repetitively applying rules of an SCFG to non-terminal symbols.

Ant-TAG starts with a given grammar with uniform distribution because it does not have any prior knowledge. At each generation, the probabilities are updated by increasing the probabilities of those rules which contributed to superior individuals. In some variants, crossover of population members is used as a local search operator.

Note that in each iteration, only the probabilities are updated, while the structure of the grammar is fixed. From the perspective of EDA–GP, ant-TAG is very similar to SG-GP [53], except that different grammar formalisms are used. We anticipate that ant-TAG may suffer from the same difficulties as SG-GP due to the fixed grammar structure, namely due to its fixed complexity, it might not be able to capture all the necessary information for constructing a good solution, hence the authors foreshadow extensions using incremental grammar learning methods. The name ant-TAG reflects its probability update mechanism, which is motivated by ACO [11], and the representation, which uses TAG grammars.

Generalised ant programming (GAP) [31] is a CFG grammar-based system. It differs from ant-TAG in that it deals with a model of variable structure.

Instead of updating only the probabilities of the grammar, and keeping the structure unchanged, GAP records the whole path which an ant has visited. In this respect, it is very much like PIPE-based EDA–GP work, especially EDP, because the probabilities in GAP are attached to the rules, and thus are conditional probabilities, representing pairwise interactions.

However, because the probabilities in GAP are attached to rules, they are conditional probabilities representing pairwise interactions, while probabilities in PIPE are independent of other nodes.

6.7.2 Non-Grammar-Based Works

Of the non-grammar-based methods, ant programming (AP) [58] appears to be the earliest attempt at using ACO to synthesise programs. It is elegant, and consistent with PIPE-based EDA–GP works. ACO search is used to explore good paths along the PIPE prototype tree, representing programs. The probability update mechanism (pheromone update policy) is the major characteristic that discriminates it from PIPE.

Other non-grammar-based works, such as ant colony programming (ACP) [12, 22], employ arbitrarily predefined graphs whose nodes could be either a GP symbol (terminal or non-terminal) or a program statement. ACO search is used to find a path representing a program. A variant, grid ant colony programming (GACP) [55], introduces a temporal index, which closely corresponds to the depth constraint of some EDA–GP grammar-based methods, so that the probability (pheromone) can have a depth reference.

6.8 Conclusion and Future Research

We have presented what we hope is a critical and comprehensive survey of EDA with GP-style tree representations, referred to as EDA–GP. We have identified two streams of work in EDA–GP, based on their model formalism. One is based on the PIPE prototype tree, while the other is based on a grammar model. PIPE-based work is relatively less computationally expensive than grammar-based work, while the latter is more flexible in terms of capturing different kinds of interactions among nodes. An alternative perspective for understanding this work – inference of probabilistic graph models – is also provided.

In addition to strictly EDA–GP methods, we have also surveyed the synthesis of programs using ACO, due to the strong similarity between EDA and ant-based approaches.

It is clear that, given the limited number of studies, the field of EDA–GP is still in its infancy. However, as tree representations, which have been widely used in GP, are suitable for a number of problems, the limited research in EDA–GP is a significant gap. This gap may result from two causes. The first is the strong connection between EDA and GA. Since EDA research stemmed from addressing problems in GA, it is natural that most studies focus on the GA side of EDA research. The second reason is the complication resulting from the complex tree structure. Suitable probabilistic models are needed to model the GP-style tree structure, but common probabilistic models are not directly applicable to EDA–GP. Because of this, we have witnessed slow progress in this field. The earliest EDA–GP work [63] dates back to 1997, and there appears to have been no subsequent work until 2001 [53]. Most EDA–GP publications have appeared in the last three or four years, which may suggest that appropriate probabilistic models are emerging. We hope that this survey may contribute to the awareness of the existing EDA–GP algorithms and thus facilitate the communication among researchers in the field.

We suggest a number of future research issues later. Some are applicable not only to EDA–GP, but more widely to EDA in general.

1. Exploring a variety of relevant model formalism and their learning methods. For example, in the field of grammar-based EDA–GP, investigating whether other forms of grammar, such as context-sensitive grammars, may be suitable for EDA–GP. Even with SCFG, there are many grammar learning methods. Only two have been studied [67, 68]. It is worth investigating the pros and cons of other grammar learning methods in the context of EDA–GP.
2. *Knowledge extraction, reuse and incorporation.* In EDA–GP, as well as EDA in general, a well-formed model is used to encode the knowledge accumulated in the course of the search. Extraction and study of the knowledge embedded in a model is an important research direction.
   The EDA model, being a probabilistic model, is more interpretable than a population. More importantly, even if it turns out to be too complex for human comprehension, it may still be possible to extract some form of knowledge that can be used by the EDA algorithm in other applications. This leads to one possible way of incorporating knowledge. If the knowledge can be extracted from other EDA models in similar applications, and represented in a standard format, such as symbolic logic, it is then perfectly possible to exploit this information. While this has not yet happened in EDA–GP, related fields, such as Bayesian networks and artificial neural networks, have explored the possibility of such knowledge extraction and reuse [19, 25, 38, 73].
   The second way to incorporate knowledge is to make use of prior human knowledge. For real world problems, domain experts usually have some level of background knowledge. The model used in EDA provides a mechanism to incorporate background knowledge. By incorporating domain dependent knowledge into EDA, we hope to achieve superior performance. In more general EDA, it has been shown in [7] that incorporating prior knowledge may improve EDA performance. In grammar-based EDA–GP, this may be done by using a carefully designed initial grammar (rather than a general grammar) to include prior knowledge. Similar methods have already been tested in GGGP [80].

3. *Parsimony pressure and noisy data.* Usually, EDA–GP has a built-in Occam's razor, which causes small individuals to be preferred. We conjecture that this parsimony pressure results from sampling bias - the more complex an individual is, the more difficult it is to produce in the sampling process, and especially, the more difficult it is to reproduce its effective code without error. It is well known in machine learning that less complicated models generalise better than more complicated ones. Hence, we expect EDA–GP to perform particularly well in learning from noisy data. This is very important for real world applications where training data are noisy and generalisation is critical. Research in this direction is promising.
4. *Incremental learning.* It would be desirable to update the model in each generation rather than learn a new model from scratch. One possible way to do this, may be to use the model of the previous generation as a prior, and then incorporate more information from the fitness distribution of the current generation into the model. This approach fits well into a Bayesian framework.
5. *Making use of negative examples.* In most EDA work, only the positive examples, i.e. high fitness individuals, are used for inferring the models. The high fitness individuals may indicate promising search areas, while the low fitness ones may tell us about the less potential areas which we need to avoid. The use of both examples may improve the efficiency of model inference, and therefore is worth further investigation.
   In the field of EDA, LEM [40] uses both positive and negative examples. However, in the field of EDA–GP, we are not aware of any work of this kind. There is no obvious reason that this option should not be explored in EDA–GP.
6. *Developing theory.* EDA eschews genetic operators and maintains a well-structured model. This provides an opportunity to develop a series of theories to characterise EDA. In conventional EC, this has proven to be very difficult because of the highly complex behaviours of genetic operators and the dynamics of the population. In EDA, these are greatly simplified, and there has been some progress in theoretical analysis, for example some EDA algorithms have been based on sound theory [43] and there are interesting attempts at characterising conventional EDA [49]. So far, there have been no theoretical analyses of EDA–GP. For EDA–GP approaches using a fixed model, such as the PIPE prototype tree model, there seem no insuperable difficulties in extending the analyses of conventional EDA. EDA–GP approaches incorporating model structure learning are clearly more complex to analyse, because of the interaction between model structure learning and probability learning. The well-known equivalence between MML measures and probability, provide some reason for hope, that it might be possible to study these two components within a common framework.
7. *Contributing to the understanding of GP.* EDA–GP is partially motivated by GP, and due to the similarity in the representations they share many

characteristics. EDA–GP has a neat probability model to replace the population of GP. This provides some potential to study the dynamics of EDA–GP which may in turn shed light on understanding GP. We believe it may assist in understanding some important questions in GP, such as what form building blocks take, and how building blocks contribute to problem solving.

Acknowledgments

The authors are grateful for the constructive comments from the OBUPM workshop at GECCO 2005, which significantly improved the quality of this chapter, and comments from colleagues, in particular Lam Bui and Ang Yang. Thanks to Erick Cantu-Paz, Kumara Sastry, and Martin Pelikan for their effort to bring the researchers in the field of EDA together to present their diverse works in one book.

The first author would also like to thank University of New South Wales for its funding (UIPRS) during his Ph.D candidature.

References

[1] H. A. Abbass, N. X. Hoai, and R. I. McKay. AntTAG: A new method to compose computer programs using colonies of ants. In *The IEEE Congress on Evolutionary Computation*, pages 1654–1659, 2002.

[2] P. J. Angeline. Genetic programming and emergent intelligence. In K. E. Kinnear, Jr., editor, *Advances in Genetic Programming*, chapter 4, pages 75–98. MIT, Cambridge, MA, 1994.

[3] P. J. Angeline. Two self adaptive crossover operations for genetic programming. In *Advances in Genetic Programming II*, pages 89–110. MIT Press, Cambridge, MA, 1995.

[4] P. J. Angeline and J. B. Pollack. Coevolving high-level representations. In C. G. Langton, editor, *Artificial Life III*, volume XVII, pages 55–71, Santa Fe, New Mexico. Addison-Wesley, Reading, MA, 1994.

[5] D. Angluin. Negative results for equivalence queries. *Mach. Learn.* 5(2):121–150, 1990.

[6] S. Baluja. Population-based incremental learning: A method for integrating genetic search based function optimization and competitive learning. Technical Report CMU-CS-94-163, Carnegie Mellon University, Pittsburgh, PA, 1994.

[7] S. Baluja. Using a priori knowledge to create probabilistic models for optimization. *Int. J. Approx. Reason.*, 31(3):193–220, 2002.

[8] W. Banzhaf, P. Nordin, R. E. Keller, and F. D. Francone. *Genetic Programming – An Introduction; On the Automatic Evolution of Computer Programs and its Applications*. Morgan Kaufmann, dpunkt.verlag, 1998.

[9] E. Bengoetxea, P. Larrañaga, I. Bloch, A. Perchant, and C. Boeres. Inexact graph matching by means of estimation of distribution algorithms. *Pattern Recognit.*, 35(12):2867–2880, 2002.

[10] R. Blanco, I. Inza, and P. Larrañaga. Learning Bayesian networks in the space of structures by estimation of distribution algorithms. *Int. J. Intell. Syst.* 18(2):205–220, 2003.

[11] E. Bonabeau, M. Dorigo, and T. Theraulaz. *From Natural to Artificial Swarm Intelligence.* Oxford University Press, New York, 1999.

[12] M. Boryczka and Z. J. Czech. Solving approximation problems by ant colony programming. In W. B. Langdon, E. Cantú-Paz, K. Mathias, R. Roy, D. Davis, R. Poli, K. Balakrishnan, V. Honavar, G. Rudolph, J. Wegener, L. Bull, M. A. Potter, A. C. Schultz, J. F. Miller, E. Burke, and N. Jonoska, editors, *GECCO 2002: Proceedings of the Genetic and Evolutionary Computation Conference*, page 133, New York. Morgan Kaufmann, Los Altos, CA, 2002.

[13] P. Bosman and D. Thierens. An algorithmic framework for density estimation based evolutionary algorithms. Technical Report Technical Report UU-CS-1999-46, Utrecht University, 1999.

[14] P. A. N. Bosman and E. D. de Jong. Grammar transformations in an eda for genetic programming. In *Special session: OBUPM - Optimization by Building and Using Probabilistic Models, GECCO*, Seattle, Washington, USA, 2004.

[15] S. F. Chen. Bayesian grammar induction for language modeling. In *Meeting of the Association for Computational Linguistics*, pages 228–235, 1995.

[16] S. F. Chen. *Buidling Probabilistic Models for Natural Language.* PhD thesis, Harvard University Press, Cambridge, MA, USA, 1996.

[17] N. Cramer. A representation for the adaptive generation of simple sequential programs. In *Proceedings of an International Conference on Genetic Algorithms and their Applications*, pages 183–187, Carnegie-Mellon University, Pittsburgh, PA, USA, 1985.

[18] N. L. Cramer. A representation for the adaptive generation of simple sequential programs. In J. J. Grefenstette, editor, *Proceedings of an International Conference on Genetic Algorithms and the Applications*, pages 183–187, Carnegie-Mellon University, Pittsburgh, PA, USA, 1985.

[19] A. S. d'Avila Garcez, K. Broda, and D. M. Gabbay. Symbolic knowledge extraction from trained neural networks: a sound approach. *Artif. Intell.* 125(1–2):155–207, 2001.

[20] P. D'haeseleer. Context preserving crossover in genetic programming. In *Proceedings of the 1994 IEEE World Congress on Computational Intelligence*, volume 1, pages 256–261, Orlando, Florida, USA. IEEE, New York, 1994.

[14] D. E. Goldberg. *Genetic algorithms in search, optimization, and machine learning.* Addison-Wesley, Reading, MA, 1989.

[22] J. Green, J. L. Whalley, and C. G. Johnson. Automatic programming with ant colony optimization. In M. Withall and C. Hinde, editors, *Proceedings of the 2004 UK Workshop on Computational Intelligence*, pages 70–77. Loughborough University, 2004.

[23] F. Gruau. On using syntactic constraints with genetic programming. In P. J. Angeline and K. E. Kinnear, Jr., editors, *Advances in Genetic Programming 2*, chapter 19, pages 377–394. MIT, Cambridge, MA, USA, 1996.

[24] P. Grunwald. A minimum description length approach to grammar inference. In *Connectionist, Statistical and Symbolic Approaches to Learning for Natural Language*, volume 1004 of *Lecture Notes in AI*, pages 203–216. Springer, Berlin Heidelberg New York, 1994.

[25] P. Haddawy. Generating Bayesian networks from probability logic knowledge bases. In *Proceedings of the Tenth Conference on Uncertainty in Articial Intelligence*, 1994.

[26] G. Harik. Linkage learning via probabilistic modeling in the ECGA. Technical Report IlliGAL Report No. 99010, University of Illinois at Urbana-Champaign, 1999.

[27] G. R. Harik, F. G. Lobo, and D. E. Goldberg. The compact genetic algorithm. *IEEE Trans. Evol. Comput.* 3(4):287–297, 1999.

[28] J. H. Holland. *Adaptation in Natural and Artificial Systems.* University of Michigan Press, 1975.

[29] H. Iba and H. de Garis. Extending genetic programming with recombinative guidance. In *Advances in Genetic Programming II*, pages 69–88. MIT, Cambridge, MA, 1996.

[30] A. K. Joshi, L. S. Levy, and M. Takahashi. Tree adjunct grammars. *J. Comput. Syst. Sci.* 10(1):136–163, 1975.

[31] C. Keber and M. G. Schuster. Option valuation with generalized ant programming. In *Proceedings of the Genetic and Evolutionary Computation Conference*, pages 74–81. Morgan Kaufmann, Los Altos, CA, 2002.

[32] B. Keller and R. Lutz. Evolving stochastic context-free grammars from examples using a minimum description length principle. In *Proceedings of Workshop on Automata Induction Grammatical Inference and Language Acquisition, ICML-97*, Nashville, Tennessee, 1997.

[33] J. R. Koza. *Genetic Programming: On the Programming of Computers by Means of Natural Selection.* MIT, Cambridge, MA, USA, 1992.

[34] W. B. Langdon and R. Poli. *Foundations of Genetic Programming.* Springer, Berlin Heidelberg New York, 2002.

[35] K. Lari and S. J. Young. The estimation of stochastic context-free grammars using the inside-outside algorithm. *Comput. Speech Lang.* 4:35–56, 1990.

[36] K. Lari and S. J. Young. Applications of stochastic context-free grammars using the inside-outside algorithm. *Comput. Speech Lang.* 5: 237–257, 1991.

[37] P. Larrañaga and J. A. Lozano. *Estimation of Distribution Algorithms: A New Tool for Evolutionary Computation.* Kluwer, Dordrecht, 2001.

[38] P. P. Le, A. Bah, and L. H. Ungar. Using prior knowledge to improve genetic network reconstruction from microarray data. *Silico Biol.* 4(27), 2004.

[39] C. Manning and H. Schütze. *Foundations of Statistical Natural Language Processing.* MIT, Cambridge, MA, 1999.

[40] R. S. Michalski. Learnable evolution model: Evolutionary processes guided by machine learning. *Mach. Learn.* 38:9–40, 2000.

[41] D. J. Montana. Strongly typed genetic programming. BBN Technical Report #7866, Bolt Beranek and Newman, Inc., 10 Moulton Street, Cambridge, MA 02138, USA, 1993.

[42] H. Müehlenbein and G. Paaß. From recombination of genes to the estimation of distributions i.binary parameters. In *Lecture Notes in Computer Science 1411: Parallel Problem Solving from Nature, PPSN IV*, pages 178–187. Springer, Berlin Heidelberg New York, 1996.

[43] H. Mühlenbein and T. Mahnig. The factorized distribution algorithm for additively decompressed functions. In *1999 Congress on Evolutionary Computation*, pages 752–759, Piscataway, NJ. IEEE Service Center, 1999.

[44] P. Nordin. A compiling genetic programming system that directly manipulates the machine code. In K. E. Kinnear, Jr., editor, *Advances in Genetic Programming*, chapter 14, pages 311–331. MIT, Cambridge, MA, 1994.

[45] M. O'Neill and C. Ryan. Grammatical evolution. *IEEE Trans. Evol. Comput.* 5(4):349–358, 2001.

[46] U.-M. O'Reilly and F. Oppacher. The troubling aspects of a building block hypothesis for genetic programming. In L. D. Whitley and M. D. Vose, editors, *Foundations of Genetic Algorithms 3*, pages 73–88, Estes Park, Colorado, USA. Morgan Kaufmann, Los Altos, CA, 1995.

[47] M. Osborne. DCG induction using MDL and parsed corpora. In J. Cussens, editor, *Proceedings of the 1st Workshop on Learning Language in Logic*, pages 63–71, Bled, Slovenia, 1999.

[48] T. K. Paul and H. Iba. Identification of informative genes for molecular classification using probabilistic model building genetic algorithm. In e. a. Kalyanmoy Deb, editor, *Proceedings of Genetic and Evolutionary Computation (GECCO 2004) (Lecture Notes in Computer Science)*, volume 3102, pages 414–425, Seattle, USA, 2004.

[49] M. Pelikan. *Bayesian optimization algorithm: From single level to hierarchy.* PhD thesis, University of Illinois at Urbana-Champaign, Urbana, IL, 2002. Also IlliGAL Report No. 2002023.

[50] M. Pelikan, D. E. Goldberg, J. Ocenasek, and S. Trebst. Robust and scalable black-box optimization, hierarchy, and Ising spin glasses. IlliGAL Report No. 2003019, Illinois Genetic Algorithms Laboratory, University of Illinois at Urbana-Champaign, Urbana, IL, 2003.

[51] F. Pereira and Y. Schabes. Inside-outside reestimation from partially bracketed corpora. In *Proceedings of the 30th conference on Association for Computational Linguistics*, pages 128–135. Association for Computational Linguistics, 1992.

[52] P. Prusinkiewicz and A. Lindenmayer. *The Algorithmic Beauty of Plants.* Springer, Berlin Heidelberg New York, 1990.

[53] A. Ratle and M. Sebag. Avoiding the bloat with probabilistic grammar-guided genetic programming. In P. Collet, C. Fonlupt, J.-K. Hao, E. Lutton, and M. Schoenauer, editors, *Artificial Evolution 5th International Conference, Evolution Artificielle, EA 2001*, volume 2310 of *LNCS*, pages 255–266, Creusot, France. Springer, Berlin Heidelberg New York, 2001.

[54] J. Rissanen. *Stochastic Complexity in Statistical Inquiry.* World Scientific, Singapore, 1989.

[55] S. A. Rojas and P. J. Bentley. A grid-based ant colony system for automatic program synthesis. In M. Keijzer, editor, *Late Breaking Papers at the 2004 Genetic and Evolutionary Computation Conference*, Seattle, Washington, USA, 2004.

[56] J. P. Rosca. Analysis of complexity drift in genetic programming. In J. R. Koza, K. Deb, M. Dorigo, D. B. Fogel, M. Garzon, H. Iba, and R. L. Riolo, editors, *Genetic Programming 1997: Proceedings of the Second Annual Conference*, pages 286–294, Stanford University, CA, USA. Morgan Kaufmann, Los Altos, CA, 1997.

[57] J. P. Rosca and D. H. Ballard. Genetic programming with adaptive representations. Technical Report TR 489, University of Rochester, Computer Science Department, Rochester, NY, USA, 1994.

[58] O. Roux and C. Fonlupt. Ant programming: or how to ants for automatic programming. In *Proceedings of the Second International Conference on Ant Algorithms (ANTS2000)*, Belgium, 2000.

[59] R. Sagarna and J. Lozano. On the performance of estimation of distribution algorithms applied to software testing. *Appl. Artif. Intell.* 19(5): 457–489, 2004.

[60] Y. Sakakibara. Learning contextfree grammars from structural data in polynomial time. *Theor. Comput. Sci.* 76:223–242, 1990.

[61] Y. Sakakibara. Efficient learning of context-free grammars from positive structural examples. *Inf. Comput.* 97(1):23–60, 1992.

[62] Y. Sakakibara. Recent advances of grammatical inference. *Theor. Comput. Sci.* 185(1):15–45, 1997.

[63] R. P. Salustowicz and J. Schmidhuber. Probabilistic incremental program evolution. *Evol. Comput.* 5(2):123–141, 1997.

[64] K. Sastry and D. E. Goldberg. Probabilistic model building and competent genetic programming. In R. L. Riolo and B. Worzel, editors, *Genetic Programming Theory and Practise*, chapter 13, pages 205–220. Kluwer, Dordecht, 2003.

[65] J. Schmidhuber. Evolutionary principles in self-referential learning. on learning now to learn: The meta-meta-meta...-hook. Diploma thesis, Technische Universitat Munchen, Germany, 1987.

[66] Y. Shan. *Program Distribution Estimation with Grammar Models.* PhD thesis, University of New South Wales, Australia, 2005.

[67] Y. Shan, R. McKay, R. Baxter, H. Abbass, D. Essam, and H. Nguyen. Grammar model-based program evolution. In *Proceedings of The Congress on Evolutionary Computation*, Portland, USA. IEEE, New York, 2004.

[68] Y. Shan, R. I. McKay, H. A. Abbass, and D. Essam. Program evolution with explicit learning: a new framework for program automatic synthesis. In *Proceedings of 2003 Congress on Evolutionary Computation*, Canberra, Australia. University College, University of New South Wales, Australia, 2003.

[69] W. M. Spears, K. A. D. Jong, T. Bäck, D. B. Fogel, and H. de Garis. An overview of evolutionary computation. In P. Brazdil, editor, *Machine Learning: ECML-93, European Conference on Machine Learning*, pages 442–459, Vienna, Austria. Springer, Berlin Heidelberg New York, 1993.

[70] A. Stolcke. *Bayesian Learning of Probabilistic Language Models.* PhD thesis, University of California, Berkeley, CA, 1994.

[71] I. Tanev. Implications of incorporating learning probabilistic context-sensitive grammar in genetic programming on evolvability of adaptive locomotion gaits of snakebot. In *Proceedings of GECCO 2004*, Seattle, Washington, USA, 2004.

[72] A. Teller and M. Veloso. PADO: A new learning architecture for object recognition. In K. Ikeuchi and M. Veloso, editors, *Symbolic Visual Learning*, pages 81–116. Oxford University Press, Oxford, 1996.

[73] G. G. Towell and J. W. Shavlik. Extracting refined rules from knowledge-based neural networks. *Mach. Learn.* 13(1):71–101, 1993.

[74] C. S. Wallace and D. M. Boulton. An information measure for classification. *Comput. J.* 11(2):185–194, 1968.

[75] C. S. Wallace and D. L. Dowe. Minimum message length and kolmogorov complexity. *Comput. J.* 42(4):270–283, 1999.

[76] C. S. Wallace and P. R. Freeman. Estimation and inference by compact coding. *J. R. Statist. Soc. Ser. B (Methodological)*, 49(3):240–265, 1987.

[77] P. Whigham. Grammatically-based genetic programming. In J. P. Rosca, editor, *Proceedings of the Workshop on Genetic Programming: From Theory to Real-World Applications*, pages 33–41, Tahoe City, California, USA, 1995.

[78] P. Whigham. Inductive bias and genetic programming. In *Proceedings of First International Conference on Genetic Algorithms in Engineering Systems: Innovations and Applications*, pages 461–466. IEE, London, 1995.

[79] P. Whigham. A schema theorem for context-free grammars. In *1995 IEEE Conference on Evolutionary Computation*, volume 1, pages 178–181, Perth, Australia. IEEE, New York, 1995.

[80] P. Whigham and F. Recknagel. Predicting chlorophyll-a in freshwater lakes by hybridising process-based models and genetic algorithms. *Ecol. Modell.* 146(1–3):243–252, 2001.

[81] M. L. Wong and K. S. Leung. Genetic logic programming and applications. *IEEE Expert*, 10(5):68–76, 1995.

[82] K. Yanai and H. Iba. Estimation of distribution programming based on Bayesian network. In *Proceedings of Congress on Evolutionary Computation*, pages 1618–1625, Canberra, Australia, 2003.

[83] M. Zlochin, M. Birattari, N. Meuleau, and M. Dorigo. Model-based search for combinatorial optimization: A critical survey. *Ann. Oper. Res.* 131:373–395, 2004.

7

Efficiency Enhancement of Estimation of Distribution Algorithms

Kumara Sastry, Martin Pelikan, and David E. Goldberg

Summary. Efficiency-enhancement techniques speedup the search process of estimation of distribution algorithms (EDAs) and thereby enable EDAs to solve hard problems in *practical* time. This chapter provides a decomposition and an overview of different efficiency-enhancement techniques for estimation of distribution algorithms. Principled approaches for designing an evaluation-relaxation, and a time-continuation technique are discussed in detail.

Key words: Efficiency enhancement, evolutionary computation, estimation of distribution algorithms, parallelization, evaluation relaxation, hybridization, time continuation, speedup

7.1 Introduction

A key challenge in genetic and evolutionary algorithm research is the design of *competent* genetic algorithms (GAs) that can solve *hard problems* quickly, reliably, and accurately. Estimation of distribution algorithms (EDAs) are one such class of competent GAs. In essence, EDAs take problems that were intractable with first-generation GAs and render them tractable, oftentimes requiring only a polynomial (usually subquadratic) number of fitness evaluations. However, for large-scale problems, the task of computing even a subquadratic number of function evaluations can be daunting. This is especially the case if the fitness evaluation is a complex simulation, model, or computation. For example, if a search problem requires over a million evaluations, and if each evaluation takes about 10 s, EDAs would take over 120 days to successfully solve the problem. This places a premium on a variety of *efficiency enhancement techniques*. In essence, while competence leads us from *intractability* to *tractability*, efficiency enhancement takes us from *tractability* to *practicality*. In addition to function evaluations, in EDAs, the probabilistic model building process can also be computationally intensive, especially with

K. Sastry et al.: *Efficiency Enhancement of Estimation of Distribution Algorithms*, Studies in Computational Intelligence (SCI) **33**, 161–185 (2006)
www.springerlink.com

increasing problem sizes, making a variety of *model-efficiency-enhancement techniques* also a necessity.

A distinct advantage of EDAs over many other evolutionary algorithms is that the probabilistic models contain useful information about problem structure that can be exploited in the principled design of various efficiency-enhancement methods. Systematically incorporating problem knowledge *mined* through the model-building process of EDAs into the design of an efficiency-enhancement technique makes it adaptive and can potentially enhance the speed-up of the method. For example, when a simple surrogate (approximate fitness function) is used as an alternative to an expensive and accurate fitness evaluation, we obtain a moderate speed-up of about 1.3[75]. On the other hand, when the probabilistic model is used to design a surrogate, we obtain a speed-up of about 50 [64]. That is, by incorporating problem knowledge contained in the probabilistic model into the design of the surrogate, we obtain about *39-fold* increase in the speed-up.

In this chapter, we present an overview of different efficiency-enhancement techniques, used to speedup not only the search process, but also the model-building process. We will also illustrate systematic and principled ways of incorporating and integrating the knowledge gained through probabilistic models in the efficiency-enhancement methods – specifically, evaluation relaxation [72], and time continuation [24] – to yield maximum speedup. Additionally, subsequent chapters will discuss in detail some of the efficiency-enhancement methods outlined here.

This chapter is organized as follows. We start with a brief outline of fundamental tradeoffs exploited by different EDA efficiency-enhancement methods and discuss four broad classes of efficiency enhancement techniques (1) Parallelization, (2) hybridization, and (3) time continuation, and (4) evaluation relaxation. We then provide examples of two principled efficiency-enhancement techniques (1) An evaluation-relaxation scheme where we build an endogenous fitness-estimate model using the probabilistic models built by EDAs – specifically, the Bayesian optimization algorithm (BOA) [62, also see chapter by Pelikan et al] – in Sect. 7.3, and (2) a time-continuation scheme where we develop a scalable mutation operator in the extended compact GA (eCGA) [37, also see chapter by Harik et al] that searches locally in the substructural neighborhood in Sect. 7.4. Summary and key conclusions are given in Sect. 7.5.

7.2 Decomposition of Efficiency Enhancement Techniques

In practical optimization problems we are often faced with limited computation resources, which brings forth different tradeoffs involving (1) time, which is the product of population size, number of generations per epoch, and the number of convergence epochs, and (2) solution quality assessment. Note that the time includes both the function-evaluation time and the EDA time (time for selection, model building, model sampling, and replacement). It should be noted that the EDA time – especially the model building, sampling, and replacement – can be very significant and sometimes comparable to – if not more than – the function-evaluation time.

One or more of the following tradeoffs are exploited by efficiency-enhancement techniques to speedup EDAs:

Quality-Duration Tradeoff: Usually, the longer we run an EDA (with a sufficient population size), the higher will be the solution quality. However, in real-world scenarios, the computational resources are often limited, which leads to a tradeoff between solution quality and the search duration. Therefore, efficiency can be gained by choosing a search procedure that maximizes solution quality given the computational resource requirements. For example, quality-duration tradeoff might result in deciding between running a single epoch of an EDA with large population, as opposed to multiple epochs of the EDA with small population.

In addition to the search process, building a high quality model in EDAs might require longer time. On the other hand, reasonably accurate models might be built in less amount of time. Therefore, efficiency in EDAs can be gained by correctly deciding model accuracies during the search process.

Accuracy-Cost Tradeoff: Oftentimes, many complex real-world optimization problems involve computationally expensive function evaluation. However, an array of cheaper fitness functions can be easily developed, but at the cost of accuracy of fitness estimation. That is, the approximate fitness functions (or surrogates) suffer from various levels of error, and typically, cheaper the fitness function, larger the error in it. This introduces a tradeoff between fitness functions that are computationally cheap, but less accurate and fitness functions that are accurate, but computationally expensive. Therefore, we have to decide on the level of solution-quality assessment accuracy required during the search process, such that high-quality solutions can be obtained at minimum computational cost.

Additionally, in EDAs, a similar tradeoff exists between the probabilistic model accuracy and the cost. Typically, high-quality models are more expensive than low-quality models and striking an appropriate balance between model accuracy and model cost can significantly improve the model-building efficiency of EDAs.

Time Budget and Resource Allocation Tradeoff: Given limited computational resource its allocation in terms of population size, run duration, and number of convergence epochs can significantly influence the efficiency of the search algorithm. Time budgeting tradeoffs are often faced when distributing the EDA process between multiple processors (to strike a balance between communication and computation times), dividing the overall search time between different variation operators of an EDA such as crossover and mutation, or dividing the search time between different local and global search methods.

Additionally, in EDAs time budgeting tradeoffs can also be faced when dividing resources between model building and model usage (in terms of exploration via model sampling and evaluation of sampled individuals).

Efficiency-enhancement techniques that exploit one or more of the aforementioned tradeoffs can be broadly classified into four categories:

Parallelization: EDAs are run on multiple processors and the computations are distributed among these processors [14]. The use of parallelization – of both the search process and the model-building process – in EDAs is discussed in detail elsewhere in this book (see chapter by Ocenasek et al).

Hybridization: Domain-specific knowledge and other techniques are coupled with EDAs to create a search bias and to accelerate the search process [16, 26, 39, 44, 54, 80]. In addition to traditional hybridization methods, prior knowledge can be incorporated in the probabilistic models of EDAs, details of which are provided elsewhere in this book (see chapter by Baluja). Additionally, the effectiveness of hybridizing EDAs with local search methods is empirically demonstrated for the spin-glass problems elsewhere in this book (see chapter by Pelikan and Hartmann).

Time continuation/utilization: Capabilities of both mutation and recombination are utilized to the fullest extent, and time budgeting issues are addressed depending on the problem type [24, 47, 73, 74, 82, 83]. In this chapter, we will illustrate a principled manner of incorporating neighborhood information, contained in the probabilistic models, with time-continuation operators to yield maximum speedup.

Evaluation relaxation: Accurate, but expensive fitness functions are replaced by less accurate, but inexpensive fitness functions (or surrogates), and thereby the total number of costly fitness evaluations is reduced [2, 8, 31, 43, 53, 64, 72, 75, 76, 81]. In this chapter, we will illustrate a principled approach for using substructural knowledge provided by probabilistic models of EDAs to develop an endogenous surrogate that can be used instead of the expensive fitness function to obtain high-quality solutions and thus provide maximum speedup. In addition to relaxing the solution-quality assessment measures, we can also relax the model-quality assessment in EDAs [66].

The speedup obtained by employing an efficiency-enhancement technique (EET) is measured in terms of a ratio of the computation effort required by an EDA when the is not used to that required when the EET is used. That is, $\eta = T_{\text{base}}/T_{\text{efficiency-enhanced}}$. The speedup obtained by employing even a single EET can potentially be significant. Furthermore, assuming that the performance of one of the above methods does not affect the performance of others, if we employ more that one EET, the overall speedup is the product of individual speedups. That is, if the speedups obtained by employing parallelization, hybridization, time continuation and evaluation relaxation be η_{p}, η_{h}, η_{t}, and η_{e} respectively, then the overall speedup obtained is

$$\eta_{\mathrm{total}} = \eta_{\mathrm{p}}\eta_{\mathrm{h}}\eta_{\mathrm{t}}\eta_{\mathrm{e}}.$$

Even if the speedup obtained by a single EET is modest, a combination of two or more EETs can yield a significant speedup. For example, if we use a parallel EDA that yields linear speedup with 100 processors, and each of the other three EETs makes EDAs 25% more efficient, then together they yield a speedup of $100 * 1.25^3 = 195.3$. That is evaluation relaxation, time continuation, and hybridization would give slightly more than 95 processors' worth of additional computation power.

Before we demonstrate principled methodologies for utilizing information from probabilistic models of EDAs for maximum efficiency enhancement, we present a brief outline of each of the four classes of efficiency-enhancement techniques.

7.2.1 Parallelization

In parallelization, EDAs are run on multiple processors and the computations are distributed among these processors [13, 14]. Evolutionary algorithms are by *nature* parallel, and many different parallelization approaches such as a simple master-slave [9, 30], coarse-grained [32, 67, 84], fine-grained [27, 28, 50, 70], or hierarchical [23, 29, 33, 48] architectures can be readily used. Regardless of how parallelization is done, the key idea is to distribute the computational load of EDAs on several processors thereby speeding-up the search process. A principled design theory exists for developing an efficient parallel GA and optimizing the key facts of parallel architecture, connectivity, and deme size [14], some of which are discussed in the next chapter. Apart from parallelizing the function evaluations, the probabilistic model building process can also be parallelized [56–58] which is also discussed in the chapter by Ocenasek et al.

7.2.2 Hybridization

In hybridization, domain-specific knowledge and other local-search techniques are coupled with evolutionary algorithms to obtain high-quality solutions in reasonable time [16, 26, 39, 44, 54, 80]. Most industrial-strength evolutionary algorithms employ some sort of local search for a number of reasons such as achieving faster convergence [12, 39, 80], repairing infeasible solutions into legal ones [42, 59], initializing the population [21, 68], and refining of solutions obtained by a GA [41]. In addition to traditional ways of hybridizing EDAs [34, 46, 60, 61, 71], prior knowledge of the search problem can be incorporated into the probabilistic models as discussed elsewhere in this book (see chapter by Baluja).

While evolutionary-algorithm practitioners have often understood that real-world or commercial applications require hybridization, there have been limited efforts in developing a principled design framework on answering critical issues such as the optimal division of labor between global and local searches (or the right mix of exploration and exploitation) [26, 79], the effect of local search on sampling [39, 40], and the optimal duration of local search [39, 45], and similar efforts are yet to be attempted for understanding and designing hybrid EDAs.

7.2.3 Time Continuation

In time continuation, capabilities of both mutation and recombination are optimally utilized to obtain a solution of as high quality as possible with a given limited computational resource [24, 47, 73, 74, 82, 83]. Time utilization (or continuation) exploits the tradeoff between the search for solutions with large population and a single convergence epoch and using a small population with multiple convergence epochs.

Early theoretical investigations indicate that when the subsolutions are of equal (or nearly equal) salience and both recombination and mutation operators have the linkage information, then a small population with multiple convergence epochs is more efficient. However, if the fitness function is noisy or has overlapping subsolutions, then a large population with single convergence epoch is more efficient [73, 74]. On the other hand, if the subsolutions of the problem are of nonuniform salience, which essentially requires serial processing, then a small population with multiple convergence epochs is more efficient [24]. While early efforts on developing adaptive continuation operators using probabilistic models of EDAs are promising [35, 47, 73], much work needs to be done to develop a principled design theory for efficiency enhancement via time continuation and to design adaptive continuation operators to reinitialize population between convergence epochs.

7.2.4 Evaluation relaxation

In evaluation relaxation, an accurate, but computationally expensive fitness evaluation is replaced with a less accurate, but computationally inexpensive fitness estimate. The low-cost, less-accurate fitness estimate can either be (1) *exogenous*, as in the case of approximate fitness functions [8, 43, 49], where external means can be used to develop the fitness estimate, or (2) *endogenous*, as in the case of *fitness inheritance* [81] where, some of the offspring fitness is estimate based on fitness of parental solutions.

Evaluation relaxation in GAs dates back to early, largely empirical work of Grefenstette and Fitzpatrick [31] in image registration [20] where significant speedups were obtained by reduced random sampling of the pixels of an image. Approximate models have since been used extensively to solve complex optimization problems in many engineering applications such as aerospace and structural engineering [8, 11, 19].

While early evaluation-relaxation studies were largely empirical in nature, design theories have since been developed to understand the effect of approximate evaluations via surrogates on population sizing and convergence time and to optimize speedups in approximate fitness functions with known variance [51, 53], in integrated fitness functions [3, 4], in simple functions of known variance or known bias [72], and in fitness inheritance [75]. While exogenous surrogates can be readily used in EDAs, the probabilistic models of EDAs can be effectively used to develop endogenous surrogates that provide significant speedup [64, 76], details of which are provided in the next section. In addition to relaxing the solution-quality assessment measures, we can also relax the model-quality assessment in EDAs. For example, we can use *sporadic* model building, where the structure of the probabilistic model is built once every few generations and the probabilities are updated every generation [66].

7.3 Evaluation Relaxation: Designing Adaptive Endogenous Surrogates

As mentioned earlier, a distinct advantage of EDAs over first-generation GAs is the availability of variable-interaction information in terms of the probabilistic models *mined* from a population of promising solutions. Therefore, we can use the probabilistic models to infer the structural form of the surrogate. This is in contrast to surrogates often used to speedup evolutionary algorithms, which are of fixed form and do not adapt to key variable interactions of the underlying search problem. In other words, with the help of probabilistic models built in EDAs, we can use the probabilistic models to decide on the form of the surrogate and use one of the system identification, estimation, or regression methods to estimate the coefficients of the surrogate.

For example, the probabilistic model of eCGA represents nonoverlapping partitions of variables. The resulting surrogate inferred from the model would then be a polynomial, whose order and terms are decided based on the substructures identified by the model, and the coefficients of the surrogate represent the partial contribution of the subsolutions to the overall fitness of the individual [76]. The surrogates designed with the information provided by the probabilistic models are quite accurate and yield substantial speedups. For example, on a class of boundedly difficult additively decomposable problems endogenous surrogates in BOA yields speedups of about 50 [64]. This is in contrast to a moderate speed-up of about 1.3 obtained by using a simple *fitness inheritance* method [75, 81].

In the remainder of this section, we illustrate the design of adaptive endogenous surrogates in the Bayesian optimization algorithm. We note that the design method can be extended to other EDAs and the key idea is to using the probabilistic model to infer the structural form of the surrogate and to use system-identification and estimation tools for computing the coefficients of the surrogate.

7.3.1 Evaluation Relaxation: Endogenous Surrogates in BOA

The Bayesian optimization algorithm (BOA) uses Bayesian networks to model candidate solutions [62, also see chapter by Pelikan et al]. The structure of the Bayesian network is encoded by a directed acyclic graph with the nodes corresponding to the variables in the modeled data set and the edges corresponding to conditional dependencies. A Bayesian network encodes a joint probability distribution given by

$$p(X) = \prod_{i=1}^{n} p(X_i|\Pi_i), \tag{7.1}$$

where $X = (X_0, \ldots, X_{n-1})$ is a vector of all the variables in the problem; Π_i is the set of parents of X_i (the set of nodes from which there exists an edge to X_i); and $p(X_i|\Pi_i)$ is the conditional probability of X_i given its parents Π_i.

The parameters of the Bayesian networks are represented by a set of conditional probability tables (CPTs) specifying a conditional probability for each variable given any instance of the variables that the variable depends on. CPTs store conditional probabilities $p(X_i|\Pi_i)$ for each variable X_i. Local structures – in the form of decision trees or decision graphs – can also be used in place of full CPTs to enable more efficient representation of local conditional probability distributions in Bayesian networks. While we describe the design of endogenous surrogate in BOA with CPTs, similar methodology can be used for BOA with decision trees and graphs.

Given the probabilistic model (in form of a Bayesian network), we can infer the form of the surrogate as an acyclic tree whose nodes correspond to the variables and the edges correspond to the marginal fitness contributions of subsolutions (or the coefficients of the surrogate). That is, for every variable X_i and each possible value x_i of X_i, an estimate of the marginal fitness contribution of a subsolution with $X_i = x_i$ must be stored for each instance π_i of X_i's parents Π_i. In the binary case, each row in the CPT is thus extended by two additional entries. Figure 7.1 shows an example of the probability model and the substructural surrogate in BOA. The substructural fitness can be estimated as

$$f_{\mathrm{est}}(X_1, X_2, \ldots, X_\ell) = \bar{f} + \sum_{i=1}^{\ell} \left(\bar{f}(X_i|\Pi_i)\right), \tag{7.2}$$

where $\bar{f}(X_i|\Pi_i)$ denotes the average fitness of solutions with X_i and Π_i. That is,

$$\bar{f}(X_i|\Pi_i) = \frac{1}{n_h} \sum_{\{j|y_j \supset x_i, \pi_i\}} f\left(y_j\right) - \bar{f}(\Pi_i), \tag{7.3}$$

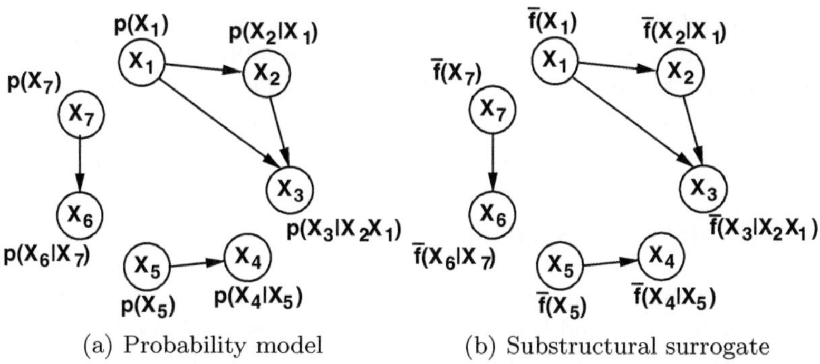

Fig. 7.1. Substructural fitness estimation model in Bayesian optimization algorithm. The estimated fitness for the model is given by $f_{\text{est}}(X_1, X_2, \ldots, X_7) = \bar{f} + \bar{f}(X_1) + \bar{f}(X_2|X_1) + \bar{f}(X_3|X_2X_1) + \bar{f}(X_5) + \bar{f}(X_4|X_5) + \bar{f}(X_7) + \bar{f}(X_6|X_7)$

where n_h is the total number of individuals that contain the schema π_i, y_j is the jth individual and $f(y_j)$ is its fitness, and $\bar{f}(\Pi_i)$ is the average fitness of all solutions with Π_i.

Similar to earlier fitness-inheritance studies, we begin with fully evaluating the initial population, and thereafter evaluating an offspring with a probability $1-p_i$. In other words, we use the endogenous surrogate to estimate the fitness of an offspring with probability p_i. One question remains as to where to obtain information for computing the coefficients of the surrogate, which is addressed in Sect 7.3.2.

7.3.2 Where to Estimate the Marginal Fitnesses From?

In the proposed method, for each instance x_i of X_i and each instance π_i of X_i's parents Π_i, we must compute the average fitness of all solutions with $X_i = x_i$ and $\Pi_i = \pi_i$. In this section, we discuss two sources for computing the coefficients of the surrogate:

1. Selected parents that were evaluated using the actual fitness function
2. The offspring that were evaluated the actual fitness function

The reason for restricting computation of the coefficients of the surrogate to selected parents and offspring is that the probabilistic model used as the basis for selecting relevant statistics represents nonlinearities in the population of parents and the population of offspring. Since it is best to maximize learning data available, it seems natural to use both populations to compute the marginal fitness of the components of the surrogate. The reason for restricting input for computing these statistics to solutions that were evaluated using the actual fitness function is that the fitness of other solutions was estimated only and it involves errors that could mislead the surrogate and propagate through generations.

We have extensively tested the proposed evaluation-relaxation scheme on a class of boundedly difficult additively decomposable problems. Before presenting the key results, we now briefly introduce facetwise models to predict the scalability and speed-up of using endogenous surrogates as an alternative to expensive fitness evaluation.

7.3.3 Scalability and Speedup

Facetwise and dimensional models can be used to analyze the scalability of and the speedup provided by endogenous surrogates in EDAs. In this section, we present the key results of the analysis and the details are given elsewhere [76].

The error introduced by the surrogate can be modeled as additive Gaussian noise with zero mean and variance $p_{\mathrm{i}}\sigma^2_{\mathrm{f,t}}$, where p_{i} is the probability of an individual receiving estimated fitness, and $\sigma^2_{\mathrm{f,t}}$ is the true fitness variance. However, this approximation is not valid for very high p_{i} values as the substructural fitness is estimated from very few individuals, which increases the error in the estimate significantly. Empirically, we observed that the noise variance becomes significantly higher than the above approximation when $p_{\mathrm{i}} \geq 0.85$. Error due to variance (as in additive Gaussian noise) increases both the population size and run duration required for EDA success [25, 38, 52, 72].

The increase in the required population size due to the use of the substructural surrogate is given by

$$n_{\mathrm{r}} = \frac{n}{n_o} = (1 + p_{\mathrm{i}})\,. \tag{7.4}$$

where n_o is the minimum population size required to obtain a solution of quality $(m-1)/m$ when the endogenous surrogate is not used. Here, m is the number of key substructures of the search problem.

The increase in the run duration due to the use of the surrogate is given by

$$t_{\mathrm{c,r}} = \frac{t_{\mathrm{c}}}{t_{\mathrm{c,o}}} = \sqrt{1 + p_{\mathrm{i}}}\,. \tag{7.5}$$

where $t_{\mathrm{c,o}}$ is the run duration – in other words, the number of generations – taken by the EDA to successfully solve the search problem when the surrogate is not used.

Using (7.4) and (7.5) and, after further simplifications and approximations, we can estimate the increase in the total number of function evaluations required to obtain a solution with at least $m-1$ out of m substructures at their optimal values as

$$n_{\mathrm{fe,r}} \approx (1 + p_i)^{1.5}\,(1 - p_i)\,. \tag{7.6}$$

Therefore, the speedup provided by using endogenous fitness-estimation model is given by the inverse of the function-evaluation ratio:

$$\eta_{\text{endogenous fitness model}} = \frac{1}{(1+p_{\mathrm{i}})^{1.5}(1-p_{\mathrm{i}})}. \tag{7.7}$$

Equation (7.6) indicates that the number of function evaluations initially increases with p_{i}, reaching a maximum at $p_{\mathrm{i}} = 0.2$. The function-evaluation-ratio model indicates that the number of function evaluations decreases with p_{i} for $p_{\mathrm{i}} > 0.2$ and reaches a minimum at $p_{\mathrm{i}} = 1$. In other words, the speedup decreases initially ($p_{\mathrm{i}} < 0.2$) and then increases reaching a maximum at $p_{\mathrm{i}} = 1$. However, as mentioned earlier, the facetwise models for the population sizing and the convergence time are not valid at very high values of p_{i}. Nevertheless, the models are suggestive and as shown in the results, we obtain maximum speedups when p_{i} is close to 1.0.

It should be noted that in our scalability and speedup analysis, we only considered the cost of actual fitness evaluation. In other words, we ignored the time complexity of selection, fitness model construction, generation of new candidate solutions, and fitness estimation. Combining these factors with the complexity estimate for the actual fitness evaluation can be used to compute the optimal proportion of candidate solutions whose fitnesses can be estimated using the endogenous surrogate. We reiterate that the proposed evaluation-relaxation scheme is beneficial when the actual fitness evaluation is expensive, in which case the above costs are indeed negligible and the models developed in this section valid.

7.3.4 Results and Discussion

We use two test functions for verifying and validating the use of the endogenous surrogate instead of costly, but accurate function-evaluation method. Our approach in verifying the models and observing if the proposed evaluation-relaxation scheme yields speedup is to consider bounding *adversarial problems* that exploit one or more dimensions of problem difficulty [25]. Particularly, we are interested in problems where substructure identification and exchange is critical for the EDA success. Specifically, we use OneMax – where the fitness function is the number of ones in the binary string – and $m-k$ deceptive trap problem [1,17,18,22].

While the optimization of the OneMax problem is easy, the probabilistic models built by EDAs such as eCGA and BOA, however, are known to be only partially correct and include spurious linkages [63]. Therefore, the speed-up results on the OneMax problem will indicate if the effect of using partially correct linkage mapping on the endogenous surrogate is significant. For an ideal surrogate developed for the OneMax problem, the average fitness of a 1 in any leaf should be approximately 0.5, whereas the average fitness of a 0 in any leaf should be approximately −0.5.

Unlike, the OneMax problem, $m-k$ deceptive problems are boundedly difficult and the accurate identification and exchange of key substructures are critical to EDA success. For the $m-k$ trap problem, $\bar{f}(X_{\mathrm{i}} = 0)$ and

$\bar{f}(X_\mathrm{i} = 1)$ depend on the state of the search because the distribution of contexts of each bit changes over time and bits in a trap are not independent. The context of each leaf also determines whether $\bar{f}(X_\mathrm{i} = 0) < \bar{f}(X_\mathrm{i} = 1)$ or $\bar{f}(X_\mathrm{i} = 0) > \bar{f}(X_\mathrm{i} = 1)$ in that particular leaf.

Figure 7.2(a) and 7.2(b) present the scalability and speedup results of the evaluation-relaxation scheme for BOA on a 50-bit OneMax, 10-4 and 10-5 deceptive trap functions. We considered a binary ($s = 2$) tournament selection without replacement. For each test problem, the following proportions of using the surrogate, p_i, were considered: 0–0.9 with step 0.1, 0.91–0.99 with step 0.01, and 0.991–0.999 with step 0.001. For each test problem and p_i value, 30 independent experiments were performed. Each experiment consisted of 10 independent runs with the minimum population size to ensure convergence to a solution within 10% of the optimum (i.e., with at least 90% correct bits) in all 10 runs. For each experiment, bisection method was used to determine the minimum population size, and the number of evaluations (excluding the evaluations done using the model of fitness) was recorded. The average of 10 runs in all experiments was then computed and displayed as a function of the proportion of candidate solutions for which fitness was estimated using the surrogate. Therefore, each point in Figs. 7.2(a) and 7.2(b) represents an average of 300 BOA runs that found a solution that is at most 10% from the optimum.

In all experiments, the number of actual fitness evaluations decreases with p_i. Furthermore, the surrogates built in BOA are applicable at high p_i values, even as high as 0.99. That is, by evaluating less than 1% of candidate solutions and estimating the fitness for the rest using the endogenous surrogate, we obtain speedup of 31 (for OneMax) to 53 (for m k-Trap). In other words, by developing and using the endogenous surrogate to estimate the fitness of

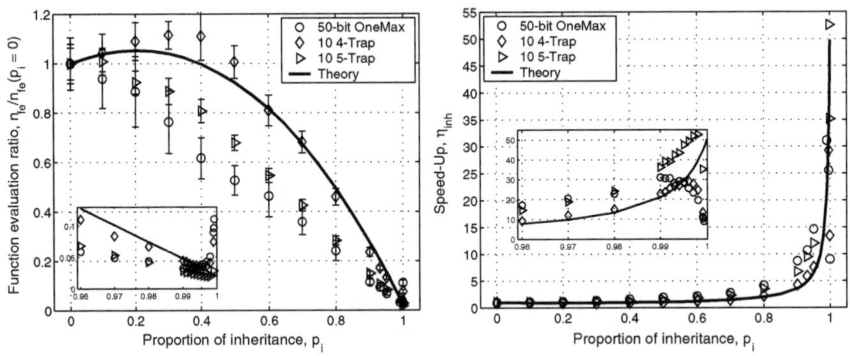

(a) Function-evaluation-ratio

(b) Speedup in BOA

Fig. 7.2. The effect of using the endogenous surrogate on the total number of function evaluations required for BOA success, and the speedup obtained by using the evaluation relaxation-scheme in BOA. The empirical results are obtained for a 50-bit OneMax, 10 4-Trap and 10 5-trap problems

99% of the individuals, we can reduce the number of actual fitness evaluation required to obtain high quality solutions by a factor of up to 53.

Overall, the results suggest that significant efficiency enhancement can be achieved through an endogenous surrogate that incorporates knowledge of important subsolutions of a problem and their partial fitnesses. The results clearly indicate that using the surrogate in EDAs can reduce the number of solutions that must be evaluated using the actual fitness function by a factor of 31–53. Consequently, if fitness evaluation is a bottleneck, there is a lot of room for improvement using endogenous surrogates in EDAs in general, and BOA, in particular. For real-world problems, the actual savings may depend on the problem being considered. However, it can be expected that developing and using the fitness-estimate model enables significant reduction in the number of fitness evaluations on many problems because deceptive problems of bounded difficulty bound a large class of important nearly decomposable problems.

The probabilistic models are not only useful for the design of surrogates, but can be exploited in other facets of efficiency enhancement as well. In the following section, we illustrate the use of probabilistic models in the principled design of time continuation operators.

7.4 Time Continuation: Mutation in EDAs

In time continuation, we investigate and decide between the fundamental tradeoff between using an evolutionary algorithm with a large population for a single convergence epoch or with a small population for multiple convergence epochs, as illustrated in Fig. 7.3. A *continuation operator* is required when using a small-population, multiple-epoch evolutionary algorithm for maintain diversity in population between *epochs*. In the ideal case, the continuation operator perturbs only bad building blocks (BBs) at the end of each convergence epoch. However, in practice, the continuation operator not only perturbs bad building blocks, but also some good ones, and a regeneration cost – or cost of reduction in solution quality between two epochs – is incurred. Since the continuation operator modifies only a few individuals to seed the population in subsequent epochs, it is assumed to be some form of mutation or local search method. Therefore, the decision making involved in time continuation can also be posed as choosing between two key genetic operators – recombination and mutation.

Goldberg [24] developed an analytical framework to optimally solve the *time continuation* problem. Early studies considered the effect of the salience structure and observed that while a large population run is preferable for problems with near-uniform salience structure, a small population run is advantageous for problems with exponential-salience structure [24, 82, 83]. More recent studies considered the effectiveness of incorporating the building-block structure into both global and local evolutionary algorithm operators and the effect of noise, salience structure and the crosstalk between different building

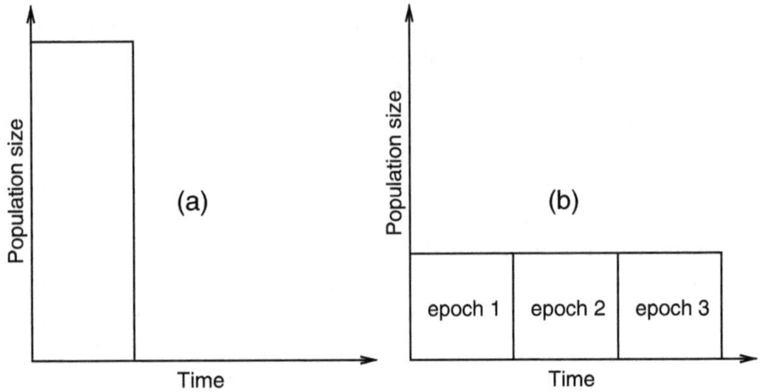

Fig. 7.3. Two scenarios of resource utilization: **(a)** Large population, single convergence epoch, and **(b)** Small population, multiple convergence epochs

blocks on time continuation [74, 77]. When both global and local operators are given the substructural information (or good neighborhood information), a small population run is beneficial for deterministic problems with near-uniform salience structure. On the other hand, for noisy problems, a large population is preferred.

One of the key challenges in time continuation is the design of effective continuation operators that searches in the correct neighborhoods. Existing mutation (or continuation) operators usually search in the local neighborhood of an individual, without taking into account the *global* neighborhood information. In genetic algorithms, mutation is usually a secondary search operator which performs random walk *locally* around a solution. On the other hand, in evolution strategies, while powerful mutation operators are used [6, 10, 36, 69, 78], the neighborhood information is still *local* around a single or few solutions. In local-search literature, while the importance of using a good neighborhood operator is often highlighted [5, 7, 15, 85, 86], no systematic methods for designing neighborhood operators that can solve a broad class of bounding problems have been developed.

However, for solving boundedly difficult problems, local neighborhood information is not sufficient, and a mutation operator which uses local neighborhoods requires $\mathcal{O}(m^k \log m)$ number of evaluations [55]. Therefore, we utilize the probabilistic models built in eCGA for automatically building *global* neighborhood (or linkage) information into the mutation operator. Unlike, adaptive mutation techniques in evolution strategies, which usually have local neighborhood information adapted over time, our method leads to a more global induction of the neighborhood.

In Sect. 7.4.1, we illustrate the design of an efficient operator that utilizes the *global* neighborhood information mined by the probabilistic models of the extended compact genetic algorithm (eCGA) [37, also see chapter by Harik et al] to search among competing subsolutions.

7.4.1 Mutation in eCGA: Searching in Substructural Neighborhood

As described in the chapter by Harik et al, eCGA builds marginal product models that yields a direct mapping of linkage groups among successful individuals. Therefore, we use the model-building procedure of eCGA to identify the key substructures of a problem. Once the linkage-groups are identified, we use an *enumerative building-block-wise mutation* operator [74] to search for the best among competing subsolutions. For example, if the model builder identifies m BBs of size k each, the eCGA continuation operator will select the best subsolution out of 2^k possible ones in each of the m partition.

That is, from a sample of randomly generated candidate solution the top solutions (as determined by the selection mechanism) are used to build probabilistic model in eCGA. The best solution in the population is used for substructural mutation: Consider the first nonmutated substructure, where the substructures are arbitrarily chosen from left-to-right, however, different schemes can be – or may required to be – chosen to decide the order of choosing substructures to mutate. For example, substructural partitions that contain most *active* variables might be mutated before those that contain less active variables. For the substructure in consideration, create $2^k - 1$ unique individuals with all possible subsolutions in the chosen partition, where k is order of the substructure. The subsolutions in other partitions are not modified. Evaluate all $2^k - 1$ individuals and retain the best for mutation of other substructures. Thus at each convergence epoch the best subsolution in each partition is chosen and the search ends after m convergence epochs, where m is the number of substructures in the problem.

Note that the performance of the above mutation can be slightly improved by using a greedy heuristic to search for the best among competing BBs, however, as shown later, the scalability of the mutation operator is determined by the population-size required to accurately identify the building blocks. Furthermore, we perform linkage identification only once in the initial generation. This offline linkage identification works well on problems with BBs of nearly equal salience. However, for problems with BBs of nonuniform salience, we would have to perform linkage identification and update BB information in regular intervals. The key idea in designing the mutation operator in other EDAs such as BOA is that the operator should effectively use the neighborhood information contained in the probabilistic models.

We now present the scalability of the enumerative BB-wise mutation operator and followed by an analysis of the efficiency enhancement provided by time continuation in eCGA.

7.4.2 Scalability of Building-Block-Wise Mutation

The scalability of the BB-wise mutation operator depends on two factors (1) the population size required to build accurate probabilistic models of the linkage groups, and (2) the total number of evaluations required by the BB-wise mutation operator to find optimal subsolutions in all the partitions. Pelikan, Sastry, and Goldberg [65] developed facetwise models for predicting the critical and maximum population-size required to correctly identify good interactions among variables. They showed that the minimum population size scales as

$$n_{\text{min}} = \Theta\left(2^k m^{1.05}\right), \tag{7.8}$$

and the maximum population size which avoids discovery of false dependencies between independent variables is given by

$$n_{\text{max}} = \Theta\left(2^k m^{2.1}\right). \tag{7.9}$$

In other words, to avoid incorrect identification of BBs, the population size should be less than n_{max}. Since we require that *all* the BBs be correctly identified in the first generation itself, the population size required should be greater than n_{min}, but less than n_{max}. That is,

$$\Theta\left(2^k m^{1.05}\right) \leq n \leq \Theta\left(2^k m^{2.1}\right). \tag{7.10}$$

Since the model building is performed only once, the total number of function evaluations scales as the population size. That is,

$$\Theta\left(2^k m^{1.05}\right) \leq n_{\text{fe},1} \leq \Theta\left(2^k m^{2.1}\right). \tag{7.11}$$

During BB-wise mutation, we evaluate $2^k - 1$ individuals for determining the best BBs in each of the m partitions. Therefore, the total number of function evaluations used during BB-wise mutation is

$$n_{\text{fe},2} = \left(2^k - 1\right) m = \mathcal{O}\left(2^k m\right). \tag{7.12}$$

From Equations 7.11 and 7.12, the total number of function evaluations scales as

$$\Theta\left(2^k m^{1.05}\right) \leq n_{\text{fe}} \leq \Theta\left(2^k m^{2.1}\right). \tag{7.13}$$

We now empirically verify the scale-up of the population size and the number of function evaluations required for successfully solving the $m - k$ deceptive trap problem in Figs. 7.4(a) and 7.4(b), respectively. For the empirical runs, we use tournament selection without replacement with a tournament size of 8. The average number of subsolutions correctly converged are computed over 30 independent runs. The minimum population size required such that $m - 1$ subsolutions converge to the correct value is determined by a bisection method [72]. The results of population-size is averaged over 30 such bisection

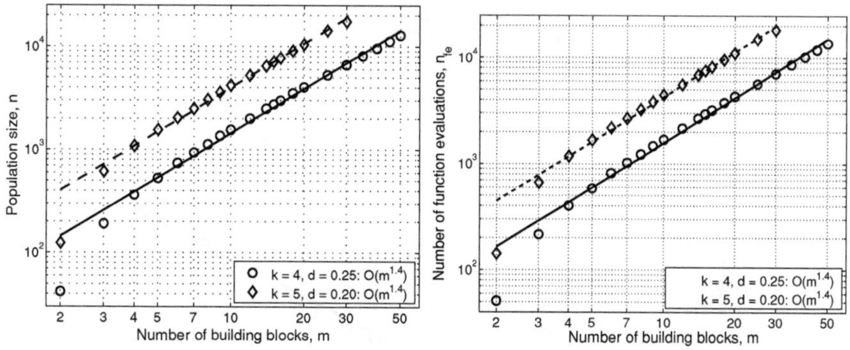

(a) Population size

(b) Number of function evaluations

Fig. 7.4. Population size (7.10) and the number of function evaluations (7.13) required by BB-wise mutation for solving $m - k$ Trap function. The results are averaged over 900 runs for the number of function evaluations and 30 bisection runs for the population size. The relative deviation for the empirical results is less than 0.2%. The population size and the number of function evaluations both scale as $\Theta(2^k m^{1.5})$

runs, while the results for the function-evaluation ratio is averaged over 900 independent runs.

In contrast to fixed mutation operators which require $\mathcal{O}(m^k \log m)$ number of function evaluations to solve additively separable GA-hard problems [55], the proposed eCGA-based BB-wise mutation operator that automatically identifies the linkage groups requires only $\mathcal{O}(2^k m^{1.5})$ (polynomial) number of evaluations.

7.4.3 Crossover vs. Mutation in eCGA

We know that eCGA scales as $\Theta\left(2^k \sqrt{k} m^{1.5} \log m\right)$ [73, also see chapter by Harik et al], and from (7.13), we know that the BB-wise mutation scales as $\Theta\left(2^k m^{1.5}\right)$ for additively separable problem of bounded difficulty. Therefore, the BB-wise mutation operator in eCGA is $\Theta\left(\sqrt{k} \log m\right)$ faster than eCGA in solving boundedly difficult additively separable problems. That is, the speedup – which is defined as the ratio of number of function evaluations required by eCGA to that required by the selectomutative GA – is given by

$$\eta = \frac{n_{\text{fe}}(\text{eCGA})}{n_{\text{fe}}(\text{BBwise Mutation})} = \Theta\left(\sqrt{k} \log m\right), \tag{7.14}$$

which is empirically verified in Fig. 7.5. The results clearly indicate the efficiency enhancement provided by the time continuation operator that automatically and adaptively searches for the subsolution neighborhood as identified by eCGA.

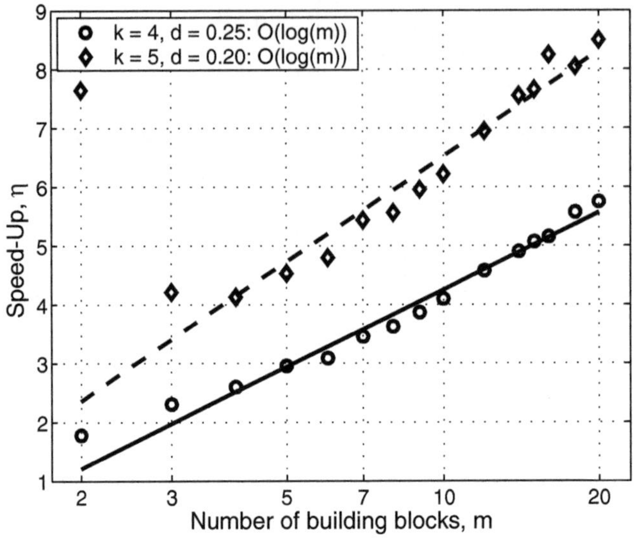

Fig. 7.5. Empirical verification of the speedup (7.14) obtained by using the probabilistic model building BB-wise mutation over eCGA for the $m - k$ Trap function. The results show that the speedup scales as $\Theta(\sqrt{k} \log m)$

Time-continuation scenarios have also been studied when dealing with noisy problems and problems with overlapping building blocks, where a competent crossover is often more efficient than a competent mutation, and therefore a large-population, single convergence epoch eCGA run is preferred [74, 77]. One of the important efforts directly motivated by this study, which is currently underway, is the design and development of *adaptive time continuation operators* that utilize the substructural models built by eCGA not only in mutation and recombination operators, but also to automatically decide between using a large population with single convergence epoch or a small population eCGA with multiple convergence epochs [47]. Simply stated, the model building is used to identify the appropriate population size regime and whether local or global operators are used. For example, for noisy problems an adaptive time continuation operator should implicitly switch from local to global search operator as the problem becomes more noisy. The decision making depends upon the type of the problem being solved, and results in significant savings even for modestly sized problems.

7.5 Summary and Conclusions

Like any industrial-strength search algorithm, practical deployment of EDAs strongly rely on one or more efficiency-enhancement techniques such as parallelization, hybridization, time continuation, and evaluation relaxation. While EDAs take problems that were intractable by first generation genetic algorithms, and render them tractable, principled efficiency-enhancement techniques take us from tractability to practicality. In this chapter, we presented an overview of various efficiency-enhancement techniques for speeding-up EDAs. We also provided two examples of principled efficiency-enhancement techniques, both of which utilize the probabilistic models built by the EDAs. The first example was an evaluation-relaxation method, where we build an endogenous substructural surrogate to estimate fitness of majority of the population, while actual fitness is computed for only a small portion of the population. The second example developed a competent mutation (or time continuation) operator in the extended compact genetic algorithm, which uses the probabilistic models and searches locally in the subsolution neighborhood.

The two examples clearly demonstrate that by systematically incorporating problem knowledge gained through the probabilistic models built in EDAs into the efficiency-enhancement technique, the speedup can be significantly enhanced. Furthermore, the overall efficiency of combining such nearly independent efficiency-enhancement techniques is multiplicative. For example, if we use a parallel EDA that yields linear speedup with 100 processors, and each of the other three EETs makes EDAs 25% more efficient, then together they yield a speedup of $100 * 1.25^3 = 195.3$. That is, evaluation relaxation, time continuation, and hybridization would give slightly more than 95 processors' worth of additional computation power.

Acknowledgments

This work was sponsored by the Air Force Office of Scientific Research, Air Force Materiel Command, USAF, under grant F49620-03-1-0129, the National Science Foundation under ITR grant DMR-03-25939 at Materials Computation Center, and the Research Award and the Research Board at the University of Missouri. The U.S. Government is authorized to reproduce and distribute reprints for government purposes notwithstanding any copyright notation thereon.

The views and conclusions contained herein are those of the authors and should not be interpreted as necessarily representing the official policies or endorsements, either expressed or implied, of the Air Force Office of Scientific Research, the National Science Foundation, or the U.S. Government.

References

[1] Ackley, D. H. (1987). *A Connectionist Machine for Genetic Hill Climbing.* Kluwer, Boston, MA

[2] Albert, L. A. (2001). Efficient genetic algorithms using discretization scheduling. Master's thesis, University of Illinois at Urbana-Champaign, General Engineering Department, Urbana, IL

[3] Albert, L. A. and Goldberg, D. E. (2001). Efficient evaluation relaxation under integrated fitness functions. *Intelligent Engineering Systems*

Through Artificial Neural Networks, 11:165–170. (Also IlliGAL Report No. 2001024)

[4] Albert, L. A. and Goldberg, D. E. (2002). Efficient discretization scheduling in multiple dimensions. *Proceedings of the Genetic and Evolutionary Computation Conference*, pages 271–278. (Also IlliGAL Report No. 2002006)

[5] Armstrong, D. E. and Jacobson, S. H. (2005). Data independent neighborhood functions and strict local optima. *Discrete Applied Mathematics*, 146(3):233–243

[6] Bäck, T. (1996). *Evolutionary Algorithms in Theory and Practice.* Oxford University Press, New York

[7] Barnes, J. W., Dimova, B., and Dokov, S. P. (2003). The theory of elementary landscapes. *Applied Mathematical Letters*, 16:337–343

[8] Barthelemy, J.-F. M. and Haftka, R. T. (1993). Approximation concepts for optimum structural design—a review. *Structural Optimization*, 5:129–144

[9] Bethke, A. D. (1976). Comparison of genetic algorithms and gradient-based optimizers on parallel processors: Efficiency of use of processing capacity. Tech. Rep. No. 197, Univeristy of Michigan, Logic of Computers Group, Ann Arbor, MI

[10] Beyer, H.-G. (1996). Toward a theory of evolution strategies: Self-adaptation. *Evolutionary Computation*, 3(3):311–347

[11] Booker, A. J., Dennis, J. E., Frank, P. D., Serafini, D. B., Torczon, V., and Trosset, M. W. (1998). A rigorous framework for optimization of expensive functions by surrogates. Technical report, National Aeronautics and Space Administration (NASA), Hampton, VA. ICASE Report No. 98-47

[12] Bosworth, J. L., Foo, N., and Zeigler, B. P. (1972). Comparison of genetic algorithms with conjugate gradient methods. ORA Tech. Report No. 00312-1-T, University of Michigan, Department of Computer and Communication Sciences, Ann Arbor, MI

[13] Cantú-Paz, E. (1997). A summary of research on parallel genetic algorithms. IlliGAL Report No. 97003, University of Illinois at Urbana-Champaign, General Engineering Department, Urbana, IL

[14] Cantú-Paz, E. (2000). *Efficient and accurate parallel genetic algorithms.* Kluwer, Boston, MA

[15] Colletti, B. W. and Barnes, J. W. (2004). Using group theory to construct and characterize metaheuristic search neighborhoods. In Rego, C. and Alidaee, B., editors, *Adaptive Memory and Evolution: Tabu Search and Scatter Search*, pages 303–329. Kluwer , Boston, MA

[16] Davis, L., editor (1991). *Handbook of Genetic Algorithms.* Van Nostrand Reinhold, New York, NY

[17] Deb, K. and Goldberg, D. E. (1992). Analyzing deception in trap functions. *Foundations of Genetic Algorithms*, 2:93–108. (Also IlliGAL Report No. 91009)

[18] Deb, K. and Goldberg, D. E. (1994). Sufficient conditions for deceptive and easy binary functions. *Annals of Mathematics and Artificial Intelligence*, 10:385–408. (Also IlliGAL Report No. 92001)
[19] Dennis, J. E. and Torczon, V. (1997). Managing approximation models in optimization. In Alexandrov, N. M. and Hussaini, M. Y., editors, *Multidisciplinary Design Optimization: State-of-the-Art*, pages 330–347, Philadelphia, PA. SIAM
[20] Fitzpatrick, J. M., Grefenstette, J. J., and Van Gucht, D. (1984). Image registration by genetic search. In *Proceedings of IEEE Southeast Conference*, pages 460–464, Piscataway, NJ. IEEE press
[21] Fleurent, C. and Ferland, J. (1994). Genetic hybrids for the quadratic assignment problem. In *DIMACS Series in Mathematics and Theoretical Computer Science*, volume 16, pages 190–206
[22] Goldberg, D. E. (1987). Simple genetic algorithms and the minimal deceptive problem. In Davis, L., editor, *Genetic algorithms and simulated annealing*, chapter 6, pages 74–88. Morgan Kaufmann, Los Altos, CA
[23] Goldberg, D. E. (1989). *Genetic Algorithms in Search Optimization and Machine Learning*. Addison-Wesley, Reading, MA
[24] Goldberg, D. E. (1999). Using time efficiently: Genetic-evolutionary algorithms and the continuation problem. *Proceedings of the Genetic and Evolutionary Computation Conference*, pages 212–219. (Also IlliGAL Report No. 99002)
[25] Goldberg, D. E. (2002). *Design of innovation: Lessons from and for competent genetic algorithms*. Kluwer, Boston, MA
[26] Goldberg, D. E. and Voessner, S. (1999). Optimizing global-local search hybrids. *Proceedings of the Genetic and Evolutionary Computation Conference*, pages 220–228. (Also IlliGAL Report No. 99001)
[27] Gorges-Schleuter, M. (1989). ASPARAGOS: A population genetics approach to genetic algorithms. *Evolution and Optimization '89*, pages 86–94
[28] Gorges-Schleuter, M. (1989). ASPARAGOS: An asynchronous parallel genetic optimization strategy. *Proceedings of the Third International Conference on Genetic Algorithms*, pages 422–428
[29] Gorges-Schleuter, M. (1997). ASPARAGOS96 and the traveling salesman problem. *Proceedings of the IEEE International Conference on Evolutionary Computation*, pages 171–174
[30] Grefenstette, J. J. (1981). Parallel adaptive algorithms for function optimization. Tech. Rep. No. CS-81-19, Vanderbilt Univeristy, Computer Science Department, Nashville, TN
[31] Grefenstette, J. J. and Fitzpatrick, J. M. (1985). Genetic search with approximate function evaluations. *Proceedings of the International Conference on Genetic Algorithms and Their Applications*, pages 112–120
[32] Grosso, P. B. (1985). *Computer simulations of genetic adaptation: Parallel subcomponent interaction in a multilocus model.* PhD thesis, University of Michigan, Ann Arbor, MI. (University microfilms no. 8520908)

[33] Gruau, F. (1994). *Neural network synthesis using cellular encoding and the genetic algorithm.* PhD thesis, L'Universite Claude Bernard-Lyon I

[34] Handa, H. (2003). Hybridization of estimation of distribution algorithms with a repair method for solving constraint satisfaction problems. *Proceedings of the Genetic and Evolutionary Computation Conference*, pages 991–1002

[35] Handa, H. (2005). The effectiveness of mutation operation in the case of estimation of distribution algorithms. *Journal of Biosystems.* (To appear)

[36] Hansen, N. and Ostermeier, A. (2001). Completely derandomized self-adaptation in evolution strategies. *Evolutionary Computation*, 9(2): 159–195

[37] Harik, G. (1999). Linkage learning via probabilistic modeling in the ECGA. IlliGAL Report No. 99010, University of Illinois at Urbana-Champaign, Urbana, IL

[38] Harik, G., Cantú-Paz, E., Goldberg, D. E., and Miller, B. L. (1999). The gambler's ruin problem, genetic algorithms, and the sizing of populations. *Evolutionary Computation*, 7(3):231–253. (Also IlliGAL Report No. 96004)

[39] Hart, W. E. (1994). *Adaptive global optimization with local search.* PhD thesis, University of California, San Diego, San Diego, CA

[40] Hart, W. E. and Belew, R. K. (1996). Optimization with genetic algorithm hybrids using local search. In Belew, R. K. and Mitchell, M., editors, *Adaptive Individuals in Evolving Populations*, pages 483–494. Addison-Wesley, Reading, MA

[41] Hartmann, A. K. and Rieger, H. (2001). *Optimization algorithms in physics*, chapter 9, pages 192–203. Wiley-VCH, Berlin

[42] Ibaraki, T. (1997). Combinations with other optimization methods. In Bäck, T., Fogel, D. B., and Michalewicz, Z., editors, *Handbook of Evolutionary Computation*, pages D3:1–D3:2. Institute of Physics Publishing and Oxford University Press, Bristol and New York

[43] Jin, Y. (2005). A comprehensive survey of fitness approximation in evolutionary computation. *Soft Computing Journal*, 9(1):3–12

[44] Krasnogor, N. (2002). *Studies on the Theory and Design Space of Memetic Algorithms.* PhD thesis, University of the West of England, Bristol, England

[45] Land, M. (1998). *Evolutionary algorithms with local search for combinatorial optimization.* PhD thesis, University of California at San Diego, San Diego, CA

[46] Lima, C. F. and Lobo, F. G. (2004). Parameter-less optimization with the extended compact genetic algorithm and iterated local search. *Proceedings of the Genetic and Evolutionary Computation Conference*, 1:1328–1339. (Also technical report cs.NE/0402047 at arXiv.org)

[47] Lima, C. F., Sastry, K., Goldberg, D. E., and Lobo, F. G. (2005). Combining competent crossover and mutation operators: A probabilistic model

building approach. *Proceedings of the 2005 Genetic and Evolutionary Computation Conference*, pages 735–742. (Also IlliGAL Report No. 2005002)

[48] Lin, S.-C., Goodman, E. D., and Punch, W. F. (1997). Investigating parallel genetic algorithms on job shop scheduling problem. *Sixth International Conference on Evolutionary Programming*, pages 383–393

[49] Llorà, X., Sastry, K., Goldberg, D. E., Gupta, A., and Lakshmi, L. (2005). Combating user fatigue in iGAs: Partial ordering, support vector machines, and synthetic fitness. *Proceedings of the Genetic and Evolutionary Computation Conference*, pages 1363–1370. (Also IlliGAL Report No. 2005009)

[50] Manderick, B. and Spiessens, P. (1989). Fine-grained parallel genetic algorithms. *Proceedings of the Third International Conference on Genetic Algorithms*, pages 428–433

[51] Miller, B. L. (1997). *Noise, Sampling, and Efficient Genetic Algorithms.* PhD thesis, University of Illinois at Urbana-Champaign, General Engineering Department, Urbana, IL. (Also IlliGAL Report No. 97001)

[52] Miller, B. L. and Goldberg, D. E. (1995). Genetic algorithms, tournament selection, and the effects of noise. *Complex Systems*, 9(3):193–212. (Also IlliGAL Report No. 95006)

[53] Miller, B. L. and Goldberg, D. E. (1996). Optimal sampling for genetic algorithms. *Intelligent Engineering Systems through Artificial Neural Networks*, 6:291–297

[54] Moscato, P. (1989). On evolution, search, optimization, genetic algorithms and martial arts: Towards memetic algorithms. Technical Report C3P 826, Caltech Concurrent Computation Program, California Institute of Technology, Pasadena, CA

[55] Mühlenbein, H. (1992). How genetic algorithms really work: Mutation and hillclimbing. *Parallel Problem Solving from Nature II*, pages 15–26

[56] Munetomo, M., Murao, N., and Akama, K. (2005). Empirical studies on parallel network construction of Bayesian optimization algorithms. *Proceedings of the Congress of Evolutionary Computation*, 1-2-3: 1234–1238

[57] Ocenasek, J. and Pelikan, M. (2003). Parallel spin glass solving in hierarchical Bayeisan optimization algorithm. *Proceedings of the 9th International Conference on Soft Computing, Mendel 2003*, pages 120–125

[58] Ocenasek, J., Schwarz, J., and Pelikan, M. (2003). Design of multithreaded estimation of distribution algorithms. *Proceedings of the Genetic and Evolutionary Computation Conference*, pages 1247–1258

[59] Orvosh, D. and Davis, L. (1993). Shall we repair? Genetic algorithms, combinatorial optimization, and feasibility constraints. *Proceedings of the Fifth International Conference on Genetic Algorithms*, page 650

[60] Pelikan, M. (2005). *Hierarchical Bayesian Optimization Algorithm: Toward a New Generation of Evolutionary Algorithm.* Springer, Berlin Heidelberg New York

[61] Pelikan, M. and Goldberg, D. E. (2003). Hierarchical BOA solves Ising spin glasses and MAXSAT. *Proceedings of the Genetic and Evolutionary Computation Conference*, pages 1271–1282. (Also IlliGAL Report No. 2003001)

[62] Pelikan, M., Goldberg, D. E., and Cantú-Paz, E. (2000). Linkage learning, estimation distribution, and Bayesian networks. *Evolutionary Computation*, 8(3):314–341. (Also IlliGAL Report No. 98013)

[63] Pelikan, M., Goldberg, D. E., and Sastry, K. (2001). Bayesian optimization algorithm, decision graphs, and Occam's razor. *Proceedings of the Genetic and Evolutionary Computation Conference*, pages 519–526. (Also IlliGAL Report No. 2000020)

[64] Pelikan, M. and Sastry, K. (2004). Fitness inheritance in the Bayesian optimization algorithm. *Proceedings of the Genetic and Evolutionary Computation Conference*, 2:48–59. (Also IlliGAL Report No. 2004009)

[65] Pelikan, M., Sastry, K., and Goldberg, D. E. (2003). Scalability of the Bayesian optimization algorithm. *International Journal of Approximate Reasoning*, 31(3):221–258. (Also IlliGAL Report No. 2001029)

[66] Pelikan, M., Sastry, K., and Goldberg, D. E. (2005). Sporadic model building for efficiency enhancement of hBOA. IlliGAL Report No. 2005026, University of Illinois at Urbana-Champaign, Urbana, IL

[67] Pettey, C. C., Leuze, M. R., and Grefenstette, J. J. (1987). A parallel genetic algorithm. *Proceedings of the Second International Conference on Genetic Algorithms*, pages 155–161

[68] Ramsey, C. L. and Grefenstette, J. J. (1993). Case-Based initialization of genetic algorithms. *Proceedings of the Fifth International Conference on Genetic Algorithms*, pages 84–91

[69] Rechenberg, I. (1973). *Evolutionsstrategie: Optimierung Technischer Systeme nach Prinzipien der Biologischen Evolution.* Frommann-Holzboog, Stuttgart

[70] Robertson, G. G. (1987). Parallel implementation of genetic algorithms in a classifier system. *Proceedings of the Second International Conference on Genetic Algorithms*, pages 140–147

[71] Sastry, K. (2001a). Efficient cluster optimization using a hybrid extended compact genetic algorithm with a seeded population. IlliGAL Report No. 2001018, University of Illinois at Urbana-Champaign, Urbana, IL

[72] Sastry, K. (2001b). Evaluation-relaxation schemes for genetic and evolutionary algorithms. Master's thesis, University of Illinois at Urbana-Champaign, General Engineering Department, Urbana, IL. (Also IlliGAL Report No. 2002004)

[73] Sastry, K. and Goldberg, D. E. (2004a). Designing competent mutation operators via probabilistic model building of neighborhoods. *Proceedings of the 2004 Genetic and Evolutionary Computation Conference*, 2:114–125. Also IlliGAL Report No. 2004006

[74] Sastry, K. and Goldberg, D. E. (2004b). Let's get ready to rumble: Crossover versus mutation head to head. *Proceedings of the 2004 Genetic*

and Evolutionary Computation Conference, 2:126–137. Also IlliGAL Report No. 2004005
[75] Sastry, K., Goldberg, D. E., and Pelikan, M. (2001). Don't evaluate, inherit. *Proceedings of the Genetic and Evolutionary Computation Conference*, pages 551–558. (Also IlliGAL Report No. 2001013)
[76] Sastry, K., Pelikan, M., and Goldberg, D. E. (2004). Efficiency enhancement of genetic algorithms building-block-wise fitness estimation. *Proceedings of the IEEE International Conference on Evolutionary Computation*, pages 720–727
[77] Sastry, K., Winward, P., Goldberg, D. E., and Lima, C. F. (2005). Fluctuating crosstalk as a source of deterministic noise and its effects on ga scalability. IlliGAL Report No. 2005025, University of Illinois at Urbana-Champaign, Urbana, IL
[78] Schwefel, H.-P. (1977). Numerische optimierung von computer-modellen mittels der evolutionsstrategie. *Interdisciplinary Systems Research*, 26
[79] Sinha, A. (2003a). Designing efficient genetic and evolutionary hybrids. Master's thesis, University of Illinois at Urbana-Champaign, Urbana, IL. Also IlliGAL Report No. 2003020
[80] Sinha, A. (2003b). A survey of hybrid genetic and evolutionary algorithms. IlliGAL Report No. 2003004, University of Illinois at Urbana-Champaign, Illinois Genetic Algorithms Laboratory, Urbana, IL
[81] Smith, R., Dike, B., and Stegmann, S. (1995). Fitness inheritance in genetic algorithms. In *Proceedings of the ACM Symposium on Applied Computing*, pages 345–350, New York, NY, USA. ACM
[82] Srivastava, R. (2002). Time continutation in genetic algorithms. Master's thesis, University of Illinois at Urbana-Champaign, General Engineering Department, Urbana, IL
[83] Srivastava, R. and Goldberg, D. E. (2001). Verification of the theory of genetic and evolutionary continuation. *Proceedings of the Genetic and Evolutionary Computation Conference*, pages 551–558. (Also IlliGAL Report No. 2001007)
[84] Tanese, R. (1989). *Distributed genetic algorithms for function optimization.* PhD thesis, University of Michigan, Ann Arbor, MI. (University microfilms no. 8520908)
[85] Vaughan, D., Jacobson, S., and Armstrong, D. (2000). A new neighborhood function for discrete manufacturing process design optimization using generalized hill climbing algorithms. *ASME Journal of Mechanical Design*, 122(2):164–171
[86] Watson, J.-P. (2003). *Empirical Modeling and Analysis of Local Search Algorithms For The Job-Shop Scheduling Problem.* PhD thesis, Colorado State University, Fort Collins, CO

8

Design of Parallel Estimation of Distribution Algorithms

Jiri Ocenasek, Erick Cantú-Paz, Martin Pelikan, and Josef Schwarz

Summary. This chapter focuses on the parallelization of Estimation of Distribution Algorithms (EDAs). More specifically, it presents guidelines for designing efficient parallel EDAs that employ parallel fitness evaluation and parallel model building. Scalability analysis techniques are employed to identify and parallelize the main performance bottlenecks to ensure that the achieved speedup grows almost linearly with the number of utilized processors. The proposed approach is demonstrated on the parallel Mixed Bayesian Optimization Algorithm (MBOA). We determine the time complexity of parallel MBOA and compare this complexity with experimental results obtained on a set of random instances of the spin glass optimization problem. The empirical results fit well the theoretical time complexity, so the scalability and efficiency of parallel MBOA for unknown spin glass instances can be predicted.

Key words: Estimation of distribution algorithms, Parallelization, Efficiency enhancement, Evolutionary computation, Bayesian networks

8.1 Introduction

Estimation of Distribution Algorithms (EDAs) [1, 2], also called Probabilistic Model-Building Genetic Algorithms (PMBGAs) [3] and Iterated Density Estimation Evolutionary Algorithms (IDEAs) [4], often require fewer fitness evaluations than standard evolutionary algorithms. However, the overall execution time is still a limiting factor that determines the size of problems that are tractable in practice.

A promising way to reduce the execution time is through parallelization. Most papers on parallel Estimation of Distribution Algorithms concentrate on parallel construction and sampling of probabilistic models. The algorithms employing parallel construction of Bayesian networks include the Parallel Bayesian Optimization Algorithm (PBOA) [5] designed for pipelined architecture, the Distributed Bayesian Optimization Algorithm (DBOA) [6] designed for message passing architecture (MPI), the Parallel Estimation of

J. Ocenasek et al.: *Design of Parallel Estimation of Distribution Algorithms*, Studies in Computational Intelligence (SCI) **33**, 187–203 (2006)
www.springerlink.com

Bayesian Network Algorithm (EBNA_{BIC}) [7] designed for MIMD architecture with shared memory, and variants of EBNA_{BIC} [8] designed both for MPI and POSIX threads. The parallel Mixed Bayesian Optimization Algorithm (MBOA) [9], which uses Bayesian networks with local structures in the form of decision trees to model and sample solutions, was first designed and simulated for multithreaded environment in [10], and the real implementation of parallel MBOA suited for MPI appeared in [11].

In contrast to prior work on parallel EDAs, we provide general guidelines for designing the whole parallel EDA, including the detailed scalability analysis of each part. Without loss of generality, the proposed techniques are demonstrated on the MBOA example.

The following section explains the main paradigms used for designing parallel EDAs. Section 8.3 briefly describes MBOA. Section 8.4 analyzes the time complexity of each MBOA part. Section 8.5 provides the scalability analysis of parallel MBOA, identifies the important parts of MBOA that should be parallelized, and derives the guidelines that can be used to design effective parallel EDAs. Section 8.6 presents the performance prediction methods. Section 8.7 analyzes the performance of parallel MBOA on a real-world problem – spin glass optimization. Finally, Sect. 8.8 provides conclusions.

8.2 Main Paradigms of EDA Parallelization

There are two main approaches to parallelize EDAs: distribute the load of evaluating fitness among several processors and parallelize the model building. This section provides a brief description of these approaches.

8.2.1 Distributed Fitness Evaluation

One of the approaches to exploit parallelism with EDAs is to emulate the parallelization methods used by classical evolutionary algorithms (EAs). Classical EAs are usually parallelized either by parallelizing only the fitness evaluation component or by using multiple populations.

Single-population master–slave EAs have a master node that stores the population and executes the EA operations of selection, crossover, and mutation. In the case of a master–slave EDA, the master executes selection and builds and samples the probabilistic model. The evaluation of fitness is distributed among several slave processors. Despite being very simple algorithms, master–slave implementations can be very efficient. The master–slave architecture is visualized in Fig. 8.1a.

Multiple-population EAs are the most sophisticated and popular type of parallel EAs. They consist of several subpopulations or demes that evolve mostly independently on different processors, but occasionally exchange individuals by using a migration operator (see Fig. 8.1b for an illustration).

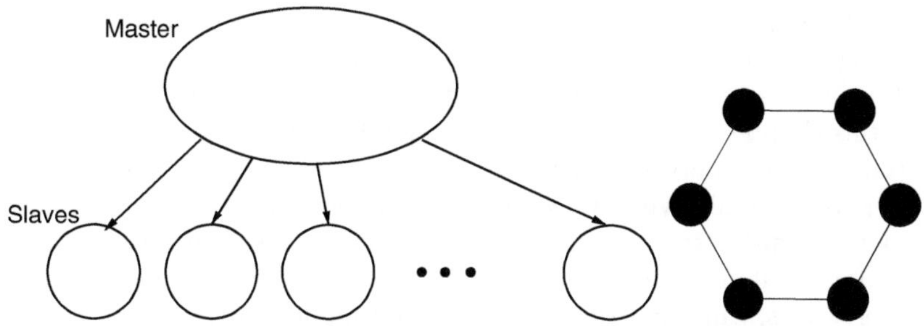

(a) Master Slave

(b) Multiple Populations

Fig. 8.1. Classical models of parallel EAs

In general, the communication requirements of parallel EAs are low, and inexpensive Beowulf clusters or Web-based computations can be practical. In fact, the behavior of GAs with spatially-distributed populations is interesting, regardless of their implementation on serial or parallel computers [12–14]. Having a spatially distributed population may have some algorithmic benefits that are independent of the efficiency gains obtained from using multiple processors (e.g., [15–18]).

The term *islands model* is easily understandable; the GA behaves as if the world was constituted of islands where populations evolve isolated from each other. On each island the population is free to converge toward different optimum. The migration operator is supposed to mix good features that emerge locally in the different demes.

For more detailed discussion of parallel fitness evaluation see [19].

8.2.2 Distributed Model Building

The most significant difference between classical genetic algorithms (GAs) and EDAs lies in the complexity of creation of the new population. In GAs, the recombination and mutation operator do not introduce significant computational costs: they are inexpensive binary and unary operators that can be performed locally in case of the island topology. In most EDAs, the building and sampling of the probabilistic model is computationally expensive. Additionally, sequential model building can be considered an N-ary operator, where N is the size of parent population, and that is why it is not straightforward to parallelize this operator.

Some algorithms – such as the Distributed Probabilistic Model-Building Genetic Algorithm (DPMBGA) [20] – are based on the straightforward parallelization of EDAs, where in each island the local model is constructed from a portion of population located on each island and migration is used to exchange locally generated individuals between islands. For simple problems this approach yields good results, especially if the class of utilized models is best suited for solving unimodal problems. Building several such models locally and migrating sampled individuals is then equivalent to using one mixture model, which is more general and better suited for solving multimodal problems.

In this chapter, we focus on models capable of learning higher-order linkage (interactions) between optimized variables. The accuracy of these models decreases rapidly with decreasing size of the parent population because the more sophisticated models involve the estimation of a large number of parameters that may not be supported adequately by a small population. In other words, sampling several distributed models constructed from fragmented parent population yields worse solutions than sampling one global model. Therefore, we focus on the parallel construction and sampling of a single probabilistic model in the pseudo-sequential manner.

8.3 Mixed Bayesian Optimization Algorithm

The Mixed Bayesian Optimization Algorithm (MBOA) [9] is based on BOA with Bayesian networks, but it uses more effective structure for representing conditional probability distributions in the form of decision trees, as proposed in the hierarchical Bayesian Optimization Algorithm (hBOA, [21]). MBOA can be also formulated for continuous domains, where it tries to find the partitioning of a search space into subspaces where the variables seem to be mutually independent. This decomposition is captured globally by the Bayesian network model with decision trees and the Gaussian kernel distribution is used locally to approximate the variable values in each resulting partition. The implementation details are described in [22].

Parallel MBOA was first simulated in the TRANSIM tool [10] and its real implementation was reported in [11].

In this chapter, we focus only on the binary domain, where MBOA can be seen as a variant of hBOA described earlier in this book.

8.4 Complexity Analysis

To parallelize any algorithm, it is important to understand time complexity of its different parts, so that one can identify the parts that are suitable for parallelization. To understand the overall complexity of MBOA, this section examines the complexity of each component of MBOA: selection, model construction, model sampling, replacement, and fitness evaluation.

8.4.1 Complexity of Selection Operator

The most commonly used selection operator in EDAs is tournament selection, where pairs of randomly chosen individuals compete to take place in the parent population that serves for model learning. Denoting the population size by N and the number of bits in solution strings by n, the number of tournaments is $O(N)$ and for each tournament we perform $O(n)$ steps to copy the winner, so the total complexity is $O(nN)$. This also holds for most other selection operators, such as truncation selection.

8.4.2 Complexity of Model Construction

As described elsewhere in this book, Bayesian networks with decision trees use one decision tree per variable. The sequential MBOA builds all the decision trees at once. It starts with empty trees and it greedily adds decision nodes. The quality of potential decision nodes is determined by the Bayesian–Dirichlet metric [23–25], which is able to determine the significance of statistical correlations between combinations of variable values in the population. This greedy algorithm picks the best candidate and never looks back to reconsider earlier choices. A necessary condition for adding a new split is that the Bayesian network remains acyclic after the split; that is, there does not exist a sequence of nodes in which each node contains its predecessor in its decision tree (as a split) and the last node is in the decision tree of the first node.

A straightforward approach to parallelizing model building in MBOA would be to distribute nodes of the Bayesian network to a number of processors, where each processor would compute the decision trees for the nodes it received. Nonetheless, this approach would require the processors to communicate after performing each split operation to avoid introducing cycles into the network. Furthermore, the work may be divided unevenly, causing ineffective utilization of the available computational resources.

Earlier, we proposed a method that solves the acyclicity problem [10, 11]. In each generation the algorithm uses a random permutation $\mathbf{o} = (o_0, o_1, \ldots, o_{n-1})$ to predetermine the topological ordering of variables in advance. That means that only the variables $X_{o_0} \cdots X_{o_{i-1}}$ can serve as splits in the binary decision tree for X_{o_i}. In this manner, no cycles can be introduced and the splits in different decision trees can thus be executed in parallel without the need for extensive communication.

Restricting the nodes to contain the dependencies on their predecessors in a particular permutation restricts the class of models that can be expressed. Consequently, the resulting model may not be as good as the one created by the standard model-building procedure. Nonetheless, experimental results indicate that this restriction does not lead to negative effects on reliability or efficiency of MBOA. Most importantly, given the constraint on the topological ordering of the nodes, each processor can create the whole decision tree asynchronously and independently of the other processors. Consequently, higher speedup is achieved by removing the communication overhead.

The code for building each decision tree can be described using this skeleton:

```
Function BuildTree(Population Pop, TargetVariable X_i,
          ListOfCandidateSplitVariables Pa): DecisionTreeNode;
Begin
  Initialize the frequency tables;                          ...O(n)
  For each Variable X_j in Pa                               ...O(n)
    Evaluate the metrics of X_j split for X_i target;       ...O(N)
  End for
  Pick up the split X_j' with the highest quality;          ...O(n)
  Pop1 := SelectIndividuals (Pop, "X_j' = 0");              ...O(N)
  Pop2 := SelectIndividuals (Pop, "X_j' = 1");              ...O(N)
  return new SplitNode(new SplitCondition("X_j'"),
                       BuildTree(Pop1, X_i, Pa\X_j'),
                       BuildTree(Pop2, X_i, Pa\X_j'));
End;
```

To express the time complexity of the whole model-building algorithm, we start with the complexity of one run of *BuildTree()* procedure, which is $O(n + nN + n + N + N) = O(n) + O(nN)$. The total number of *BuildTree()* calls is $O(2^h)$, where h is the average height of final decision tree, but note that the population size N is decreased exponentially in the recursion:

$$\sum_{i=1}^{h} 2^h(O(n) + O(nN/2^h)) \approx \sum_{i=1}^{h} 2^h O(nN/2^h) = \sum_{i=1}^{h} O(nN) = O(hnN) \tag{8.1}$$

To be precise, the time spent on initializing the frequency tables and the time spent on picking-up the best split is not compensated by this exponential decrease of population size, since it does not depend on N. However, in practice we can neglect these terms because, due to the need for a reliable statistical support for each decision node, the recursion stops with a sufficiently large number of individuals in a node to neglect the frequency initialization and split selection. The final complexity of building the entire decision tree is thus $O(hnN)$ and the complexity of building the entire probabilistic model composed of n decision trees is $O(hn^2N)$.

Because the sample size must be at least linearly proportional to the number of model parameters and because the decision trees are usually nearly balanced, the tree height is usually upper bounded by $O(\log N)$. Hence, the time complexity of model building can be rewritten as $O(n^2 N \log N)$. This time complexity also holds for hBOA.

8.4.3 Complexity of Model Sampling

Model sampling generates $O(N)$ individuals of length n, where each variable value is generated by traversing down a decision tree to a leaf. On average, it takes $O(h) = O(\log N)$ decisions before a leaf is reached. Thus, the overall complexity of model sampling is $O(nN \log N)$.

8.4.4 Complexity of Replacement Operator

MBOA uses restricted tournament replacement (RTR) to replace a part of the target population by generated offspring. RTR was proposed in [26] and its code can be specified as follows:

```
For each offspring individual                                ...O(N)
   For a*N randomly chosen individuals from target ...O(a*N)
      Compute the distance between chosen individual
                                  and offspring individual; ...O(n)
   End for
   If fitness of offspring individual is higher than the
                           fitness of nearest chosen individual then
      Replace the nearest chosen individual
                                   by offspring individual; ...O(n)
   End if
End for
```

The overall time complexity is then $O(N(anN + n)) = O(anN^2)$, where the coefficient a determines the percentage of randomly chosen individuals in the target population that undergo the similarity comparison with each offspring. The greater the a, the stronger the pressure on diversity preservation. Note that the complexity of RTR exceeds the complexity of most other existing replacement operators. Nonetheless, the benefits of effective diversity maintenance usually overshadow the computational effort.

8.4.5 Complexity of Fitness Evaluation

Clearly, the time complexity of fitness evaluation depends on the application.

In the experimental part of this chapter, we use the problem of determining ground states of random two-dimensional $\pm J$ Ising spin glasses, which is described in the chapter "Searching for Ground States of Ising Spin Glasses with Hierarchical BOA and Cluster Exact Approximation". The spin glass problem was chosen because of its interesting properties that make it a challenging benchmark for any optimization method. The evaluation time of one spin glass configuration is linearly proportional to the number of bonds, which is linearly proportional to the number of spins $O(n)$. For N individuals the evaluation time grows with $O(nN)$. This complexity also holds for many other decomposable problems for which each variable participates in at most a constant number of subproblems (for example, cluster optimization in physics).

Optionally, MBOA can use a simple hill climber to improve the fitness of each individual. The hill climber tries to flip each bit and chooses the change with the highest fitness increase. This step is repeated until no flip with a positive outcome exists.

In our application, empirically, the number of successful acceptances per individual is $O(\sqrt{n})$. After each acceptance, it takes $O(1)$ time to recompute the outcomes of neighboring flips and $O(n)$ time to pick the best flip for the next change. The total complexity of the hill climber for the whole population is thus $O(nN\sqrt{n}) = O(n^{1.5}N)$. The algorithm for picking the best flip for the next change can be implemented even more effectively using bookkeeping, so a total complexity of $O(Nn \log n)$ can be achieved. In our analysis, we consider the case without heuristics, but even in that case the complexity of evaluating a solution with the hill climber and the complexity of evaluating a solution without the hill climber are nearly the same.

8.5 Scalability Analysis

The overall time complexity of one generation of MBOA can be computed by summing up the complexity of selection (weighted by c_1), model building (weighted by c_2), model sampling (weighted by c_3), replacement (weighted by c_4), and fitness evaluation (weighted by c_5)

$$c_1 O(nN) + c_2 O(n^2 N \log(N)) + c_3 O(nN \log(N)) + c_4 O(anN^2) + c_5 O(nN). \tag{8.2}$$

To develop scalable and efficient parallel MBOA it is necessary to identify the main parts that are candidates for parallelization.

Apparently, one of the most complex parts of sequential MBOA is the construction of the probabilistic model. In Sect. 8.4.2, we outlined an algorithm for asynchronous construction of decision trees. To ensure that the load on all processors is approximately equal, the overall construction of an n-node Bayesian network can be partitioned into at most $P = n/2$ equally complex subtasks (if the first processor builds the tree for X_{o_0} and $X_{o_{n-1}}$, the second processor builds the tree for X_{o_1} and $X_{o_{n-2}}$, etc.). As the first scenario of our analysis, we consider only this parallelization of probabilistic model construction and keep the remaining parts sequential.

With P processors, the overall time complexity of parallel MBOA is

$$c_1 O(nN) + \frac{1}{P} c_2 O(n^2 N \log(N)) + c_3 O(nN \log(N)) + c_4 O(anN^2) + c_5 O(nN).$$

The proportion between the time spent in the sequential part and the parallel part of MBOA is given by

$$\frac{c_1 O(nN) + c_3 O(nN \log(N)) + c_4 O(anN^2) + c_5 O(nN)}{\frac{1}{P} c_2 O(n^2 N \log(N))}. \tag{8.3}$$

To obtain an algorithm that effectively utilizes available computational resources, this proportion should approach zero as n grows.

8.5.1 Scalability for a Fixed Number of Processors

We first analyze scalability in the case of constant P and increasing n. This is the typical scenario when the computational resources are fixed but the problem to be solved is very large. The detailed analysis of the terms in fraction (8.3) for a constant P and $n \to \infty$ gives us the following suggestions for design of parallel MBOA:

- The terms with c_1 and c_5 are negligible for scalability:
$$\lim_{n\to\infty} \frac{c_1\mathrm{O}(nN) + c_5\mathrm{O}(nN)}{\frac{1}{P}c_2\mathrm{O}(n^2 N\log(N))} = 0. \tag{8.4}$$
In another words, neither the selection operator nor the population evaluation have to be implemented in parallel. Of course, this outcome is valid only for fitness functions of complexity $\mathrm{O}(n)$. For fitness functions of quadratic complexity, the scalability will depend on the absolute values of constants c_5, c_2 and on the problem-dependent relation between n and N. Theoretically, if the population size grows nearly linearly with problem size ($N \propto n$) as suggested by [27], it is still possible to keep the fitness evaluation sequential, because the $\log N$ term in the denominator dominates. Nevertheless, in practical applications we suggest parallel evaluation of problems with quadratic and higher complexity. For the parallelization of fitness evaluation, the techniques mentioned in Sect. 8.2.1 can be used; detailed guidelines for effectively distributing fitness evaluations can be found in [19].
- The sampling of the model does not have to be performed in parallel, since for all possible assignments to constants c_2, c_3 and P,
$$\lim_{n\to\infty} \frac{c_3\mathrm{O}(nN\log N)}{\frac{1}{P}c_2\mathrm{O}(n^2 N\log N)} = 0 \tag{8.5}$$
always holds.
- The restricted tournament replacement has not to be performed in parallel if
$$\lim_{n\to\infty} \frac{c_4\mathrm{O}(anN^2))}{\frac{1}{P}c_2\mathrm{O}(n^2 N\log N)} = 0. \tag{8.6}$$
In this case, the scalability highly depends on the problem-dependent relation between n and N. Theoretically, even if the population size grows linearly with the problem size ($N \propto n$), the above fraction approaches zero because the $\log N$ term in the denominator dominates. However, in practical applications we suggest, RTR to be performed in parallel.

8.5.2 Scalability for an Increasing Number of Processors

So far we have analyzed the scalability of sequential MBOA for a fixed number of processors. Now we will analyze how the scalability changes if the number of available processors scales up with n. In this case the execution time is reduced by an order of n. By assuming $P \propto n$, we obtain from (8.3)

$$\frac{c_1 \mathrm{O}(nN) + c_3 \mathrm{O}(nN \log(N)) + c_4 \mathrm{O}(anN^2) + c_5 \mathrm{O}(nN)}{c_2 \mathrm{O}(nN \log(N))}. \tag{8.7}$$

We see that the selection operator and the simple evaluation of the population (terms with constants c_1 and c_5) can still be implemented sequentially, but it does not hold for fitness evaluation with quadratic and higher complexity any more. The decision about implementation of model sampling strongly depends on the required speedup. If sequential model sampling is performed, then the speedup is saturated at c_2/c_3. RTR must be necessarily implemented in parallel, because for fixed c_4, c_2, and a, the numerator always dominates the denominator.

8.6 Performance Prediction

This section applies the described approach to the parallelization of MBOA on two-dimensional spin glasses (see the chapter "Searching for Ground States of Ising Spin Glasses with Hierarchical BOA and Cluster Exact Approximation").

8.6.1 Fitting Complexity Coefficients

We performed a series of experiments on random instances of 2D Ising spin glass benchmarks of size 100, 225, 400, 625, 900 for population sizes $N = 500$, $N = 1{,}000$, $N = 1{,}500$, $N = 2{,}000$, $N = 4{,}000$, $N = 6{,}000$ and $N = 8{,}000$. We measured separately the duration of each part of the sequential MBOA in order to determine the coefficients c_1, c_2, c_3, c_4, and c_5. The fitted coefficients are stated in Table 8.1. We observed that for larger problems the fit is in agreement with the empirical data, but for smaller problems the measured time is lower than that expected from the theoretical complexity. This can be explained by the effects of cache memory and the assumptions of asymptotic bounds.

8.6.2 Using Complexity Coefficients

The obtained coefficients can be used to predict the speedup of MBOA with parallel model building. Given the problem size n, the population size N, and the number of processors P, we get:

$$S = \frac{c_1 nN + c_2 n^2 N \log(N) + c_3 nN \log(N) + c_4 anN^2 + c_5 nN}{c_1 nN + \frac{1}{P} c_2 n^2 N \log(N) + c_3 nN \log(N) + c_4 anN^2 + c_5 nN}. \tag{8.8}$$

Table 8.1. The resulting values of coefficients c_1, c_2, c_3, c_4, c_5

MBOA part	Coefficient	Estimated value	R^2 value
Selection	c_1	8.73E−09	0.978
Model building	c_2	1.00E−07	0.979
Model sampling	c_3	1.58E−07	0.934
Replacement (RTR)	$c_4 * a$	2.18E−10	0.989
Evaluation	c_5	1.34E−07	0.918

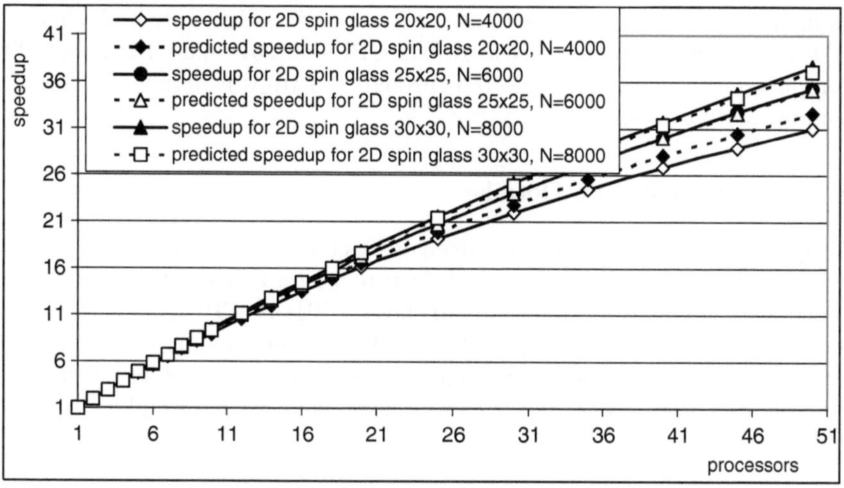

Fig. 8.2. The comparison of the speedup predicted from the numerical model and the speedup computed from the empirical data measured on sequential MBOA solving 2D Ising spin glass instances of size 20×20, 25×25, and 30×30. Population size was scaled approximately linearly with the problem size

For each Ising spin glass size 100, 225, 400, 625, 900 and each population size $N = 500$, $N = 1{,}000$, $N = 1{,}500$, $N = 2{,}000$, $N = 4{,}000$, $N = 6{,}000$, and $N = 8{,}000$ we choose 10 random benchmark instances and average the duration of each MBOA part. The coefficients hold for one generation of MBOA performed on Intel Pentium-4 at 2.4 GHz.

Figure 8.2 shows how the predicted speedup changes for increasing P and compares it with the speedup computed from the measured duration of each part of sequential MBOA. We considered three different sizes of spin glass instances 20×20, 25×25, and 30×30 and we linearly increased the population size with problem size ($N = 4{,}000, 6{,}000, 8{,}000$). The predicted speedup fits nicely the empirical speedup, especially for large problems. Additionally, it can be seen that – in the idealized case without communication – it is possible to use a large number of processors (more than $P = 50$) without observing any significant speedup saturation.

8.7 Experiments

The speedup presented in Fig. 8.2 was calculated using the time measurements from sequential MBOA and it assumed that the model building is ideally parallelized with no communication cost or delays. However, in the real world, communication significantly influences the achieved speedup. This section verifies the proposed approach using an actual MPI implementation of MBOA on a cluster of PCs. Two-dimensional spin glass problem instances are used in all tests.

All experiments were done at the Asgard cluster, which is an Intel Pentium III Beowulf cluster located at the Swiss Federal Institute of Technology (ETH) in Zürich. Asgard consists of 502 CPUs on 251 dual-CPU nodes. The computational nodes are connected through 100 Mbps Ethernet switches and communicate via MPI.

8.7.1 Implementation Details

Based on the design guidelines given in Sect. 8.5, we implemented parallel MBOA. We use a master–slave architecture where CPU_0 acts as the master controlling the distribution of jobs to slaves $CPU_1 \dots CPU_m$. Due to the dynamic workload assignment, parallel MBOA does not require a homogeneous cluster of workstations and it is adaptable to various parallel architectures.

The operation of the master can be illustrated by this simplified pseudocode:

```
Set t := 0;
Randomly generate Population(0) of size N;
Evaluate Population(0);
While termination criteria are not satisfied do
  Parents(t) := Tournament_selection(Population(t));
  Broadcast Parents(t) to CPU_1...CPU_m;
  Generate random permutation o = (o_0...o_{n-1});
  Broadcast the permutation o to CPU_1...CPU_m;
  For i:= 0 to n-1 do
    Wait for any idle slave processor CPU_j;
    Send job i to CPU_j;
  End for
  Gather all decision trees T_0...T_{n-1} from CPU_1...CPU_m;
  For all individuals in Offspring(t) do
    For i:= 0 to n-1 do
      Traverse the tree T_i conditionally on X_{o_0}...X_{o_{i-1}};
      Generate X_{o_i} according to the reached leaf in T_i;
    End for
  End for
  Evaluate Offspring(t);
```

```
   Population(t+1) := RTR(Population(t),Offspring(t));
   Set t := t+1;
End while
```

The operation of slave processors can be illustrated as follows:

```
While termination criteria are not satisfied do
   Receive Parents(t) from CPU_0;
   Receive permutation o = (o_0...o_{n-1}) from CPU_0;
   While job-queue in CPU_0 is not empty do
      Receive next job number i from CPU_0;
      T_i := BuildTree(Parents(t), X_{o_i}, X_{o_0}...X_{o_{i-1}});
   End while
   Send all constructed decision trees to CPU_0;
End while
```

In the above code, each job corresponds to one call of *BuildTree()* defined in Sect. 8.4.2. Note that not all $(m + 1)$ processors are necessary; one processor can be spared because the processes CPU_0 and CPU_1 do not overlap much and can be physically mapped to the same processor. For the spin glass problem we decided to keep the fitness evaluation sequential, as suggested in Sect. 8.4.5.

8.7.2 Results

The results shown in Fig. 8.3 confirm high efficiency of parallel MBOA. First we measured separately the speedup of parallel model construction. The model construction appears to be successfully parallelized; for example, for 50 processors and population size $N = 8{,}000$, we achieved a speedup of 45. Moreover, the speedup grows when the population size is increased, because the proportion between decision trees construction and decision trees communication increases. This is good news for solving large problems where very large populations are usually needed. The results thus confirm that model building can be effectively parallelized even for large parallel computers with a distributed memory and relatively slow communication.

Then, we measured the speedup of the entire algorithm (including sequential selection, model sampling, fitness evaluation, and replacement). Figure 8.3 shows that the speedup of the whole algorithm is also acceptable for a small number of processors, but it easily gets saturated and for a large number of processors it is hard to achieve speedup better than 20. This indicates that the additional parallelization of the remaining sequential parts might be necessary to achieve a better speedup in this setting. Nonetheless, note that the achieved speedup gets better as the computational complexity of MBOA increases due to the problem size or the population size; consequently, the speedup should be much better for problems where parallel implementation is necessary.

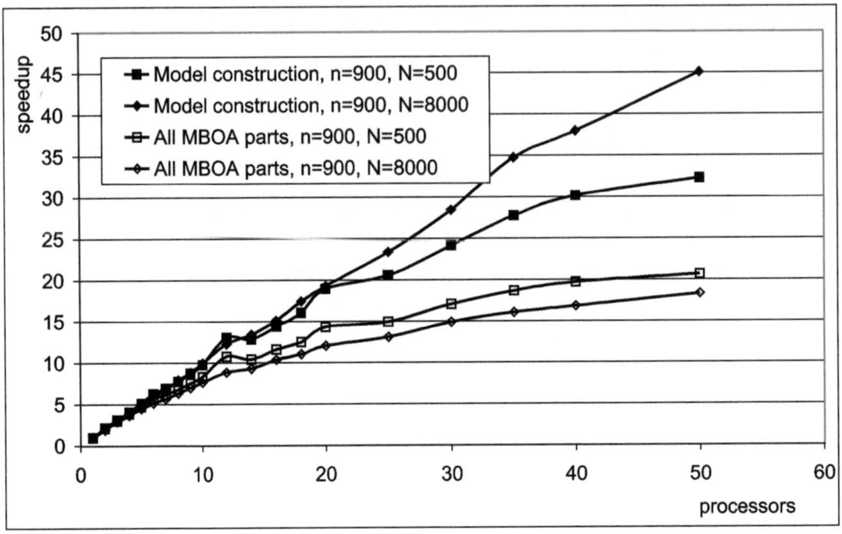

Fig. 8.3. Speedup of the whole parallel MBOA and the speedup of model building part (including true MPI communication) on random spin glass instance with 30×30 spins

8.8 Conclusions and Future Work

In this chapter we derived guidelines for designing efficient parallel estimation of distribution algorithms (EDAs). Especially, we focused on the scalability analysis that identifies the parts of EDAs that are best candidates for parallelization.

The detailed scaleup analysis was demonstrated on the example of the Mixed Bayesian Optimization Algorithm (MBOA). We fitted the complexity of MBOA to experimental data obtained by solving random instances of the spin glass optimization problem. The empirical results fit well the theoretical model. Consequently, the scalability and algorithmic efficiency of parallel MBOA can be predicted accurately; for example, the expected speedup can be estimated from the problem size and the number of processors.

In the experimental part of this chapter, we designed and implemented parallel MBOA on a Beowulf cluster of 502 processors. Parallel MBOA parallelizes the construction of the Bayesian network with decision trees for the selected population of parents. Parallel MBOA was tested on the challenging problem of finding ground states of two-dimensional $\pm J$ Ising spin glasses. The empirical results indicate that parallel MBOA can effectively utilize even a large number of processors.

Acknowledgments

The authors would like to thank the members of the Computational Laboratory ETH Zürich for valuable comments and discussions. Our experiments were run at the Asgard Beowulf cluster at ETH Zürich. Pelikan was also supported by the Research Board at the University of Missouri and the Research Award at the University of Missouri, St. Louis.

References

[1] Mühlenbein H, Paass G (1996) From Recombination of Genes to the Estimation of Distributions: I. Binary Parameters. *Lecture Notes in Computer Science,* 1141, pp. 178–187

[2] Larranaga P, Lozano JA (2002) *Estimation of Distribution Algorithms. A new Tool for Evolutionary Computation.* Kluwer Academic, Dordrecht, pp. 57–100

[3] Pelikan M, Goldberg DE, Lobo F (1999) *A survey of optimization by building and using probabilistic models,* IlliGAL Report No. 99018, University of Illinois at Urbana-Champaign, Illinois Genetic Algorithms Laboratory, Urbana, IL

[4] Bosman PAN, Thierens D (1999) *An algorithmic framework for density estimation based evolutionary algorithms.* Utrecht University Technical Report UU-CS-1999-46, Utrecht

[5] Ocenasek J, Schwarz J (2000) The Parallel Bayesian Optimization Algorithm, In: *Proceedings of the European Symposium on Computational Inteligence,* Physica, Kosice, Slovak Republic, pp. 61–67

[6] Ocenasek J, Schwarz J (2001) *The Distributed Bayesian Optimization Algorithm for combinatorial optimization, EUROGEN 2001 – Evolutionary Methods for Design, Optimisation and Control,* Athens, Greece, CIMNE, pp. 115–120

[7] Lozano JA, Sagarna R, Larrañaga P (2002) Parallel Estimation of Distribution Algorithms. In: P. Larrañaga, and J. A. Lozano (eds.): *Estimation of Distribution Algorithms. A New Tool for Evolutionary Computation,* Kluwer Academic, Dordrecht, pp. 129–145

[8] Mendiburu A, Miguel-Alonso J, Lozano JA (2004) *Implementation and performance evaluation of a parallelization of estimation of Bayesian networks algorithms.* Technical Report EHU-KAT-IK-04-04. Department of Computer Architecture and Technology, The University of the Basque Country

[9] Ocenasek J, Schwarz J (2002) *Estimation of Distribution Algorithm for mixed continuous discrete optimization problems, In: 2nd Euro-International Symposium on Computational Intelligence,* Kosice, Slovakia, IOS, Amsterdam, pp. 227–232

[10] Ocenasek J, Schwarz J, Pelikan M (2003) Design of Multithreaded Estimation of Distribution Algorithms. In: Cantú-Paz et al. (Eds.): *Genetic and Evolutionary Computation Conference – GECCO 2003.* Springer, Berlin Heidelberg New York, pp. 1247–1258

[11] Ocenasek J, Pelikan M (2003) Parallel spin glass solving in hierarchical Bayesian optimization algorithm. In: *Proceedings of the 9th International*

Conference on Soft Computing, Mendel 2003, Brno University of Technology, Brno, Czech Republic, pp. 120–125
[12] Gordon VS, Whitley D (1993) Serial and parallel genetic algorithms as function optimizers. In: S. Forrest, editor, *Proceedings of the Fifth International Conference on Genetic Algorithms*, Morgan Kaufmann, San Mateo, CA, pp. 177–183
[13] Hart WE (1994) *Adaptive Global Optimization with Local Search.* PhD thesis, University of California, San Diego
[14] Punch WF (1998) How effective are multiple programs in genetic programming. In: JR Koza, W Banzhaf, K Chellapilla, K Deb, M Dorigo, DB Fogel, MH Garzon, DE Goldberg, H Iba, RL Riolo, (ed.), *Genetic Programming 98*, Morgan Kaufmann, San Francisco, pp. 308–313
[15] Mühlenbein H (1991) Evolution in time and space – The parallel genetic algorithm. In: GJE Rawlins, (ed.), *Foundations of Genetic Algorithms*, Morgan Kaufmann, San Mateo, CA, pp. 316–337
[16] Whitley D, Starkweather T (1990) Genitor II: A distributed genetic algorithm. *Journal of Experimental and Theoretical Artificial Intelligence*, 2:189–214
[17] Davidor Y (1993) The ECOlogical framework II: Improving GA performance at virtually zero cost. In: S Forrest, (ed.), *Proceedings of the Fifth International Conference on Genetic Algorithms*, Morgan Kaufmann, San Mateo, CA, pp. 171–176
[18] Calégari PR (1999) *Parallelization of Population-Based Evolutionary Algorithms for Combinatorial Optimization Problems.* Unpublished doctoral dissertation, École Polytechnique Fédérale de Lausanne (EPFL)
[19] Cantú-Paz E (2000) Efficient and Accurate Parallel Genetic Algorithms, Kluwer Academic, Boston, MA
[20] Hiroyasu T, Miki M, Sano M, Shimosaka H, Tsutsui S, Dongarra J (2003) Distributed Probabilistic Model-Building Genetic Algorithm, In: Cantú-Paz et al. (Eds.): *Genetic and Evolutionary Computation Conference – GECCO 2003.* Springer, Berlin Heidelberg New York, pp. 1015–1028
[21] Pelikan M, Goldberg DE, Sastry K (2000) *Bayesian Optimization Algorithm, Decision Graphs, and Occam's Razor,* IlliGAL Report No. 2000020, University of Illinois at Urbana-Champaign, Illinois Genetic Algorithms Laboratory, Urbana, IL
[22] Ocenasek J (2002) *Parallel Estimation of Distribution Algorithms.* PhD. Thesis, Faculty of Information Technology, Brno University of Technology, Brno, Czech Republic, pp. 1–154
[23] Cooper GF, Herskovits EH (1992). A Bayesian method for the induction of probabilistic networks from data. *Machine Learning*, 9, pp. 309–347
[24] Heckerman D, Geiger D, Chickering M (1994) *'Learning Bayesian networks: The combination of knowledge and statistical data'.* Technical Report MSR-TR-94-09, Microsoft Research, Redmond, WA

[25] Chickering DM, Heckerman D, Meek C (1997) *A Bayesian approach to learning Bayesian networks with local structure.* Technical Report MSR-TR-97-07, Microsoft Research, Redmond, WA

[26] Harik G (1995) Finding multimodal solutions using restricted tournament selection. In: *Sixth International Conference on Genetic Algorithms* (ICGA-95), pp. 24–31

[27] Pelikan M, Sastry K, Goldberg DE (2002). Scalability of the Bayesian optimization algorithm. *International Journal of Approximate Reasoning,* 31(3), pp. 221–258

9

Incorporating a priori Knowledge in Probabilistic-Model Based Optimization

Shumeet Baluja

Summary. Recent studies have examined the effectiveness of using probabilistic models to guide the sample generation process for searching high dimensional spaces. Building complex dependency networks that can account for the interactions between parameters are often used; however, they may necessitate enormous amounts of sampling. In this chapter, we demonstrate how a priori knowledge of parameter dependencies, even incomplete knowledge, can be incorporated to efficiently obtain accurate models that account for parameter interdependencies. This is achieved by effectively putting priors on the network structures that are created. These more accurate models yield improved results when used to guide the sample generation process for search. We demonstrate the results on a variety of graph coloring problems, and examine the benefits of a priori knowledge as problem difficulty increases.

9.1 Introduction

Within the past few years, there has been increased interest in using probabilistic modeling for combinatorial optimization. Unlike hillclimbing methods, which operate by sampling solutions neighboring the current solution, probabilistic methods explicitly maintain statistics about the search space by creating models of the good solutions found so far. These models are sampled to generate the next query points to be evaluated. The high-performing sampled solutions are then used to update the model, and the cycle is continued. Comprehensive survey papers of this literature are available [28, 31, 39–41].

Some of the first work in using probabilistic modeling for optimization, such as PBIL [1] and BSC [43], used extremely simple models. In these models, no inter-parameter dependencies were taken into account; each bit was generated independently. Although this simple probabilistic model was used, PBIL was successful when compared to a variety of standard genetic algorithms and hillclimbing algorithms on numerous benchmark and real-world problems [2, 3, 18]. A more theoretical analysis of PBIL can be found in [16, 17, 22, 25, 27]. An analysis of PBIL in the Univariate Marginal Distribution framework is given in [33]. Despite the successes, limitations to the PBIL

S. Baluja: *Incorporating a priori Knowledge in Probabilistic-Model Based Optimization*, Studies in Computational Intelligence (SCI) **33**, 205–222 (2006)
www.springerlink.com

algorithm were described in [12]: PBIL and BSC may not perform as well as pair-wise operators, such as those found in genetic algorithms [10, 15, 23], when tested on problems explicitly designed with a high degree of interdependence between parameters. More complex models in the form of probabilistic networks were introduced to overcome the limitations of models that assumed each parameter was independent. Although these models provided a more accurate representation of the high evaluation solutions, they also required more samples in order to be used effectively. To reduce the amount of required data, studies were conducted with networks that modeled only a subset of the possible dependencies [4, 5, 11].

In this chapter, we show how a priori knowledge of the problem or of the search space can be used to direct the creation of the probabilistic networks. The interactions of variables in the objective function can be more accurately ascertained from the sampled points if knowledge of the problem is incorporated. This helps to overcome the drawbacks of limited sample sizes by ensuring that the modeled dependencies are reflective of real dependencies in the problem and not merely spurious correlations in the sampled solutions. We empirically demonstrate that by creating more accurate models, we improve the quality of the final solutions found through the search.

In the next section, we describe the simple tree-based model that we will use as the basis for this study; this work was originally presented in [4]. Section 9.3 gives an introduction to how a priori knowledge can be incorporated into model creation. Section 9.4 empirically demonstrates its effectiveness on a variety of graph coloring problems. Finally, Section 9.5 closes the paper with conclusions and suggestions for future work.

9.2 Probabilistic Models

In the simplest model-based optimization techniques, such as PBIL or BSC, all parameters are examined independently. Here we will give a brief background into the probabilistic models that we use to overcome the independence assumption. The probabilistic models attempt to capture dependencies, or more specifically mutual information, between the parameters to determine which parameters are dependent upon each other. These dependencies are used to generate the new candidate solutions. The reader is referred to texts by Pearl [38] and Jensen [24] for an introduction to probabilistic modeling and Bayesian networks.

The overall structure of our approach is similar to PBIL. After evaluating each member of the current generation, the best members of that population are used to update a probabilistic model from which the next generation's population will be generated. From the set of solutions evaluated in each generation, the best samples are added into a dataset, termed S. Rather than recording the individual members of S, our algorithm maintains a sufficient set of statistics in an array A. For models that use pair-wise interactions, this contains a number $A[X_i = a,\ X_j = b]$ for every pair of variables X_i and X_j and every combination of binary assignments to a and b. $A[X_i = a,\ X_j = b]$ is as an estimate of how many recently generated "good" bit-strings (from S) have bit $X_i = a$ and bit $X_j = b$. To give more weight to recently generated bit-strings, the contributions of bit-strings that were previously added to the dataset are decayed. All $A[X_i = a,\ X_j = b]$ are initialized to some constant C_{init} before the first iteration of the algorithm; this causes the algorithm's first set of bit-strings to be generated from the uniform distribution. See Fig. 9.1 for the algorithm with bit-strings; extension to real-valued optimization have been explored [7, 14, 29, 42].

The values of $A[X_i = a,\ X_j = b]$ at the beginning of an iteration may be thought of as specifying a prior probability distribution over "good" bit-strings. As the algorithm progresses, we only select the top members of the population to contribute to the probabilistic model. Although arbitrarily complex probabilistic models can be used, we use a simple one that is capable of capturing pair-wise dependencies: optimal dependency trees.

Given a dataset, S, of previously generated good bit-strings, we try to model a probability distribution $P(X) = P(X_1, \ldots, X_n)$ of bit-strings of length n, where $X_1, \ldots, X_n$ are variables corresponding to the values of the bits. We try to learn a simplified model $P'(X_1, \ldots, X_n)$ of the empirical

For all bits i and j and all binary assignments to a and b, initialize $\mathbf{A}[\mathrm{x_1} = a, \mathrm{x_j} = b]$ to $\mathrm{C_{init}}$.
Repeat until Termination Condition is met:

1. Generate probabilistic model base on A. See Figure 9.2.
2. Stochastically generate K bit-strings based on the probabilistic model. Evaluate these bit-strings.
3. Multiply all entries in A by decay factor α from (0, 1).
4. Choose the best M of the K bit-strings generated in step 2. For each bit-sting V of these M, add 1.0 to every $\mathbf{A}[\mathrm{x_i} = a, \mathrm{x_j} = b]$ such that V has $\mathrm{x_i} = a$ and $\mathrm{x_j} = b$.

CONSTANTS (Values used in this study)

C$_{\mathrm{init}}$: Constant used to initialize matrix **A**-Number of examples "seen" at initialization (1000).
K: Number of samples generated in each iteration. This is the population size (200).
M: Number of best samples (from the K generated) that are used to update the statistics (4).
α: How much to decay the effect of older examples (1.99).

Fig. 9.1. Outline for using a probabilistic model. The values in the parenthesis are those that will be used in the experiments presented later in this paper probability distribution $P(X_1, \ldots, X_n)$ entailed by the bit-strings in S. We restrict our model $P'(X_1, \ldots, X_n)$ to the following form:

$$P'(X_1, \ldots, X_n) = \prod_{i=1}^{n} P(X_i|\text{Parent}_{X_i}), \tag{9.1}$$

where Parent_{X_i} is X_i's single "parent" variable (the variable on which X_i will be conditioned). We require that there be no cycles in these "parent-of" relationships: formally, there must exist some permutation $m = (m_1, \ldots, m_n)$ of $(1, \ldots, n)$ such that $(\text{Parent}_{X_i} = X_j) \Rightarrow m(i) < m(j)$ for all i. (The "root" node, X_R, will not have a parent node; however, this case can be handled with a "dummy" node X_0 such that $P(X_\text{R}|X_0)$ is by definition equal to $P(X_\text{R})$.) In other words, we restrict P' to factorizations representable by Bayesian networks in which each node (except X_R) has one parent, i.e., tree-shaped graphs.

A method for finding the optimal model within these restrictions is given in [9]. A complete weighted graph G is created in which every variable X_i is represented by a corresponding vertex V_i, and in which the weight W_{ij} for the edge between vertices V_i and V_j is set to the mutual information $I(X_i, X_j)$ between X_i and X_j:

$$I(X_i, X_j) = \sum_{a,b} P(X_i = a, X_j = b) \cdot \log \frac{P(X_i = a, X_j = b)}{P(X_i = a) \cdot P(X_j = b)} \tag{9.2}$$

The empirical probabilities of the form $P(X_i = a)$ and $P(X_i = a, X_j = b)$ are computed directly from S for all combinations of i, j, a, and b (a and b are binary assignments to X_i and X_j). Once these edge weights are computed, the maximum spanning tree of G is calculated, and this tree determines the structure of the network used to model the original probability distribution. Since the edges in G are undirected, a decision must be made about the directionality of the dependencies with which to construct P'; however, all such orderings conforming to the restrictions described earlier model identical distributions. Among all trees, this algorithm produces a tree that minimizes the Kullback–Leibler divergence, $D(P||P')$, between P (the true empirical distributions exhibited by S) and P' (the distribution modeled by the network):

$$D(P||P') = \sum_{X} P(X) \log \frac{P(X)}{P'(X)}. \tag{9.3}$$

As shown in [9], this produces the tree-shaped network that maximizes the likelihood of S (this means that of all the tree shaped networks, this is the most likely to have generated S). This tree generation algorithm, summarized in Fig. 9.2, runs in time $O(n^2)$, where n is the number of bits in the solution encoding.

The arcs which remain in the maximum spanning tree represent the dependencies to be modeled. Since it is a tree, each variable will be conditioned

Generate an optimal dependency tree:

- Set the root to an arbitrary bit X_{root}
- For all other bits X_i, set bestMatchingsBitInTree [X_i] to X_{root}.
- While not all bits have been added to the tree:
  - Of all the bits not yet in the tree, pick bit X_{add} with the maximum mutual information I (X_{add}, bestMatchingBitInTree [X_{add}]), using **A** (which contain sufficient statistics for **S**) to estimate the relevant probability distributions.
  - Add X_{add} to tree, with bestMatchingBitInTree[X_{add}] as parent.
  - Fro each bit X_{out} not in the tree, if I(X_{out}, bestMatchingBitInTree[X_{out}]) < I(X_{out}, X_{add}) then set bestMatchingBitInTree [X_{out}] = X_{add}.

Fig. 9.2. Procedure for generating the dependency tree

on exactly one other variable (its parent). The exception to this is the root of the tree, which is set according to its unconditional probabilities. With the generated dependency tree that models $P(X_1, \ldots, X_n)$, we use it to generate $\boldsymbol{K}$ new bit-strings. Each bit-string is generated in $O(n)$ time during a depth-first traversal of the tree. Each bit-string is then evaluated. The best $\boldsymbol{M}$ of these bit-strings are selected and effectively added to S by updating the counts in $\boldsymbol{A}$. Based on the updated $\boldsymbol{A}$, a new dependency tree is created, and the cycle is continued.

9.2.1 Discussion of Related Models

Another extension to PBIL that captured pair-wise dependencies was termed Mutual Information Maximization for Input Clustering (MIMIC) [11]. MIMIC used a greedy search to generate a chain in which each variable is conditioned on the previous variable. The first variable in the chain, X_1, is chosen to be the variable with the lowest unconditional entropy $H(X_1)$. When deciding which subsequent variable X_{i+1} to add to the chain, MIMIC selects the variable with the lowest conditional entropy $H(X_{i+1}|X_i)$. While MIMIC was restricted to a greedy heuristic for finding chain-based models, the algorithm described in this paper uses a broader class of models, trees, and finds the optimal model in the class.

Example dependency graphs, shown in Fig. 9.3, illustrate the types of probability models learned by PBIL, a dependency-chain algorithm similar to MIMIC, and our dependency tree algorithm. We use Bayesian network notation for our graphs: an arrow from node X_p to node X_c indicates that X_c's probability distribution is conditionally dependent on the value of X_p. These models were learned while optimizing a noisy version of a two-color graph coloring problem (shown in Fig. 9.3a) in which there is a 0.5 probability of adding 1 to the evaluation function for every edge constraint satisfied

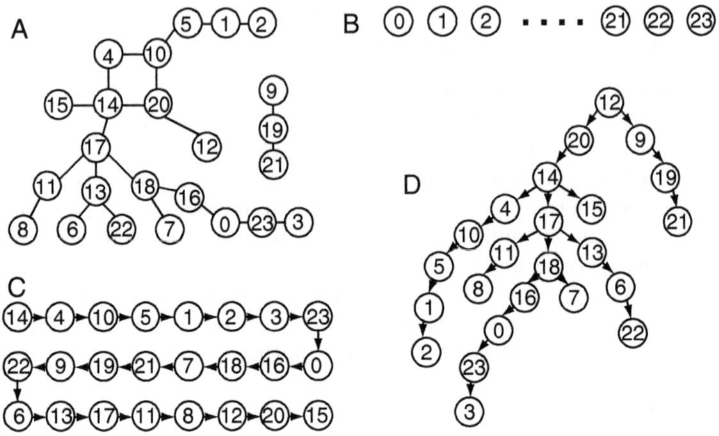

Fig. 9.3. A: The underlying graph in a noisy two color graph coloring problem. B: The empty dependency graph used by PBIL. C: The graph learned by our implementation of the dependency chain algorithm. D: The graph learned by our dependency tree algorithm.

by the candidate solution. Note that the dependency tree algorithm is able to discover the underlying structure of the graph, in terms of which bits are dependent on each other (as shown in Fig. 9.3D).

The clear next step after modeling pair-wise dependencies is modeling higher-order dependencies. The need for this has been demonstrated in [6]. However, generating models which are capable of representing higher-order dependencies may be computationally expensive. The hope is that the expense of generating the models will be offset by the savings obtained by the smaller number of function evaluations that will be required due to the more accurate modeling. A large amount of work has been done exploring different models to use. The Factorized Distribution Algorithm (FDA) [32–35] uses a fixed model throughout the search, with the model being specified by an expert. The FDA algorithm is designed to work with problems that are decomposable into independent parts. This work has been extended to incorporate learning with low complexity networks and Junction-Trees [36, 37]. The Bayesian Optimization Algorithm (BOA) and related work [13, 30, 39, 40] is the closest method to the optimization techniques presented here. The model used in BOA is able to represent arbitrary dependencies. When general Bayesian Networks are used for modeling, the scoring function used to determine the quality of the network plays a vital role in finding accurate networks. The quality of networks can be assessed through a variety of measures. For example, both Minimum Description Length and Bayesian Dirichlet metrics have been explored in [40]. The models that are found by BOA are similar to those used in FDA; however, BOA is designed to learn the models as the search progresses. Because the models used by BOA are general Bayesian Networks, it is clear that a priori information can be incorporated [19, 20]. This is the focus of the next section.

9.3 Incorporating a priori Knowledge

In this section, we describe how to incorporate information into the process of learning the probabilistic model. The method is general and can be used in MIMIC [11], COMIT [5], BOA [39] or any other algorithm which has a learning component in the model generation process. Although the a priori information that is available for a problem is often high level and specifies complex dependencies, we show how the knowledge can be used even when simple probabilistic models are employed.

To this point, we have restricted the probabilistic models that we have examined to dependency trees that only model pair-wise interactions. This was done to mitigate the need for a large number of samples that arises when higher-order dependencies are modeled. Nonetheless, in some cases more complex models are required. Additionally, even when simple models such as trees are used, by using a priori information to constrain the number of possible trees that are considered, the samples can be more effectively used since they must only select trees from a reduced set. In this paper, we use a priori knowledge about the function to be optimized to constrain the arcs that are modeled in the probabilistic models. This technique is applicable to optimization procedures regardless of whether a multiply connected Bayesian network is used or a simple dependency tree is employed.

As an introductory example, the potential for using a priori knowledge is clearly demonstrated in problems in which the dependencies are evident, such as graph coloring. Consider the graph coloring problem as shown in Fig. 9.3A. In this simple problem, it is clear that the color of node 5 should be dependent upon the color of nodes 1 and 10, and that the color of node 10 should be dependent on the colors of nodes 4, 5, and 20. There are several ways to incorporate this information into a probabilistic model. The first is to employ a model that captures more than pair-wise dependencies. For example, if we allowed arbitrary dependencies, we could create models with more than a single parent; thereby mimicing the graph structure shown in Fig. 9.3A. Although this would require maintaining more than pair-wise statistics, only a subset of these higher-order statistics would be required since we could specify the dependencies to be modeled from our knowledge of the underlying graph structure. The second approach is to select the model from a family of low complexity models (such as the set of all trees – as we have described in Sect. 9.2) but to allow the arcs to be selected only from the subset of those that exist in the graph. Continuing with the same example, the allowed parents for node 5 would be either node 10 or node 1, but not both (since that would violate the tree property).

Throughout the remainder of this paper, we concentrate on the second approach described above: constraining the tree structures that can be created. This approach has the benefit of not requiring exact information of which dependencies must be modeled; although we do not specify the specific parents of each of the nodes, we restrict the possibilities. This method also has benefits as problem sizes increase. Modeling higher-order dependencies, even when the structure of the network is known, requires a large number of samples. In the graph coloring task, this problem becomes significant when the connectivity of the graph increases.

One of many ways to implement constraints on the dependency graph is to impose a prior over network structures in which the prior likelihood of a network decreases exponentially with the number of arcs in the network that do not correspond to edges in a pre-specified set. With the optimal dependency trees, such a prior can be simply implemented. We need only subtract a penalty term from the mutual information between any two variables that are not connected by an edge (E) in the pre-specified preferred set (S) and run the maximum spanning tree algorithm on these modified weights instead. The modified mutual information calculation is shown in (9.4).

$$I'(X_i, X_j) = \begin{cases} if(E_{i,j} \in S): & \sum_{a,b} P(X_i = a, X_j = b) \cdot \log \frac{P(X_i=a,X_j=b)}{P(X_i=a)\cdot P(X_j=b)} \\ \\ if(E_{i,j} \notin S): & \sum_{a,b} P(X_i = a, X_j = b) \cdot \log \frac{P(X_i=a,X_j=b)}{P(X_i=a)\cdot P(X_j=b)} - \alpha_{i,j} \end{cases} . \tag{9.4}$$

As shown above, the penalty, α, does not need to be constant, and can vary per dependency arc. The severity of the penalty provides a means to convey confidence in the a priori information. The more confident we are that an arc should not be modeled, the larger the penalty can be.

For simplicity, however, we do not use a complex penalty setting. Instead, the penalty term is constant for every arc. In the experiments presented in this paper, a sufficiently severe penalty term was used to ensure that arcs in the pre-specified set were always favored over arcs not in the set. This simple penalty procedure was chosen to ensure that the focus of the experiments remain on demonstrating that improvements in the final search result were obtainable by incorporating a priori information into the probabilistic models. Nonetheless, we do not suggest that this will work well on all problems; in problems in which the information should be regarded only as a preferred dependency instead of one that must be enforced, a less severe penalty may yield improved results.

9.4 Empirical Demonstration

In this section, we will use the graph coloring (vertex coloring) problem to demonstrate the effectiveness of incorporating a priori knowledge into model creation. In each of the graph coloring problems, there are n nodes in a partially connected graph to be assigned one of c colors. The goal is to assign each of the connected nodes different colors. In these problems, the graphs are not planar, and a solution in which all of the constraints are met is not guaranteed to exist. The evaluation, to be maximized, is the number of constraints (connected nodes that have different colors) that are met. Graph coloring is an ideal problem to show the benefits of a priori knowledge. Without a priori knowledge, an optimization technique would simply use the evaluations of all of the previously evaluated solutions to create a new sample point to evaluate. With a priori knowledge, the underlying graph structure can be used to constrain the set of dependencies that are allowed so that only dependencies that have corresponding edges in the underlying graph are allowed in the dependency tree.

To examine the benefits of using a priori knowledge, we vary the problem difficulty along three axes. The first is by simply varying the number of nodes, n, in the graph. As the problem size grows, we expect that providing knowledge of which dependencies are important to model will grow in importance. Along the second axis, we vary the number of connections in the graph. As the number of connections grows, the problem becomes more difficult. Finally, we vary the number of colors, c, that can be assigned to each node. The number of colors has an interesting effect on the problem difficulty. As the number of colors grows, the easier it is to find a color that does not violate constraints. However, using a dependency tree with a binary solution-string encoding is challenging with more than two colors. The solution is encoded as follows: each node is assigned $\log_2 c$ bits. The bits specify which color is assigned to the node. With only two colors, each color can be represented as a single bit. The tree structure maps very well onto the model since it is setup to capture pair-wise dependencies between bits. However, with more than two colors, the color of each node is represented with more than a single bit. To represent the dependency between nodes, it is unclear which bits should be dependent on each other. As will be shown, we do not specify which bits should be dependent on each other, only which sets of bits *may* be dependent (i.e., all the bits that represent node A can be dependent on any of the bits that represent node B); this is enough information to provide benefits to the optimization procedure.

Note that the results in the section are not intended to represent a comparison of different optimization algorithms. For more comprehensive comparisons between optimization methods, such as genetic algorithms, probabilistic optimization methods, and hillclimbing methods, the reader is referred to [2, 4, 5, 12, 18]. For the experiments presented in this section, we keep the probabilistic modeling algorithms as simple as possible to concentrate our examination on the effects of incorporating knowledge into the models. We have not included operators such as mutation, local hillclimbing or any of the numerous heuristics that can be used in conjunction with optimization techniques to create a general purpose optimization tool [15, 26].

In the first set of experiments, we attempted graph coloring problems with two colors. We tried five sizes of problems, with 100, 500, 1,000, 1,500 and 2,000 nodes. We also varied the connectivity at each size – each node was randomly connected to 2, 5, 10 or 20 nodes (a node's color must be different from the nodes directly connected to it). For each problem and connectivity combination (20 in total), we randomly created 20 instantiations which had different connectivity structures. In total, 400 two-color problems were tried with three algorithms:

1. ***NO:*** No dependencies are modeled.
2. ***FULL:*** Dependency arcs can be selected from any node to any other node.
3. ***GRAPH:*** a priori knowledge of the graph was used; arcs in the dependency tree were only chosen from nodes that are actually connected in the underlying graph.

Table 9.1 shows the mean number of constraints satisfied by each of the different algorithms. The second set of three result columns show the number of trials (out of 20) that the ***FULL*** satisfied more constraints than ***NO***, ***GRAPH*** satisfied more constraints than ***NO***, and ***GRAPH*** satisfied more constraints than ***FULL***. The last set of three results columns shows which of the averages are significantly different at the $p = 0.01$ confidence level; this is measured by a pair-wise t-test. Based on the results shown in Table 9.1, several points are of interest.

- It is never the case that all of the constraints are met. This is not surprising, since the connectivity of the graph is chosen randomly, and there is no guarantee that there is a solution that will satisfy all the constraints.
- When the problem sizes are small (100 nodes), there is little difference between the algorithms. Although the differences in performance are significant in a small number of cases, the absolute differences are quite small.
- At 500 nodes, the procedures which incorporate dependency modeling (***FULL*** & ***GRAPH***) perform significantly better than not modeling any dependencies. However, there is little difference between the performance of using a priori knowledge (***GRAPH***) and not using any a priori knowledge (***FULL***).
- As the problem size increases (1,000, 1,500, 2,000 nodes), both the ***FULL*** algorithm and the ***GRAPH*** algorithm perform better than the ***NO*** algorithm in nearly every instantiation of every problem size.
- For the large problem sizes (1,000, 1,500, 2,000 nodes), the ***GRAPH*** algorithm, which incorporates knowledge of the graph structure, performs better than the ***FULL*** algorithm for every size problem examined (at the $p = 0.01$ significance level).

The results largely match our expectations. As the problem size increases, the benefit of incorporating a priori knowledge increases. In almost all of

Table 9.1. Results on graph coloring problem with two colors

		Mean constraints satisfied with different probabilistic models			No. of trials Model A better than Model B (out of 20)			Paired t-tests on means $p = 0.01$ confidence level		
No. of nodes (no. of bits)	Connectivity	No arcs (NO)	All arcs (FULL)	a priori knowl. arcs (GRAPH)	Full > no	Graph > no	Graph > full	No & full	No & graph	Full & graph
100 (100)	2	168	170	170	16	14	5	*		
	5	358	363	363	18	17	8	*	*	
	10	656	662	660	17	12	7			
	20	1,222	1,228	1,227	13	11	6			
500 (500)	2	794	842	846	20	20	14	*	*	
	5	1,726	1,778	1,792	20	20	16	*	*	*
	10	3,183	3,238	3,242	20	19	13	*	*	
	20	5,967	6,023	6,044	17	19	14	*	*	
1000 (1000)	2	1,382	1,527	1,646	20	20	20	*	*	*
	5	3,127	3,239	3,370	20	20	20	*	*	*
	10	5,895	6,015	6,145	20	20	20	*	*	*
	20	11,269	11,378	11,521	20	20	19	*	*	*
1500 (1500)	2	1,894	2,017	2,358	20	20	20	*	*	*
	5	4,415	4,511	4,814	20	20	20	*	*	*
	10	8,429	8,533	8,813	20	20	20	*	*	*
	20	16,320	16,467	16,700	19	20	20	*	*	*
2000 (2000)	2	2,405	2,496	3,011	20	20	20	*	*	*
	5	5,661	5,739	6,192	20	20	20	*	*	*
	10	10,927	11,043	11,411	20	20	20	*	*	*
	20	21,323	21,455	21,793	19	20	20	*	*	*

the problems, ***GRAPH*** performed significantly better than ***NO*** (17/20); additionally, ***GRAPH*** performed better than ***FULL*** in 13/20 problems – and consistently in the larger problems. On the smaller problems (for example when the no. of nodes = 100,500), the ***FULL*** algorithm was able to discover an underlying structure on the graph that improved its performance over ***NO*** in most cases. It is interesting to note that with a connectivity of 20 when the number of nodes equaled 100 and 500 the differences between all three algorithms were usually very small. These problems may have been too constrained for there to be any significant advantage to be seen.

In the next set of experiments, we use the same problem sizes in terms of the number of bits; however, we increase the number of possible colors to

4. Note that this also doubles the size of the solution encoding in terms of the number of bits that are required to represent the solution. The results are shown in Table 9.2, as with the previous set of results, the results represent 400 problem instantiations. In comparison to the previous experiments with two colors, there are several interesting points to note:

- In the first problem (no. of nodes = 100 and connectivity = 2), all of the algorithms were able to solve the problem optimally.
- As with the previous experiments with two colors, there is little performance difference in the smallest problems (no. of nodes = 100).

Table 9.2. Results on graph coloring problem with four colors

		Mean constraints satisfied with different probabilistic models			No. of trials Model A better than Model B (out of 20)			Paired t-tests on Means $p = 0.01$ confidence level		
No. of nodes (no. of bits)	Connectivity	No arcs (NO)	All arcs (FULL)	a priori knowledge arcs (GRAPH)	Full > no	Graph > no	Graph > full	No & full	No & graph	Full & graph
100 (200)	2	200	200	200	0	0	0			
	5	484	487	486	16	14	9	*		
	10	918	922	920	14	10	8			
	20	1,747	1,751	1,751	14	16	9			
500 (1,000)	2	883	920	977	20	20	20	*	*	*
	5	2,103	2,141	2,221	20	20	20	*	*	*
	10	4,082	4,136	4,198	18	20	20	*	*	*
	20	7,976	8,057	8,099	20	20	19	*	*	*
1,000 (2,000)	2	1,652	1,672	1,847	19	20	20	*	*	*
	5	4,004	4,030	4,174	18	20	20	*	*	*
	10	7,861	7,890	8,024	18	20	20	*	*	*
	20	15,514	15,546	15,655	15	20	20		*	*
1,500 (3,000)	2	2,422	2,433	2,680	16	20	20	*	*	*
	5	5,898	5,918	6,115	18	20	20	*	*	*
	10	7,799	7,812	8,048	15	20	20		*	*
	20	23,063	23,092	23,210	18	20	20	*	*	*
2,000 (4,000)	2	3,190	3,193	3,500	11	20	20		*	*
	5	7,799	7,812	8,048	15	20	20		*	*
	10	15,438	15,446	15,636	14	20	20		*	*
	20	30,604	30,636	30,781	16	20	20		*	*

- In every problem above 100 nodes, the ***GRAPH*** algorithm was significantly better than the ***NO*** algorithm – working better in each problem instantiation (320/320) of each problem size and complexity.
- In every problem above 100 nodes, the ***GRAPH*** algorithm was significantly better than the ***FULL*** algorithm – working better in almost each problem instantiation (319/320) of each problem size and complexity.
- As the problem sizes increased (2,000 nodes), the ***FULL*** algorithm was not significantly better than the ***NO*** algorithm, but the ***GRAPH*** algorithm remained better than both ***FULL*** and ***NO***.

The results are promising. First, it should be noted that although the colors of each node were represented by two bits, and we did not specify on which of the bits the dependencies should be modeled, this did not hurt the ability of ***GRAPH*** in making improvements over ***NO***. For all the problems, we only specified which sets of bits were allowed for inclusion into the probabilistic model. In almost each of the problem instantiations above 100 nodes (319/320 instantiations) ***GRAPH*** performed better than ***FULL***, and in all of the instantiations ***GRAPH*** performed better than ***NO***. For the largest problem sizes that we examined (2,000 nodes), the ***FULL*** algorithm was unable to find a good probabilistic model; the ***GRAPH*** algorithm performed consistently better.

Although the differences are not significant with small problem of 100 nodes, one interesting trend may be interesting to explore further: the ***FULL*** algorithm slightly outperforms the ***GRAPH*** algorithm on a number of small problem instantiations. For the smaller problem sizes, the ***FULL*** algorithm can find a set of dependencies to model that achieve results at least comparable to the ***GRAPH*** algorithm. It is possible that the ***FULL*** algorithm may find a dependency outside the graph structure that captures a logical dependency that ***GRAPH*** was not allowed to model. As the problem size increases, however, the chances of finding these outside of the underlying graph structure decreases; thereby allowing the ***GRAPH*** algorithm to outperform the ***FULL*** algorithm for larger problems.

In summary, although it is usually accepted that using a probabilistic model to guide exploration helps optimization procedures to find better solutions, as the problem sizes get large, having enough samples to accurately determine the correct dependencies may be impractical. By using problem specific knowledge, in this case the underlying graph structure, we were able to narrow down the number of potential dependencies from which to choose. For example, in the four-color problem with 2,000 nodes, the total number of dependencies that can be modeled is 4,000*3,998 (15,992,000)[1]; this is the number of dependencies from which the ***FULL*** algorithm can select in order

[1] This assumes that the bits representing the color of a node cannot be dependent on other bits (including itself) representing the same node. In this counting, A→B, B→A are counted individually.

to build the dependency tree. By using the underlying graph structure, we can significantly reduce the number of connections that we must consider. For the 20 connectivity case, there are at most 320,000 connections ((4,000 * 20 * 2)*2), or 2% of the total connections. For the two connectivity case there are only 32,000 ((4,000 * 2 * 2)*2) connections, or 0.2% of the total connections. Here, the number of connections considered for inclusion in the model scales with the complexity of the underlying graph structure.

9.5 Conclusions and Extensions

This study has demonstrated the effectiveness of incorporating a priori knowledge into the probabilistic models that are used to guide search. The knowledge was used to pre-select the set of arcs that were allowed for inclusion in the optimal dependency trees. We have also demonstrated that the knowledge that is included does not need to be exact. In all of the experiments conducted, we limited the set of edges that could be included in the tree; however, the exact tree was never specified – and was automatically generated during the search. In this study, the knowledge that was incorporated into the optimization process was straight-forward, the underlying graph is a clear source of information. In other domains, finding a good source of domain information may be less intuitive. Nonetheless for using a priori knowledge in real domains, it is important to find good sources; incorrect information has the potential to yield results worse than using a full or even no probabilistic model.

Although not explored in this paper, using a priori knowledge can be used in conjunction with other problem-specific heuristics. For example, if we are trying to solve a well studied optimization problem, for example Satisfiability, Bin Packing, Jobshop, VLSI Layout, etc. there already exist a wide variety of specialized stochastic heuristics to address these problems. We would like to be able to use the probabilistic modeling techniques in conjunction with these specialized procedures. Here, the probabilistic models are used to generate points with which to initialize search with the specialized heuristics. For example, once we make several runs with the specialized heuristic, we can model the better points found during these runs with the probabilistic models. These probabilistic models can then be used to generate new initialization points. In this manner, the probabilistic models "wrap-around" the specialized search heuristics [4]. Similar to the manner in which a priori information was incorporated into the probabilistic models in this study, we can incorporate any a priori information in the models created in this approach. Note that the wrapper approach may also be employed for computational benefits, irrespective of whether a specialized search algorithm is used. In the wrapper approaches, the probabilistic model may be created much less frequently, which can provide benefits in terms of speed since model creation is a computationally intensive procedure (even for the trees, it is $O(n^2)$ where n is the number of bits in the parameter encoding, for more complex networks the expense can be much greater [8]).

We have not attempted to propose a complete optimization "system" in this paper; there is a vast amount of literature available on heuristics that can be used in conjunction with the algorithms proposed here. Extensions to the algorithms proposed here will include such heuristics as elitist selection, mutation operators, adaptive operator probabilities, and domain-dependent operators. Future research should also examine the effects of using non-uniform penalty settings. The magnitude of the penalty can be used as a means to convey the confidence in the a priori information including whether the information is mandatory or a suggestion. Another direction for future research is to examine the convergence properties of probabilistic optimization techniques with a priori knowledge. Convergence studies have been conducted with optimization with fixed networks, which may be viewed as an extreme form of the knowledge incorporated in this paper [44].

In previous papers, we have shown that the performance of optimization algorithms consistently improves as the accuracy of their statistical models increases. In [4] we showed that trees generally performed better than chains, and chains generally performed better than models which assumed all variables were independent, such as those used in PBIL. The accuracy of the models can be improved through either using more complex models or by ensuring that the models that are created are more representative of the structure of the underlying search space. Unfortunately, when we move toward models in which variables can have more than one parent variable, the problem of finding an optimal network with which to model a set of data becomes NP-complete [8]. The methods presented in this paper provide a means to reduce the set of probabilistic models that must be considered – whether pair-wise or higher-order dependencies are included. The incorporation of a priori information improves the accuracy of the models that are created given a limited number of samples. As shown, improved accuracy in the models leads to improved search results.

References

[1] Baluja, S. (1994), "Population-Based Incremental Learning: A Method for Integrating Genetic Search Based Function Optimization and Competitive Learning," Technical Report CMU-CS-94-163, Carnegie Mellon University, Pittsburgh, PA

[2] Baluja, S. (1995), "An Empirical Comparison of Seven Iterative and Evolutionary Heuristics for Static Function Optimization," Technical Report CMU-CS-95-193, Carnegie Mellon University, Pittsburgh, PA

[3] Baluja, S. and Caruana, R. (1995), "Removing the Genetics from the Standard Genetic Algorithm," Proceedings of the Twelfth International Conference on Machine Learning, 1995, Prieditis, A. and Russel, S. (Eds.), Morgan Kaufmann, San Mateo, CA, pp. 38–46

[4] Baluja, S. and Davies, S. (1997), "Using Optimal Dependency Trees for Combinatorial Optimization: Learning the Structure of the Search

Space," Proceedings of the Fourteenth International Conference on Machine Learning, 1997, Fisher, D.H. (Ed.), Morgan Kaufmann, San Mateo, CA, pp. 30–38

[5] Baluja, S. and Davies, S. (1998), "Fast Probabilistic Modeling for Combinatorial Optimization," Proceedings of the Fifteenth National Conference on Artificial Intelligence (AAAI-98), AAAI Press, pp. 469–476

[6] Bosman, P. and Thierens, D. (1999), "Linkage Information Processing in Distribution Estimation Algorithms," Proceedings of the Genetic and Evolutionary Computation Conference 1999, Banzhaf, W., Daida, J., Eiben, A.E., Garzon, M.H., Honavar, V., Jakiela, M. (Eds.), Morgan Kaufmann, San Mateo, CA, pp. 60–70

[7] Bosman, P. and Thierens, D. (2000), "Continuous Iterated Density Estimation Evolutionary Algorithms within the IDEA Framework", Technical Report UU-CS-2000-15, Utrecht University

[8] Chickering, D., Geiger, D., and Heckerman, D. (1995), "Learning Bayesian Networks: Search Methods and Experimental Results," Preliminary Papers of the Fifth International Workshop on Artificial Intelligence and Statistics, pp. 112–128

[9] Chow C. and Liu, C. (1968), Approximating discrete probability distributions with dependence trees. IEEE Transactions on Information Theory, 14:462–467

[10] De Jong, K. (1975), Analysis of Behavior of a Class of Genetic Adaptive Systems. Ph.D. Thesis. University of Michigan, Ann Arbor, MI

[11] De Bonet, J., Isbell, C., and Viola, P. (1997), "MIMIC: Finding Optima by Estimating Probability Densities," Advances in Neural Information Processing Systems 9, Mozer, M.C., Jordan, M.I., and Petsche, T. (Eds.). The MIT Press, Cambridge, pp. 424–431

[12] Eshelman, L.J., Mathias, K.E., and Schaffer, J.D. (1996), "Convergence Controlled Variation," in Proceedings of the Foundations of Genetic Algorithms 4, Morgan Kaufmann, San Mateo, CA, pp. 203–224

[13] Etxeberria, R. and Larrañaga P. (1999), "Global Optimization Using Bayesian networks," Proceedings of the Second Symposium on Artificial Intelligence pp. 332–339

[14] Gallagher, M., Fream, M., and Downs, T. (1999), "Real-Valued Evolutionary Optimization Using a Flexible Probability Density Estimator," Proceedings of the Genetic and Evolutionary Computation Conference, 1999, Banzhaf, W., Daida, J., Eiben, A.E., Garzon, M.H., Honavar, V., and Jakiela, M. (Eds.), Morgan Kaufmann, San Mateo, CA, pp. 840–846

[15] Goldberg, D.E. (1989), Genetic Algorithms in Search, Optimization and Machine Learning.Addison-Wesley, Reading, MA.

[16] González, C., Lozano, J.A., and Larrañaga, P. (2000), "Analyzing the PBIL Algorithm by Means of Discrete Dynamical Systems", Complex Systems 12(4), 465–479

[17] González, C., Lozano, J.A., and Larrañaga, P. (2001), "The Convergence Behavior of the PBIL Algorithm: A Preliminary Approach,"

Proceedings of the Fifth International Conference on Artificial Neural Networks and Genetic Algorithms, Springer, Berlin, Heildelberg, New York, pp. 228–231

[18] Greene, J.R. (1996), "Population-Based Incremental Learning as a Simple Versatile Tool for Engineering Optimization," Proceedings of the First International Conf. on EC and Applications, pp. 258–269

[19] Heckerman, D. (1996), "A Tutorial on Learning with Bayesian Networks," Technical Report MSR-TR-95-06. Microsoft

[20] Heckerman, D., Geiger, D., and Chickering, D. (1995), "Learning Bayesian Networks: The Combination of Knowledge and Statistical Data," Machine Learning 20:197–243

[21] Hertz, J., Krogh A., and Palmer R.G. (1991), Introduction to the Theory of Neural Computing. Addison-Wesley, Reading, MA

[22] Hohfeld, M. and Rudolph, G. (1997), "Towards a Theory of Population-Based Incremental Learning," Proceedings of the Fourth IEEE Conference on Evolutionary Computation, 1997.IEEE Press, Piscataway, NJ, pp. 1–5

[23] Holland, J.H. (1975), Adaptation in Natural and Artificial Systems.University of Michigan Press, Ann Arbor, MI

[24] Jensen, F.V. (2001), Bayesian Networks and Decision Graphs. Springer, Berlin, Heidelberg, New York

[25] Juels, A. (1996), Topics in Black-box Combinatorial Optimization. Ph.D. Thesis, University of California, Berkeley

[26] Keane, A (2000), The Options Design Exploration System: Reference Manual and Users Guide, available from: http://www.soton.ac.uk/~ajk/options/welcome.html

[27] Kvasnica, V., Pelikan, M., and Pospical, J. (1996), "Hill Climbing with Learning (An Abstraction of Genetic Algorithm)," Neural Network World (Czech Republic), 6(5): 773–796

[28] Larrañaga, P., Etxeberria, R., Lozano, A., and Peña, J. (1999), Optimization by learning and simulation of Bayesian and Gaussian networks. Technical Report EHU-KZAA-IK-4/99, Department of Computer Science and Artificial Intelligence, University of the Basque Country

[29] Larrañaga, P., Etxeberria, R., Lozano, A., and Peña, J. (2000), "Optimization in continuous domains by learning and simulation of Gaussian networks," Optimization By Building and Using Probabilistic Models Workshop in the GECCO-2000 Conference, pp. 201–204

[30] Larrañaga, P., Etxeberria, R., Lozano, J.A., and Peña, J.M. (2000), "Combinatorial Optimization by Learning and Simulation of Bayesian Networks", Proceedings of the Conference in Uncertainty in Artificial Intelligence, Morgan Kauffman, San Mateo, CA, pp. 343–352

[31] Larrañaga, P. and Lozano, J. (2001), Estimation of Distribution Algorithms. A New Tool for Evolutionary Computation, Kluwer, Dordrecht, The Netherlands

[32] Mahnig, T. and Muhlenbein, H. (2001), "Mathematical Analysis of Optimization Methods Using Search Distributions," Optimization By Building and Using Probabilistic Models Workshop in the GECCO-2000 Conference, pp. 205–208
[33] Muhlenbein, H. (1997), "The Equation for Response to Selection and Its Use for Frediction," Evolutionary Computation, 5(3): 303–346
[34] Muhlenbein, H. and Mahnig, T. (1999), "Convergence Theory and Applications of the Factorized Distribution Algorithm," Journal of Computing and Information Technology, 7, pp. 19–32
[35] Muhlenbein, H., Mahnig, T. and Rodriguez, A.O. (1999), "Schemata, Distributions and Graphical Models in Evolutionary Optimization," Journal of Heuristics, 5, pp. 215–247
[36] Ochoa, A., Muehlenbein, H., and Soto, M. (2000), "Factorized Distribution Algorithms using Bayesian Networks of Bounded Complexity," Optimization By Building and Using Probabilistic Models Workshop in the GECCO-2000 Conference, pp. 212–215
[37] Ochoa A., Soto M., Santana R., Madera J.C., and Jorge N. (1999), "The Factorized Distribution Algorithm and The Junction Tree: A Learning Perspective," Proceedings of the Second Symposium on Artificial Intelligence, pp. 368–377
[38] Pearl, J. (1988) Probabilistic Reasoning in Intelligent Systems. Morgan Kaufmann, San Mateo, CA
[39] Pelikan, M., Goldberg, D.E., Lobo, F. (2002), "A Survey of Optimization by Building and Using Probabilistic Models," Computational Optimization and Applications, 21(1): 5–20
[40] Pelikan, M., Sastry, K., and Goldberg, D.E. (2001), "Evolutionary Algorithms + Graphical Models = Scalable Black-Box Optimization," Technical Report 2001029, Illinois Genetic Algorithm Laboratory, University of Illinois at Urbana-Champaign
[41] Pelikan, M., Goldberg, D.E., and Tsutsui, S. (2001), "Combining the Strengths of the Bayesian Optimization Algorithm and Adaptive Evolution Strategies," Technical Report 2001023, Illinois Genetic Algorithm Laboratory, University of Illinois at Urbana-Champaign
[42] Sebag, M., and Ducoulombier, A. (1998), "Extending Population-Based Incremental Learning to Continuous Search Spaces," Parallel Problem Solving from Nature V, pp. 418–427
[43] Syswerda, G. (1993), "Simulated Crossover in Genetic Algorithms," Foundations of Genetic Algorithms 2, Whitley, D.L. (Ed.), Morgan Kaufmann, San Mateo, CA, pp. 239–255
[44] Zhang Q. and Muehlenbein H. (1999), "On Global Convergence of FDA with Proportionate Selection," Proceedings of the Second Symposium on Artificial Intelligence, pp. 340–343

10

Multiobjective Estimation of Distribution Algorithms

Martin Pelikan, Kumara Sastry, and David E. Goldberg

Summary. Many real-world optimization problems contain multiple competing objectives and that is why the design of optimization techniques that can scalably discover an optimal tradeoff between given objectives (Pareto-optimal solutions) represents an important challenge. This chapter discusses estimation of distribution algorithms (EDAs) that address this challenge. The primary focus is on scalability on discrete multiobjective decomposable problems and the multiobjective hierarchical BOA (mohBOA); other approaches to the multiobjective EDA design are discussed briefly.

Key words: Multiobjective optimization, evolutionary computation, estimation of distribution algorithms, multiobjective decomposable problems

10.1 Introduction

Many real-world optimization problems contain multiple competing objectives. For example, in engine design, one objective may be to maximize performance whereas the other objective may be to maximize fuel efficiency. Nonetheless, most optimization algorithms are applicable to only single-objective problems. One of the approaches to multiobjective optimization is to create a new objective function that is defined as a weighted sum of the competing objectives; a single-objective optimizer can then be used. The second approach is to modify the optimizer to consider all objectives and discover an optimal tradeoff (Pareto-optimal front) between these objectives.

Since it is often impossible to choose appropriate weights without trying many settings by trial and error, designing optimization algorithms that can scalably and reliably discover the set of all Pareto-optimal solutions represents an important challenge [6, 7]. This chapter discusses multiobjective estimation of distribution algorithms (moEDAs), which address this challenge within the EDA framework.

M. Pelikan et al.: *Multiobjective Estimation of Distribution Algorithms*, Studies in Computational Intelligence (SCI) **33**, 223–248 (2006)
www.springerlink.com

The chapter starts with an introduction to multiobjective optimization in Sect. 10.2. Section 10.3 discusses multiobjective decomposable problems and analyzes the scalability of several multiobjective genetic and evolutionary algorithms and EDAs on this class of problems. Section 10.4 describes the multiobjective hierarchical Bayesian optimization algorithm (mohBOA), which is shown to solve difficult multiobjective decomposable problems scalably and reliably. Section 10.5 then reviews and discusses other multiobjective EDAs. Finally, Sect. 15.7 summarizes and concludes the chapter.

10.2 Multiobjective Optimization and Pareto Optimality

In multiobjective optimization, the task is to find a solution or solutions that are optimal with respect to multiple objectives. For example, one may want to optimize a telecommunication network to both maximize its performance as well as minimize its cost. In many problems, objectives compete and improving the value of one objective comes at the expense of the other objectives. This section introduces multiobjective optimization and the basic procedure of NSGA-II, which illustrates this class of optimization problems.

10.2.1 Basic Approaches and Pareto Optimality

There are two basic approaches to solving multiobjective optimization problems:

(1) Weigh the objectives in some way, yielding a single-objective problem where the objective consists of a weighted sum of all objectives.
(2) Find the Pareto-optimal front, which is defined as the set of solutions that can be improved with respect to any objective only at the expense of their quality with respect to at least one other objective; for example, the performance of an engine on the Pareto-optimal front could be improved only at the expense of its fuel consumption and the other way around (see Fig. 10.1).

Pareto optimality can be easily explained using the concept of *dominance.* We say that a candidate solution A dominates a candidate solution B if A is better than B with respect to at least one objective but A is not worse than B with respect to all other objectives. For example, in telecommunication network optimization, network A dominates network B if A is better than B with respect to both performance as well as cost. The *Pareto-optimal front* is then a subset of all candidate solutions that are not dominated by any other candidate solution.

The primary advantage of finding the Pareto-optimal front as opposed to finding the optimum to a single-objective problem created by weighing the objectives is that sometimes it is difficult or impossible to weigh the objectives appropriately to find satisfactory solutions. Furthermore, finding

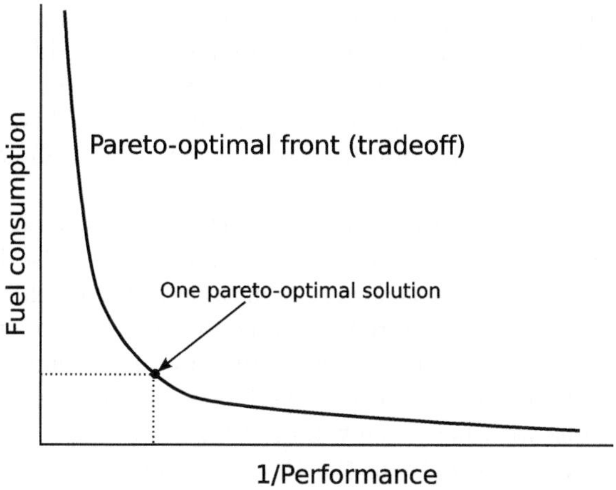

Fig. 10.1. Pareto-optimal front for two-objective engine design

the Pareto-optimal front reveals the tradeoff between the objectives, which can be used to decide which of the solutions on this front is best for each particular problem instance.

This chapter focuses on EDAs for finding the set of all Pareto-optimal solutions because the application of EDAs or other advanced evolutionary algorithms to any single-objective problem is straightforward. The task is to find the set of all Pareto-optimal solutions or to cover this set evenly if storing all solutions is intractable.

An overview of multiobjective genetic and evolutionary algorithms and their comparison can be found in Deb [7] and Coello Coello [6]. Section 10.3 describes the basic procedure of NSGA-II [9], which is one of the most popular multiobjective genetic algorithms.

10.2.2 NSGA-II

To apply genetic and evolutionary algorithms to multiobjective problems, it is common to modify selection and replacement operators. NSGA-II [9] uses *nondominated crowding selection*, which makes tournaments among different solutions and decides on the winner of each tournament using two criteria: dominance and crowding distance.

NSGA-II starts by ranking solutions based on dominance. The ranking starts by assigning rank 1 to the set of solutions that are not dominated by any other solution in the population. Next, solutions that are not dominated by any of the remaining solutions are assigned rank 2. That is, all solutions with rank 2 are dominated by at least one solution with rank 1, but are not dominated by others in the population. The sorting and ranking process continues until all solutions are ranked by assigning increasing ranks to those solutions that are not dominated by any of the remaining, unranked solutions.

After nondominated sorting, we are left with subsets of population with different ranks. Solutions with a given rank are not dominated by solutions that have the same or higher rank but they are dominated by at least one solution with a lower rank. Therefore, with respect to Pareto optimality, solutions with lower ranks should be given priority. See Fig. 10.2 for an illustration of the ranking procedure of NSGA-II.

Apart from finding solutions in the Pareto front, it is also essential to achieve good coverage or spread of solutions in the front. The diversity of solutions in the objective space is usually maintained with a niching mechanism. NSGA-II uses a niching procedure based on crowding distance, which depends on the density of solutions in the neighborhood of each solution. The higher the crowding distance of the solution, the less dense the neighborhood of the solution. Solutions with higher crowding distance are favored to ensure that the population spreads evenly on the Pareto-optimal front. The pseudocode for computing the crowding distance is outlined in Fig. 10.3.

To compare quality of two solutions, their ranks are compared first. If the ranks of the two solutions differ, the solution with the lower rank is better regardless of the crowding distance. If the ranks of the two solutions are equal, the solution with a greater crowding distance wins. If both the ranks as well as the crowding distances are equal, the winner is determined randomly. A pseudocode for the comparison of two solutions is shown in Fig. 10.4. This comparison procedure can be used in any standard selection operator, such as tournament or truncation selection. NSGA-II uses binary tournament selection.

Section 10.3 discusses multiobjective decomposable problems and proposes a simple combination of NSGA-II and EDAs. Then, the section presents initial

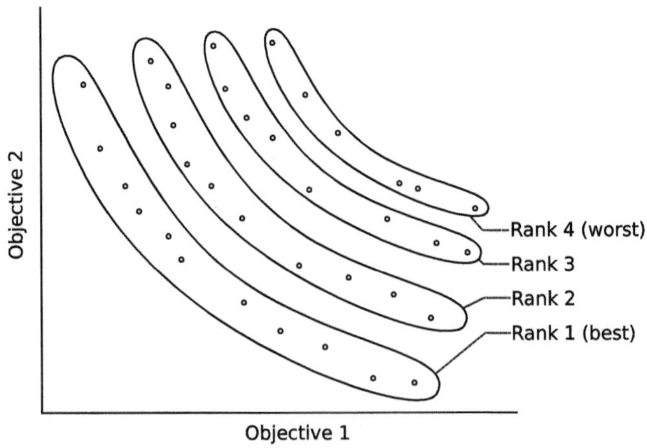

Fig. 10.2. Illustration of the ranking procedure of NSGA-II for a two-objective minimization problem

```
crowding-distance-assignment(P)
   for each rank r (nondominated sorting)
      P' = select solutions with rank r from P;
      N = size(P');
      for each solution X in P'
          d(X)=0;
      for each objective m
          Q = sort P' using m-th objective;
          d(Q(1))=infinity;
          d(Q(N))=infinity;
          for i=2 to N-1
             d(Q(i))=d(Q(i))+Q(i+1).m-Q(i-1).m;
   return d;
```

Fig. 10.3. Crowding distance assignment in NSGA-II. For a solution `X`, `X.m` denotes the value of `m`th objective for `X`. `Q(i)` denotes `i`th candidate solution in population `Q`

```
compare(A,B)
   if (rank(A)<rank(B)) then better(A,B)=A;
   if (rank(A)>rank(B)) then better(A,B)=B;
   if (rank(A)=rank(B))
      then if (crowding(A)>crowding(B))
              then better(A,B)=A;
           if (crowding(A)<crowding(B))
              then better(A,B)=B;
           if (crowding(A)=crowding(B))
              then better(A,B)=random(A,B);
```

Fig. 10.4. Nondominated crowding selection in NSGA-II

experiments with various multiobjective optimizers on several multiobjective decomposable problems and discusses the poor scalability of all algorithms on the tested problems.

10.3 Initial Experiments

Since much research in EDAs focused on the design of robust, scalable, and efficient EDAs for decomposable problems of bounded difficulty, the assumption of bounded-order decomposability appears to be a good starting point for the design of robust and scalable multiobjective EDAs. The purpose of this section is to describe a class of decomposable problems for testing multiobjective optimization techniques and to discuss performance of several multiobjective genetic algorithms and EDAs on these problems.

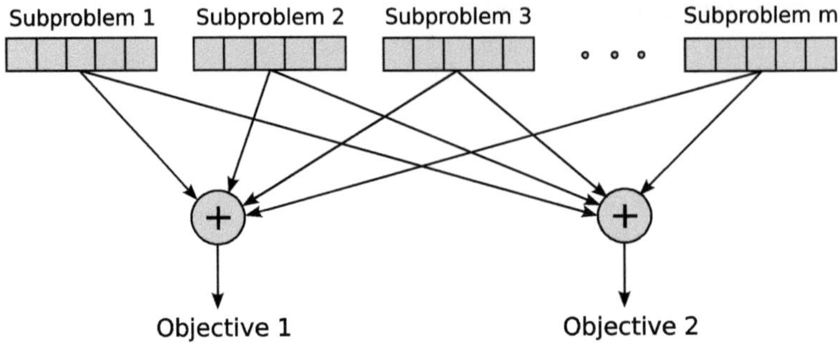

Fig. 10.5. A two-objective decomposable problem of m subproblems of order 5

The section starts by discussing multiobjective decomposable problems. The section then introduces a class of multiobjective decomposable problems that can test whether a multiobjective optimizer can scalably find a diverse set of representatives on the Pareto-optimal front of multiobjective decomposable problems of bounded difficulty. Initial experiments with multiobjective EDAs and other multiobjective evolutionary algorithms are then presented and discussed. The results indicate that when there are many subproblems for which the multiple objectives compete, niching in multiobjective evolutionary algorithms becomes overwhelmed, leading to poor, exponentially scaled performance.

10.3.1 Multiobjective Decomposable Problems

In the study on multiobjective decomposable problems, we make several important assumptions. First of all, we assume that the problem can be additively decomposed into subproblems of bounded order and that the decomposition is identical for all objectives. We also assume that out of m subproblems, $m_{\mathrm{d}} \leq m$ subproblems compete in at least one objective, which means that the optimal solution to the subproblem is not the same for all objectives.

Figure 10.5 illustrates this concept with a two-objective problem with m subproblems of order 5 and two objectives. In the remainder of this chapter we assume two objectives, although the results of the chapter can be generalized to any fixed number of objectives in a straightforward manner.

An example of a multiobjective decomposable problems is a two-objective onemax-zeromax [5, 30, 33], where the first objective counts the ones in the input string whereas the second objective counts the zeros (see Sect. 10.3.3); the task is to maximize both objectives.

10.3.2 Initial Approach to Multiobjective EDAs

The most intuitive way to extend EDAs to find Pareto-optimal solutions is to replace selection and replacement operators of standard EDAs by those of multiobjective evolutionary algorithms, such as NSGA-II [9] or SPEA2 [40]. This approach was taken in two extensions of BOA [28] and hierarchical BOA (hBOA) [26,30]: The multiobjective hierarchical BOA (mhBOA) of Khan [16,17], who combined hBOA with NSGA-II, and the multiobjective mixed BOA (mmBOA) of Laumanns et al. [18], who combined mixed BOA [23] with SPEA2 [40].

In this work, we use the selection operator of NSGA-II, but similar results can be expected for other multiobjective selection techniques. That is, nondominated crowding with binary tournament selection is used to select of promising solutions (see Sect. 10.2.2).

For replacement, two operators are evaluated (1) Elitist replacement of NSGA-II [9] and (2) restricted tournament selection [15]:

1. *Elitist replacement.* The first replacement operator is the elitist replacement of NSGA-II, where the selected population and the population of new candidate solutions are first evaluated with the nondominated crowding evaluation of NSGA-II, and then the best candidate solutions according to nondominated ranking form the next population.
2. *Restricted tournament selection.* The second replacement operator is the restricted tournament selection [15]. Also in this case, the selected population and the population of new candidate solutions are first evaluated with the nondominated crowding evaluation of NSGA-II. Then, each new candidate solution X is incorporated into the original population using the following three steps (1) select a random subset W of size w from the original population, (2) find the solution Y in W that is most similar to X, and (3) make a tournament between X and Y where X replaces Y if it is better than Y according to the nondominated ranking comparison of NSGA-II. The parameter w is called window size, and a good rule of thumb for this parameter is $w = \min\{n, N/20\}$, where n is the problem size and N is the population size [29].

To perform initial experiments, the nondominated crowding selection and the two replacement schemes were incorporated into the extended compact genetic algorithm (ECGA) [12], which is an advanced EDA based on multivariate probabilistic models. The multiobjective variant of ECGA will be referred to as *mECGA*. Since ECGA is described in detail in another chapter of this book, we omit any discussion of this algorithm.

ECGA can automatically identify and exploit appropriate problem decomposition to scalably solve decomposable problems of bounded difficulty [12, 32]. That is why mECGA should not suffer from ineffective recombination of standard variation operators of genetic algorithms, providing a good starting point for the design of robust and scalable optimizers for multiobjective decomposable problems.

Additionally, the multiobjective selection and replacement schemes were incorporated into UMDA [25], which represents a simple EDA with a univariate probabilistic model that assumes independence of all variables.

10.3.3 Test Problems

There were four primary objectives in the design of test problems used in initial experiments:

1. *Scalability.* Test problems should be scalable, that is, it should be possible to increase problem size.
2. *Decomposability.* Objective functions should be decomposable into subproblems of bounded order.
3. *Known solution.* Test problems should have a known Pareto-optimal front in order to be able to verify the results.
4. *Linkage learning.* Some test problems should require the optimizer to be capable of linkage learning [10, 15], that is, of identifying and exploiting interactions between decision variables to provide effective exploration.

Two test problems are used in this initial set of experiments: Onemax-zeromax, and trap_k-invtrap_k:

(1) *Onemax-zeromax.* The first test problem is inspired by [5]. It consists of two objectives (1) onemax and (2) zeromax. Onemax is defined as the sum of bits in the input binary string $X = (X_1, X_2, \ldots, X_n)$:

$$\text{onemax}(X) = \sum_{i=1}^{n} X_i \tag{10.1}$$

The task is to maximize the function and thus the optimum of onemax is in the string of all ones. See Fig. 10.6 to visualize onemax for 5-bit strings. Zeromax is defined as the number of positions containing a 0:

$$\text{zeromax}(X) = n - \text{onemax}(X) \tag{10.2}$$

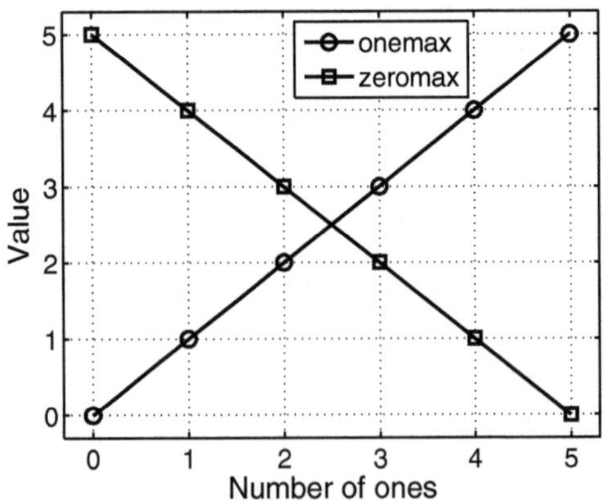

Fig. 10.6. Onemax-zeromax for 5 bits

The task is to maximize the function and thus the optimum of zeromax is in the string of all zeros. See Fig. 10.6 to visualize zeromax for 5-bit strings.

Onemax and zeromax are conflicting objectives; in fact, *any* modification that increases one objective decreases the other objective. In onemax-zeromax, any binary string is located on the Pareto-optimal front.

(2) *Trap$_k$-invtrap$_k$*. The second test problem is inspired by [16]. It consists of two objectives: (1) trap of order k and (2) inverse trap of order k. String positions are first (before running the optimizer) divided into disjoint subsets or partitions of k bits each (string length is assumed to be a multiple of k). The partitioning is fixed during the entire optimization run, but the algorithm is not given information about the partitioning in advance. Bits in each partition contribute to trap of order k using a trap function [1, 8] defined as

$$\text{trap}_k(u) = \begin{cases} 1 & \text{if } u = k \\ (1-d)\left(1 - \frac{u}{k-1}\right) & \text{otherwise} \end{cases} \tag{10.3}$$

where u is the number of ones in the input string of k bits. The task is to maximize the function and thus the optimum of order-k trap is in the string of all ones. See Fig. 10.7 to visualize order-5 trap for one block of 5 bits and $d = 0.2$.

Traps deceive the algorithm away from the optimum if interactions between the bits in each partition are not considered [4, 27, 35]. That is why standard crossover operators of genetic algorithms – such as uniform, one-point, and two-point crossover – fail to solve traps unless the bits in each partition are located close to each other in the chosen representation; in fact, standard crossover operators require exponentially scaled population sizes to solve traps [35]. Mutation operators require $O(n^k \log n)$

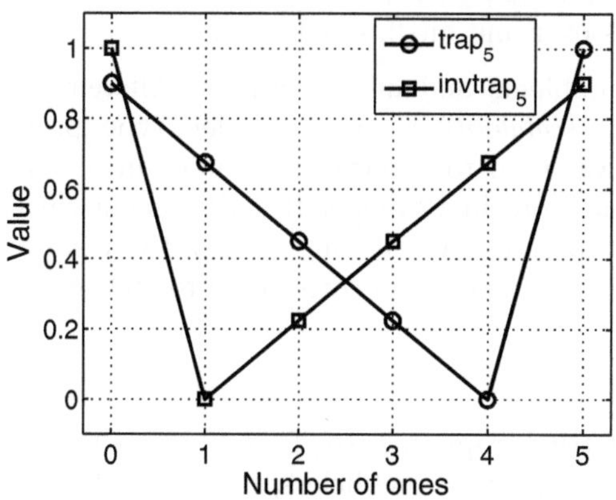

Fig. 10.7. Trap$_5$-invtrap$_5$ for 5 bits

evaluations to solve order-k traps and, therefore, are also highly inefficient in solving traps of moderate to large order.

Inverse trap of order k is defined using the same partitions as trap of order k, but the basis function, which is applied to each partition, is modified as follows:

$$\text{invtrap}_k(u) = \begin{cases} 1 & \text{if } u = 0 \\ (1-d)\frac{u-1}{k-1} & \text{otherwise} \end{cases} \qquad (10.4)$$

The task is to maximize the function and thus the optimum of order-k inverse trap is in the string of all zeros. Therefore, in trap_k-invtrap_k the two objectives compete in every partition of the problem decomposition.

10.3.4 Description of Experiments

The primary focus of our experiments is to analyze *scalability* of different multiobjective EDAs and other evolutionary algorithms on the aforementioned multiobjective decomposable problems. Algorithm performance is measured in terms of the minimum number of function evaluations required to find and maintain at least one copy of all representative Pareto-optimal solutions. In some experiments, we relax the requirements by considering all Pareto-optimal solutions that have equal values of both objectives equivalent; in this scenario, we require the algorithm to find one representative for each combination of values of the two objectives that lies on the Pareto-optimal front.

Three recombination operators are included in the initial set of experiments:

- UMDA recombination [25], which uses a probabilistic model with no interactions to model and sample solutions,
- ECGA recombination [12], which uses a probabilistic model that is a product of multivariate marginal probabilities, and
- two-point crossover and bit-flip mutation.

For all test problems and all algorithms, different problem sizes were examined to study *scalability*. For each problem type, problem size and algorithm, bisection was used to determine a minimum population size to cover the Pareto-optimal front in 10 out of 10 independent runs. To reduce noise, the bisection method was ran 10 times. Thus, the results for each problem type, problem size, and algorithm correspond to 100 successful runs.

10.3.5 Results

Although we have tried trap_k-invtrap_k for $k = 3$, 4, and 5, for brevity, we only show results for $k = 3$ here. Nonetheless, the results for other values of k are qualitatively similar and those for $k = 3$ are representative of the behavior of the multiobjective evolutionary algorithms.

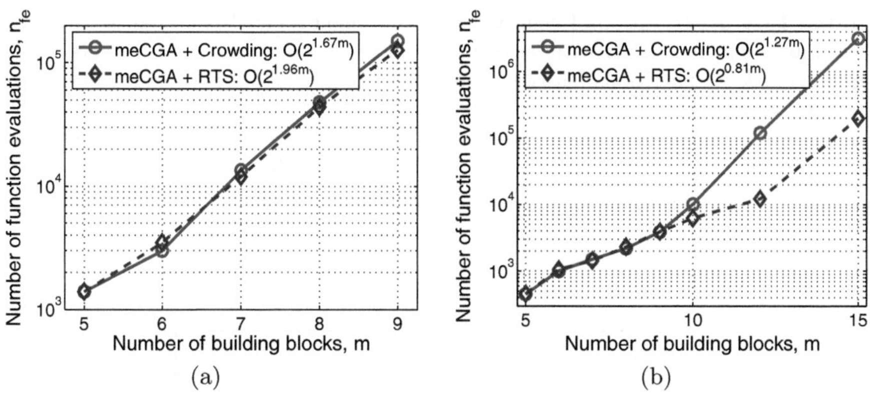

Fig. 10.8. Scalability of mECGA with crowding and with RTS for the trap$_3$-invtrap$_3$ with problem size. Here, we plot the minimum number of function evaluations required to search and maintain at least one copy of **(a)** all the 2^m solutions in the Pareto-optimal front, and **(b)** only the $m+1$ solutions in the Pareto-optimal front with different objective-value pairs where we treat the gentotypically (and phenotypically) different Pareto-optimal solution with the same values in both objectives to be equivalent

Figure 10.8a, shows the scalability of mECGA with problem size for trap$_3$-invtrap$_3$. We plot the minimum number of function evaluations required to allocate at least one copy of all solutions in the Pareto-optimal front. As shown in the figure, all algorithms scale up exponentially. The scale-up does not improve even if we relax the requirement to finding only those $m+1$ Pareto-optimal solutions with different objective-value pairs as shown in Fig. 10.8b. That is, even if we consider genotypically (and phenotypically) distinct solutions that have the same value in both objectives to be equivalent, all algorithms scale exponentially. This is despite the linkage information, which should be identified by mECGA, and the tight linkage assumption for NSGA-II. Additionally, the scalability does not improve if the niching or speciation is performed in the objective space (as in the elitist replacement of NSGA-II) or in the variable space (as in restricted tournament selection).

Therefore, the exponential scale-up is not due to incorrect linkage identification and ineffective mixing [11, 36, 39], but because the niching mechanism gets quickly overwhelmed due to the exponential growth in the number of Pareto-optimal solutions. Furthermore, the distribution of the 2^m solutions in the Pareto-optimal front is not uniform. There are exponentially as many solutions in the middle of the front than at the edges (see Table 10.1). That is, there is only one solution – a binary string with all 0s and all 1s – at each extreme of the Pareto-optimal front. In contrast, there are $\binom{m}{m/2} \approx \mathcal{O}(e^m)$ genotypically different solutions in the middle of the Pareto-optimal front with same values in both objectives.

Table 10.1. Distribution of genotypically and phenotypically different solutions in the Pareto-optimal front with same values in both objectives. $n_{1,\text{BBs}}$ refers to the number of k-bit partitions (substructures) with 1s and $n_{0,\text{BBs}}$ is the number of k-bit partitions with 0s

$n_{1,\text{BBs}}$	0	1	$\cdots$	i	$\cdots$	m
$n_{0,\text{BBs}}$	m	$m-1$	$\cdots$	$m-i$	$\cdots$	0
# solutions	1	m	$\cdots$	$\binom{m}{i}$	$\cdots$	1

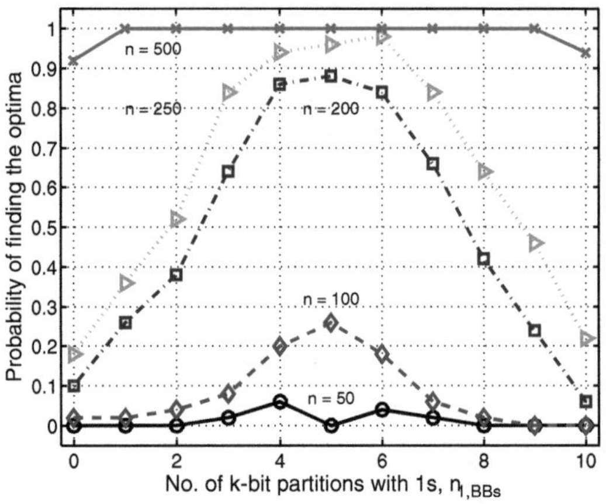

Fig. 10.9. Probability of finding and maintaining different solutions on the Pareto-optimal for the 10-3 deceptive trap and inverse trap problem as a function of population size.

This highly nonlinear distribution of solutions in the Pareto-front has two effects on the niching mechanism used in MOEAs in general, and MOEDAs in particular:

- Since the extremes of the Pareto-optimal front (maximizing most partitions or substructures with respect to one particular objective) have exponentially fewer representatives than the middle part, it takes exponentially longer time, or exponentially larger population size [10, 36] to search and maintain the solutions at the extremes of the Pareto-optimal front. When the population size is fixed, the probability of maintaining a solution in the middle of the Pareto-optimal front is higher than doing so in extremes of the front, as shown in Fig. 10.9.
- Since there are multiple points that are genotypically and phenotypically different, but lie on the same point on the Pareto-optimal front (the solutions have same values in both objectives), some of them vanish over time due to drift. The drift affects both the solutions in the middle as well as the ones near the extremes of Pareto-optimal front.

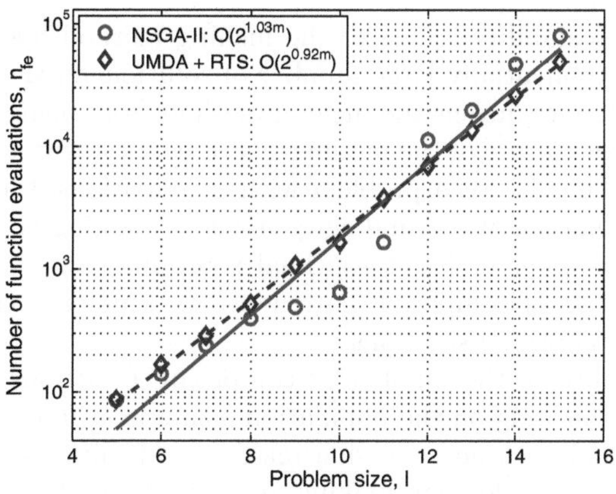

Fig. 10.10. Scalability of NSGA-II and UMDA on the onemax-zeromax problem. Both algorithms with two different niching methods scale-up exponentially with the problem size

10.3.6 Overwhelming the Niching Method

To better illustrate how competition in all the partitions of a decomposable problem can overwhelm a nicher, we use the onemax-zeromax problem. We specifically choose the onemax-zeromax problem to isolate the effects of linkage identification or lack there of from those of the niching methods on the scalability of the MOEAs. Unlike trap_k-invtrap_k, linkage identification is not necessary for the onemax-zeromax problem. Furthermore, both onemax and zeromax are GA-easy problems, which can be solved by a simple selectorecombinative GA with uniform crossover and tournament selection in nearly linear time [13, 26, 38].

Nonetheless, Fig. 10.10 shows that MOEAs scale-up exponentially even on onemax-zeromax. The results clearly indicate that the niching methods – both those that work in the parameter space (RTS) as well as those that work in the objective space (crowding) – get overwhelmed due to the exponentially large number of solutions in the Pareto-optimal front. Additionally, the results also show that even if the requirement is relaxed by treating all candidate solutions that lie on the same point in the Pareto-optimal front to be equivalent, the scale-up does not improve. Finally, the results suggest that in decomposable problems, if all or majority of partitions compete in the two objectives, then the niching method fails to maintain good coverage, leading to the exponential scale-up.

10.3.7 Circumventing the Burden on the Niching Method

The initial results clearly indicate that MOEDAs scale-up exponentially with problem size regardless of the niching method used. We also demonstrated that the exponential scalability is due to the niching method being overwhelmed because of the exponentially large number of solutions in the Pareto-optimal front. One way to circumvent the niching method from being overwhelmed is to control the growth rate of the number m_{d} of partitions that compete in the two objectives. That is, for a problem with m partitions, the two objectives compete in only m_{d} partitions and agree in the remaining $m - m_{\mathrm{d}}$ partitions. Since the total number of Pareto-optimal solutions is $n_{\mathrm{opt}} = 2^{m_{\mathrm{d}}}$, by controlling the number of competing partitions, we implicitly control the total number of Pareto-optimal solutions.

The growth-rate of the number of competing partitions should be such that the effect of niching on the population sizing is at most as strong as the effect of the model accuracy, decision making and initial supply. The effect of the model accuracy, decision making and initial supply on the population sizing of EDAs is given by [29, 32]:

$$N_{\mathrm{eda}} \propto c_1 \cdot 2^k \cdot m \log m. \tag{10.5}$$

The effect of the niching method on the population-sizing of GAs was modeled by Mahfoud [20] and is reproduced below:

$$N_{\mathrm{niching}} \propto \frac{\log\left[\left(1-\gamma^{1/t}\right)/n_{\mathrm{opt}}\right]}{\log\left[\left(n_{\mathrm{opt}}-1\right)/n_{\mathrm{opt}}\right]} \approx c_2 \cdot 2^{m_{\mathrm{d}}}, \tag{10.6}$$

where t is the number of generations we need to maintain all the niches. While Mahfoud derived the population-sizing estimate for fitness-sharing, it is generally applicable to other niching methods and MOEAs as well [16, 31].

To circumvent the niching method from being overwhelmed, we require $N_{\mathrm{eda}} \geq N_{\mathrm{niching}}$; that is,

$$c_2 \cdot 2^{m_{\mathrm{d}}} \geq c_1 \cdot 2^k \cdot m \log m. \tag{10.7}$$

The above equation can be approximated by neglecting $\log_2\left(c_1 \log m / c_2\right)$, obtaining a conservative estimate of the maximum number of competing partial solutions that circumvent the niching mechanism from being overwhelmed:

$$m_{\mathrm{d}} \approx k + \log_2(m). \tag{10.8}$$

The above growth rate is compared to empirical results for different values of k as a function of total number of partial solutions in the problem and the results are shown in Fig. 10.11. As shown in Fig. 10.12, the results indicate that once the growth of the number competing partitions is controlled, MOEDAs scale-up polynomially with problem size, even on the onemax-zeromax problem.

10.4 Multiobjective Hierarchical BOA (mohBOA)

Section 10.3 argued that in order to solve multiobjective decomposable problems with straightforward extensions of evolutionary algorithms and EDAs,

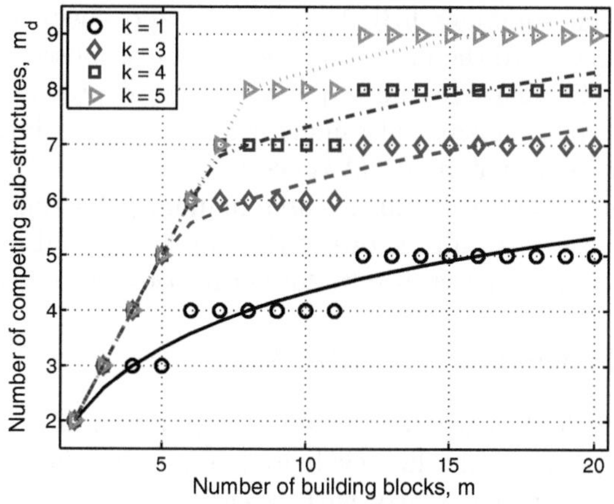

Fig. 10.11. The growth rate of number of partial solutions that compete in the two objectives for different values of k as a function of total number of partial solutions

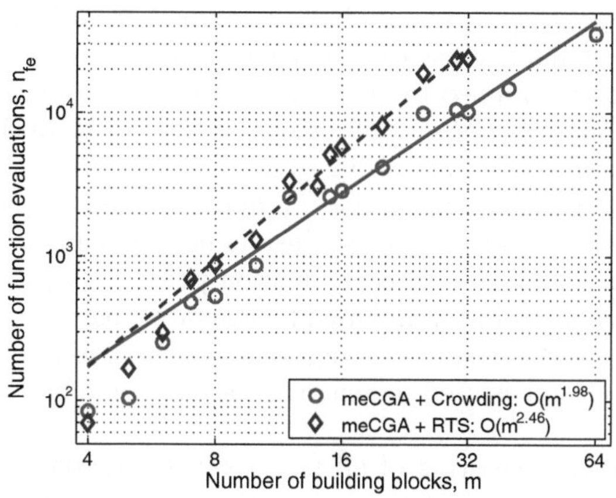

Fig. 10.12. The scalability of mECGA with the crowding mechanism of NSGA-II and RTS for both onemax-zeromax and trap$_3$-invtrap$_3$ problems. The growth rate of the number of partitions that compete in the two objectives for a given problem size is controlled as given by (10.8)

the number of competing partitions must be controlled to grow at most as a logarithm of problem size. This leads to an important question: Is it possible to scalably find representative solutions of the Pareto-optimal front even when the number of competing partitions grows faster than a logarithm of problem size? For example, is it possible to scalably solve multiobjective decomposable problems where all subproblems compete (that is, $m_{\mathrm{d}} = m$)?

This section answers the above question by proposing a multiobjective EDA that can solve even problems with a large number of competing partitions. Specifically, the section introduces the multiobjective hierarchical BOA (mohBOA), which combines hBOA, NSGA-II, and clustering in the objective space. Compared to the multiobjective hBOA proposed by Khan [16, 17], mohBOA adds clustering in the objective space to provide a polynomially scalable solution to problems with many competing partitions.

The basic procedure of mohBOA is described first. Next, the k-means clustering algorithm is described, which is used in mohBOA to ensure equal supply of candidate solutions for all regions of the Pareto-optimal front. Finally, the section presents experimental results for verification of mohBOA scalability.

10.4.1 Basic Procedure of mohBOA

Like hBOA, mohBOA generates the initial population of candidate solutions at random. The population is first evaluated. Similarly as in other evolutionary algorithms, each iteration starts with selection. However, instead of using standard selection methods, mohBOA first uses the nondominated crowding of NSGA-II to rank candidate solutions and assign their crowding distances. The ranks and crowding distances then serve as the basis for applying standard selection operators. For example, binary tournament selection can then be used where the winner of each tournament is determined by the ranks and crowding distances obtained from the nondominated crowding.

To ensure equal coverage of all regions of the Pareto-optimal front, after selecting the population of promising solutions, k-means clustering [19] in the objective space (space of fitness values) is applied to this population to obtain a specified number of clusters. Some clusters may remain empty; empty clusters are not considered in the recombination phase. A separate probabilistic model is built for each cluster and used to generate a part of the offspring population. To encourage an equal coverage of the entire Pareto-optimal front, the model for each cluster is used to generate the same number of new candidate solutions.

The population consisting of all newly generated solutions is then combined with the original population to create the new population of candidate solutions. As in the initial experiments, we use two methods to combine the two populations (1) the elitist replacement based on the nondominated crowding of NSGA-II, and (2) the restricted tournament replacement (RTS) [15] based on the nondominated crowding comparison operator. The pseudocode of the multiobjective hBOA is shown in Fig. 10.13.

10.4.2 Making Niching Easier with Clustering

Given a set X of N points, k-means clustering [19] splits X into k clusters or subsets with approximately same variance. The algorithm proceeds by updating a set of k cluster centers where each center defines one cluster. The cluster centers can be initialized randomly but more advanced algorithms can also be used to initialize the centers.

Each iteration consists of two steps. In the first step, each point in X is attached to the closest center (ties can be resolved arbitrarily). In the second

```
multiobjective-hBOA(N, k, objectives)
  t := 0;
  generate initial population P(0) of size N;
  evaluate P(0);
  while (not done) {
    rank members of P(t) using nondom. crowding;
    select S(t) from P(t) based on the ranking;
    cluster S(t) into k clusters;
    build Bayesian network with local structures
       for each cluster;
    create O(t) of size N by sampling the model
       for each cluster to generate N/k solutions;
    evaluate O(t);
    combine O(t) and P(t) to create P(t+1);
    t:=t+1;
  }
```

Fig. 10.13. Pseudocode of the multiobjective hBOA

step, cluster centers are recomputed so that each center is the center of mass of the points attached to it. The algorithm terminates when all points in X remain in the same cluster after recomputing cluster centers and reassigning the points to the newly computed centers. Points attached to each cluster center define one cluster. The numbers of points in different clusters can differ significantly if points in X are not distributed uniformly and some clusters may even become empty. Sometimes it is necessary to rerun k-means several times and use the result of the best run.

As was argued earlier, in decomposable multiobjective problems where the objectives compete in a number of problem partitions, using traditional selection and replacement mechanisms necessitates exponentially scaled populations to discover the entire Pareto-optimal front. The reason for this behavior is that the niches on the extremes of the Pareto-optimal front (maximizing most partitions with respect to one particular objective) can be expected to be exponentially smaller than the niches in the middle. To alleviate this problem, it is necessary to process different parts of the Pareto-optimal front separately and allocate a sufficiently large portion of the population to each part of the Pareto-optimal front. Of course, if the number of Pareto-optimal solutions grows exponentially, we could never find all those points with a polynomially sized population. However, it is still possible to find at least one representative solution for each combination of objective-function values that lies in the Pareto-optimal front, considering all solutions with the same values of both objectives equivalent.

It is important to note that other multiobjective evolutionary algorithms, such as NSGA-II and SPEA2, also include mechanisms that attempt to deal with a good coverage of a wide Pareto-optimal front. However, these mechanisms are insufficient for some multiobjective decomposable problems because they result in creating exponentially large niches in the middle of the Pareto-optimal front while eliminating extremes. As was argued earlier, this overwhelms the nichers and leads to poor scalability, which was shown both theoretically and empirically in Sect. 10.3.

Allocating comparable space to each part of the Pareto-optimal front can be ensured by using clustering in the objective space as suggested by Thierens and Bosman [37]. X thus consists of m-dimensional vectors where m is the number of objectives. To reduce the number of iterations until the creation of reasonable clusters, the cluster centers can be initialized by ordering points according to one objective and assigning the ith center to $(N/(2k)+i[N/k])$th point in this ordering. Since k-means clustering divides the points into clusters of approximately same variance, by forcing each cluster to produce an equal number of new candidate solutions, regular coverage of the Pareto-optimal front can be ensured even for difficult decomposable multiobjective problems.

10.4.3 Experiments

Experimental design is similar to that in Sect. 10.3.4. Solutions with the same values of both objectives are considered equivalent and thus the goal is to find at least one representative for each combination of objective-function values on the Pareto-optimal front.

To focus only on the effects of different recombination and replacement strategies, the number of clusters in k-means clustering was set to the number of unique solutions on the final Pareto-optimal front. If the number of clusters cannot be approximated in advance, it can be obtained automatically using for example the Bayesian information criterion (BIC) [34].

Bisection is used to determine a sufficient population size to find representatives of all Pareto-optimal solutions in 10 independent runs. Each run of mohBOA and UMDA is allowed to proceed for $5n$ generations, whereas each run of NSGA-II is allowed to run for $20n$ generations due to the less effective recombination. To alleviate the effects of noise, ten bisection runs are performed for each combination of the algorithm, problem and problem size, and the results are averaged.

Figure 10.14 shows the growth of the number of evaluations with problem size for onemax-zeromax. The results indicate that clustering in the objective space is necessary for a scalable solution for onemax-zeromax. Furthermore, the results show that here RTS based on nondominated crowding performs better than the elitist replacement of NSGA-II. Finally, the results indicate that UMDA with RTS solves onemax-zeromax in a low-order polynomial number of evaluations.

Figure 10.15 shows the results on trap$_5$-invtrap$_5$. The results show that, as expected, trap$_5$-invtrap$_5$ necessitates not only clustering in the objective space like onemax-zeromax but also effective identification and exploitation of interactions between different problem variables also called linkage learning. That is why standard crossover and UMDA fail to solve this problem

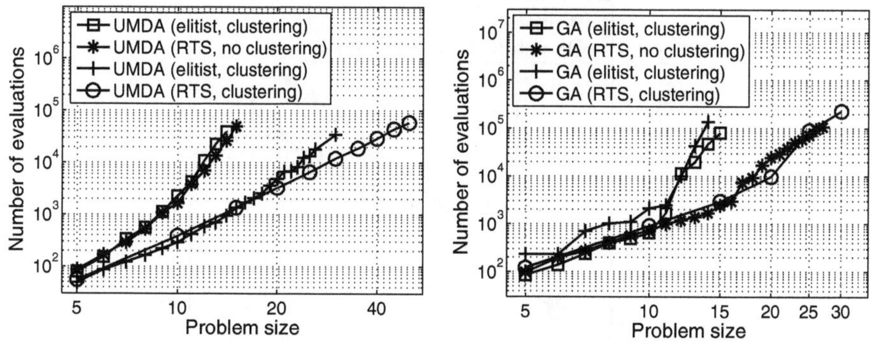

(a) UMDA on onemax-zeromax (b) GA on onemax-zeromax

Fig. 10.14. Results on onemax-zeromax indicate that k-means clustering in the objective space leads to a dramatic improvement in performance for both UMDA and GA (NSGA-II). Furthermore, they indicate that RTS performs better than the elitist replacement of NSGA-II and that multiobjective UMDA with RTS is capable of solving onemax-zeromax in low-order polynomial time

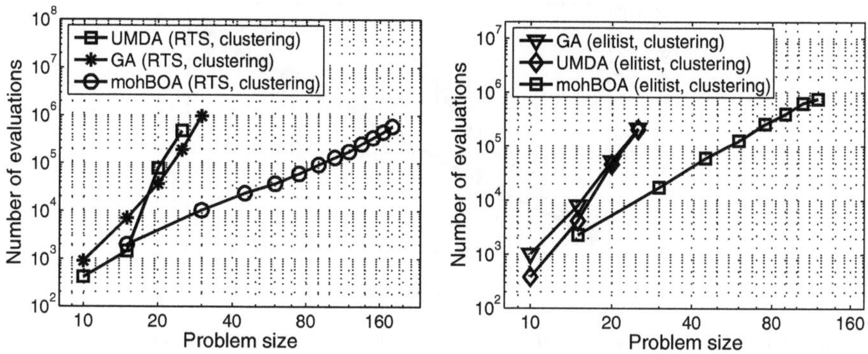

(a) RTS, $trap_5$-$invtrap_5$ (b) Elitist replacement, $trap_5$-$invtrap_5$

Fig. 10.15. Results on $trap_5$-$invtrap_5$ indicate that for some multiobjective decomposable problems it is necessary to also identify and exploit interactions between interacting string positions or decision variables

efficiently and become intractable already for relatively small problems. The algorithm mohBOA with RTS and clustering in the objective space provides best performance and scales up polynomially with problem size. Again, RTS outperforms the elitism (see Fig. 10.16).

Figure 10.17 shows that the performance of UMDA on onemax-zeromax does not change much if the distance metric in RTS is based on the objectives (as opposed to measuring the Hamming distance between binary strings). The figure also confirms that using RTS in the objective space is still not capable of ensuring scalable performance if clustering in the objective space is not

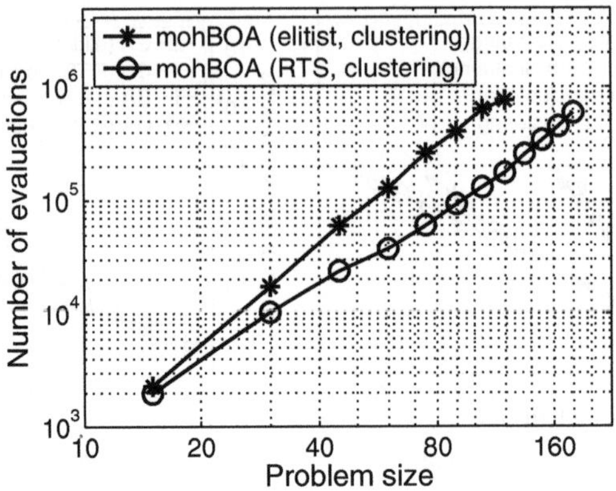

Fig. 10.16. Results on $trap_5$-$invtrap_5$ also indicate that in mohBOA, RTS outperforms the elitist replacement of NSGA-II

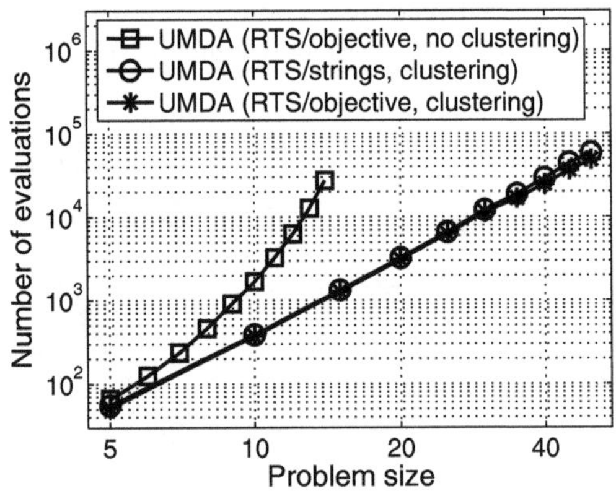

Fig. 10.17. The results of multiobjective UMDA with RTS using a distance metric in the objective space on onemax-zeromax indicate that clustering in the objective space cannot be replaced with this variant of RTS and that the choice of metric in RTS does not significantly affect performance

used, indicating that it is insufficient to incorporate niching via replacement based on the distribution of solutions no matter whether the niching method is based on the candidate solutions themselves or their objective values.

10.5 Overview of Other Multiobjective EDAs

So far, this chapter focused on solving multiobjective decomposable problems and the scalability of moEDAs on this class of problems. Nonetheless, a number of multiobjective EDAs were proposed in the past and this section attempts to provide their overview.

10.5.1 Multiobjective Mixture-Based IDEAs

Thierens and Bosman [37] proposed several variants of the multiobjective mixture-based iterated density estimation algorithm (mMIDEA), which combines EDAs, truncation selection based on dominance, and clustering.

In all variants of mMIDEA, for each candidate solution in the population, selection starts by determining the number of other solutions in the population that dominate this candidate solution. Then, truncation selection is used that selects the top $\lfloor \tau N \rfloor$ candidate solutions (solutions dominated by fewer other solutions are given preference), where the truncation threshold $\tau < 1$.

Clustering is then used to split the population into multiple subpopulations. A separate probabilistic model is then built for each subpopulation and these models are sampled to generate new candidate solutions. The number of candidate solutions generated by each model is equal to the number of candidate solutions in the corresponding subpopulation. Consequently, the clustering mechanism of mMIDEA does not provide a specific mechanism to ensure equal coverage of the Pareto-optimal front if the number of representatives in some parts of the front is much larger than the number of representatives in some other parts.

The proposed mMIDEAs considered several types of probabilistic models for both discrete and continuous problems. For discrete variables, a mixture of univariate distributions and a mixture of tree distributions were used. For continuous variables, a mixture of univariate Gaussian models and a mixture of multivariate Gaussian factorizations were used.

mMIDEAs were tested on the multiobjective 0/1 knapsack and several continuous multiobjective functions. Experimental results showed good performance of the proposed methods, including those with simple univariate models.

10.5.2 Multiobjective mixed BOA

Laumanns [18] incorporated the selection and replacement operators of SPEA2 [40] into mixed BOA (mBOA) [23], which extends BOA with decision graphs to solve problems with both discrete as well as continuous variables. The proposed method will be referred to as the multiobjective mBOA (mmBOA).

The algorithm mmBOA maintains a population of candidate solutions and an archive, which is yet another population. Initially, the population is generated randomly and the archive is empty. Each iteration starts by filling the archive with nondominated solutions from the current population and the current archive. The size of the archive is always equal to the population size. If the archive gets too big, it is truncated to have the same size as the population. If the archive is smaller than the population, mmBOA adds dominated individuals from the archive and the population. Then, the iteration proceeds by applying binary tournament selection to the archive population, selecting a population of parents. A Bayesian network with decision graphs is then built for the selected population and sampled to generate the offspring population. The population is then replaced by the offspring.

The algorithm mmBOA was tested on the multiobjective knapsack where it was shown to dominate NSGA-II, SPEA, and SPEA2 in most instances [18].

10.5.3 Multiobjective hBOA

Khan [16, 17] proposed multiobjective BOA (mBOA) and multiobjective hBOA (mhBOA) by combining BOA and hBOA with the selection and replacement operators of NSGA-II [9]. Tests on challenging decomposable multiobjective problems indicated that without identifying and exploiting interactions between different string positions, some decomposable problems become intractable using standard variation operators (crossover and mutation). On the other hand, mBOA and mhBOA could solve decomposable and hierarchical problems relatively efficiently.

The algorithm mohBOA presented in this chapter is an extension of Khan's mhBOA, which incorporates the clustering mechanism into mhBOA.

10.5.4 Multiobjective Real-Coded BOA

Ahn [2] combined the real-coded BOA (rBOA) [3], the selection procedure of NSGA-II, and adaptive sharing and crowding, creating the multiobjective rBOA (MrBOA).

MrBOA modifies the selection procedure of the single-objective rBOA. The selection procedure uses the ranking scheme of NSGA-II as the primary selection criterion. Additionally, the selection procedure applies adaptive sharing to discriminate solutions with the best rank. For the remaining solutions, the crowding mechanism of NSGA-II is used to provide a measure for comparing individuals with the same rank. Truncation selection is then used to select the best solutions based on the ranking, adaptive sharing, and crowding. New candidate solutions are created by building a factorized Gaussian mixture model of the selected solutions and sampling the model.

In comparison with the continuous variants of m$\mathbb{MIDEA}$ and NSGA-II, MrBOA was shown to provide competitive or better results on a number of newly designed and standard multiobjective test problems [2].

10.6 Summary and Conclusions

This chapter discussed multiobjective decomposable problems and their difficulty. The chapter argued that if there are many competing subproblems, standard approaches to multiobjective optimization with evolutionary algorithms fail to scale up polynomially.

The chapter then presented the multiobjective hierarchical BOA (mohBOA), which is shown to scalably solve multiobjective decomposable problems with a large number of competing subproblems. The algorithm mohBOA is capable of effective recombination by building and sampling Bayesian networks with decision trees, and it thus significantly outperforms multiobjective evolutionary algorithms with standard variation operators on problems that necessitate effective linkage learning. Furthermore, mohBOA is shown to scalably solve even multiobjective decomposable problems with many competing objectives by employing clustering.

Finally, the chapter reviewed other multiobjective EDAs to provide a starting point for pursuing other important topics and approaches in the design of multiobjective EDAs.

EDAs can be combined with multiobjective genetic and evolutionary algorithms in a straightforward manner. Since EDAs can solve many problems intractable with standard variation operators of genetic and evolutionary algorithms, multiobjective extensions of EDAs should solve many difficult problems intractable with state-of-the-art multiobjective genetic and evolutionary algorithms. Real-world applications are expected to soon confirm this hypothesis as was the case in the research on single-objective EDAs; nonetheless, extensive testing on the boundary of the design envelope clearly indicates that multiobjective hBOA and other multiobjective EDAs provide a powerful class of multiobjective optimization algorithms and can be expected to provide tractable solutions for many previously intractable multiobjective problems.

Acknowledgments

This work was sponsored by the National Science Foundation under NSF CAREER grant ECS-0547013, the Air Force Office of Scientific Research, Air Force Materiel Command, USAF, under grant F49620-03-1-0129, the National Science Foundation under ITR grant DMR-99-76550 (at Materials Computation Center), and ITR grant DMR-0121695 (at CPSD), and the Dept. of Energy under grant DEFG02-91ER45439 (at Fredrick Seitz MRL). The U.S. Government is authorized to reproduce and distribute reprints for government purposes notwithstanding any copyright notation thereon.

This work was also partially supported by the Research Award and the Research Board at the University of Missouri. Most experiments were done using the hBOA software developed by Martin Pelikan and David E. Goldberg at the University of Illinois at Urbana-Champaign.

References

[1] Ackley, D. H. (1987). An empirical study of bit vector function optimization. *Genetic Algorithms and Simulated Annealing*, pages 170–204

[2] Ahn, C.-W. (2005). *Theory, Design, and Application of Efficient Genetic and Evolutionary Algorithms.* PhD thesis, Gwangju Institute of Science and Technology, Gwangju, Republic of Korea

[3] Ahn, C. W., Ramakrishna, R. S., and Goldberg, G. (2004). Real-coded Bayesian optimization algorithm: Bringing the strength of BOA into the continuous world. *Proceedings of the Genetic and Evolutionary Computation Conference (GECCO-2004)*, pages 840–851

[4] Bosman, P. A. N. and Thierens, D. (1999). Linkage information processing in distribution estimation algorithms. *Proceedings of the Genetic and Evolutionary Computation Conference (GECCO-99)*, I:60–67

[5] Chen, J.-H. (2004). *Theory and applications of efficient multi-objective evolutionary algorithms.* PhD thesis, Feng Chia University, Taichung, Taiwan

[6] Coello Coello, C. A., Veldhuizen, D. A. V., and Lamont, G. B. (2001). *Evolutionary algorithms for solving multi-objective problems.* Kluwer

[7] Deb, K. (2001). *Multi-objective optimization using evolutionary algorithms.* Wiley, Chichester, UK

[8] Deb, K. and Goldberg, D. E. (1991). Analyzing deception in trap functions. IlliGAL Report No. 91009, University of Illinois at Urbana-Champaign, Illinois Genetic Algorithms Laboratory, Urbana, IL

[9] Deb, K., Pratap, A., Agarwal, S., and Meyarivan, T. (2002). A fast and elitist multiobjective genetic algorithm: NSGA-II. *IEEE Transactions on Evolutionary Computation*, 6(2)

[10] Goldberg, D. E. (2002). *The design of innovation: Lessons from and for competent genetic algorithms*, volume 7 of *Genetic Algorithms and Evolutionary Computation.* Kluwer

[11] Goldberg, D. E., Deb, K., and Thierens, D. (1993). Toward a better understanding of mixing in genetic algorithms. *Journal of the Society of Instrument and Control Engineers*, 32(1):10–16

[12] Harik, G. (1999). Linkage learning via probabilistic modeling in the ECGA. IlliGAL Report No. 99010, University of Illinois at Urbana-Champaign, Illinois Genetic Algorithms Laboratory, Urbana, IL

[13] Harik, G., Cantú-Paz, E., Goldberg, D. E., and Miller, B. L. (1999). The gambler's ruin problem, genetic algorithms, and the sizing of populations. *Evolutionary Computation*, 7(3):231–253

[15] Harik, G. R. (1995). Finding multimodal solutions using restricted tournament selection. *Proceedings of the International Conference on Genetic Algorithms (ICGA-95)*, pages 24–31

[15] Harik, G. R. and Goldberg, D. E. (1996). Learning linkage. *Foundations of Genetic Algorithms*, 4:247–262

[16] Khan, N. (2003). Bayesian optimization algorithms for multiobjective and hierarchically difficult problems. Master's thesis, University of Illinois at Urbana-Champaign, Urbana, IL

[17] Khan, N., Goldberg, D. E., and Pelikan, M. (2002). Multi-objective Bayesian optimization algorithm. IlliGAL Report No. 2002009, University of Illinois at Urbana-Champaign, Illinois Genetic Algorithms Laboratory, Urbana, IL

[18] Laumanns, M. and Ocenasek, J. (2002). Bayesian optimization algorithms for multi-objective optimization. *Parallel Problem Solving from Nature*, pages 298–307

[19] MacQueen, J. B. (1967). Some Methods for Classification and Analysis of Multivariate Observations. In *Proceedings of the 5th Symposium on Mathematics, Statistics and Probability*, pages 281–297, Berkeley. University of California Press

[20] Mahfoud, S. W. (1994). Population size and genetic drift in fitness sharing. *Foundations of Genetic Algorithms*, 3:185–224. (Also IlliGAL Report No. 94005)

[25] Mühlenbein, H. and Paaß, G. (1996). From recombination of genes to the estimation of distributions I. Binary parameters. *Parallel Problem Solving from Nature*, pages 178–187

[26] Mühlenbein, H. and Schlierkamp-Voosen, D. (1993). Predictive models for the breeder genetic algorithm: I. Continuous parameter optimization. *Evolutionary Computation*, 1(1):25–49

[23] Ocenasek, J. and Schwarz, J. (2002). Estimation of distribution algorithm for mixed continuous-discrete optimization problems. In *2nd Euro-International Symposium on Computational Intelligence*, pages 227–232

[29] Pelikan, M. (2005). *Hierarchical Bayesian optimization algorithm: Toward a new generation of evolutionary algorithms*. Springer, Berlin Heidelberg New York

[30] Pelikan, M. and Goldberg, D. E. (2001). Escaping hierarchical traps with competent genetic algorithms. *Proceedings of the Genetic and Evolutionary Computation Conference (GECCO-2001)*, pages 511–518

[26] Pelikan, M. and Goldberg, D. E. (2003). A hierarchy machine: Learning to optimize from nature and humans. *Complexity*, 8(5):36–45

[27] Pelikan, M., Goldberg, D. E., and Cantú-Paz, E. (1999). BOA: The Bayesian optimization algorithm. *Proceedings of the Genetic and Evolutionary Computation Conference (GECCO-99)*, I:525–532

[28] Pelikan, M. and Mühlenbein, H. (1999). The bivariate marginal distribution algorithm. *Advances in Soft Computing—Engineering Design and Manufacturing*, pages 521–535

[29] Pelikan, M., Sastry, K., and Goldberg, D. E. (2002). Scalability of the Bayesian optimization algorithm. *International Journal of Approximate Reasoning*, 31(3):221–258

[30] Pelikan, M., Sastry, K., and Goldberg, D. E. (2005). Multiobjective hBOA, clustering, and scalability. *Proceedings of the Genetic and Evolutionary Computation Conference (GECCO-2005)*, pages 663–670

[31] Reed, P. (2002). *Striking the balance: Long-term groundwater monitoring design for multiple conflicting objectives*. PhD thesis, University of Illinois, Urbana, IL

[32] Sastry, K. and Goldberg, D. E. (2004). Designing competent mutation operators via probabilistic model building of neighborhoods. *Proceedings*

of the Genetic and Evolutionary Computation Conference (GECCO-2004), pages 114–125

[33] Sastry, K., Pelikan, M., and Goldberg, D. E. (2005). Limits of scalability of multiobjective estimation of distribution algorithms. *Proceedings of the Congress on Evolutionary Computation*, pages 217–2224

[34] Schwarz, G. (1978). Estimating the dimension of a model. *The Annals of Statistics*, 6:461–464

[35] Thierens, D. (1995). *Analysis and design of genetic algorithms.* PhD thesis, Katholieke Universiteit Leuven, Leuven, Belgium

[36] Thierens, D. (1999). Scalability problems of simple genetic algorithms. Evolutionary Computation, 7(4):331–352

[37] Thierens, D. and Bosman, P. A. N. (2001). Multi-objective mixture-based iterated density estimation evolutionary algorithms. *Morgan Kaufmann*, pages 663–670

[38] Thierens, D. and Goldberg, D. (1994). Convergence models of genetic algorithm selection schemes. *Parallel Problem Solving from Nature*, pages 116–121

[39] Thierens, D. and Goldberg, D. E. (1993). Mixing in genetic algorithms. *Proceedings of the International Conference on Genetic Algorithms (ICGA-93)*, pages 38–45

[40] Zitzler, E., Laumanns, M., and Thiele, L. (2001). SPEA2: Improving the strength Pareto evolutionary algorithm. Technical Report 103, Swiss Federal Institute of Technology (ETH) Zürich

11

Effective and Reliable Online Classification Combining XCS with EDA Mechanisms

Martin Butz, Martin Pelikan, Xavier Llorà, and David E. Goldberg

Summary. Learning Classifier Systems (LCSs), such as XCS and other accuracy-based classifier systems, evolve a distributed problem solution online. During the learning process, rule quality is assessed iteratively using techniques based on gradient-descent, while the rule structure is evolved using selection and variation operators of evolutionary algorithms. While using standard variation operators suffices for solving some problems, it does not assure an effective evolutionary search in many difficult problems that contain strong interactions between features. Specifically, it was shown that standard crossover operators can frequently disrupt important combinations of features, which often results in poor performance. This chapter describes how advanced EDAs can be integrated into XCS in order to ensure effective exploration even for problems in which features strongly interact and standard variation operators lead to poor XCS performance. In particular, the chapter incorporates the model building and sampling techniques from BOA and ECGA into XCS. The chapter shows that the two proposed algorithms ensure that the solution is found efficiently and reliably. The results presented in this chapter thus suggest that the research on combining standard LCSs with advanced EDAs holds a big promise and represents an important area for future research on LCSs and EDAs.

Key words: Learning classifier systems, Hierarchical problem solving, Evolutionary computation, Bayesian networks, Classification, Reinforcement learning, Gradient-descent, Scalability, Decomposable problems

11.1 Introduction

The accuracy-based classifier system XCS may be considered the most advanced learning classifier system (LCS) to-date. The system learns distributed problem solutions both in classification (or concept learning) as well as in reinforcement learning. XCS is designed to learn online a maximally accurate and maximally general problem solution represented by a set of rules, called the population of classifiers. The population of classifiers is evolved using an accuracy-based fitness measure, and selection and variation (crossover and mutation) operators of genetic algorithms (GAs). XCS has been successfully

M. Butz et al.: *Effective and Reliable Online Classification Combining XCS with EDA Mechanisms*, Studies in Computational Intelligence (SCI) **33**, 249–274 (2006)
www.springerlink.com

applied to a wide variety of real-world classification problems [1–3], yielding results comparable to the most common machine learning algorithms.

Empirically, it was shown that crossover can improve XCS's performance, particularly in problems with strong fitness guidance towards the solution and many irrelevant attributes [12, 13]. However, from the research on GAs, we know that crossover can be advantageous, when recombining attributes effectively, but also disruptive, when destroying important subsets of correlated attributes or substructures [21, 38]. When the mixing effects overshadow the disruptive effects, GAs ensure reliable convergence to the optimum in quadratic or subquadratic number of fitness evaluations [21]. On the other hand, when the disruptive effects overshadow the mixing effects, GAs require populations of size that grows exponentially with the number of attributes [38]. It is also known that mutation alone can yield inefficient search when attributes interact strongly [30]. Specifically, mutation may require time proportional to $n^k \log n$ where n is problem size and k is the order of interactions [30].

As in GAs, also XCS needs to process interdependent subsets of features – often referred to as building blocks (BBs) – effectively [6]. In problems with strong interactions between attributes, XCS with standard or no recombination operators active often shows poor learning performance [6].

This chapter discusses the importance of processing interdependent subsets of features effectively in XCS. Additionally, the chapter describes two methods that combine XCS and EDAs, which are shown to be capable of automatic identification and processing of interdependent subsets of features in XCS. Specifically, crossover in XCS is replaced by probabilistic model building and sampling using either marginal product models (MPMs), also used in the extended compact GA (ECGA) [22], or Bayesian networks (BNs), also used in the Bayesian optimization algorithm (BOA) [35].

The result consists of two LCSs that yield efficient and reliable performance on problems that cannot be efficiently solved with XCS with standard variation operators. In fact, the results show that the two discussed methods achieve performance similar to that with an informed crossover operator, which exploits provided feature-subset information for optimal recombination. Thus, we create the first competent LCSs, XCS/ECGA and XCS/BOA, that detect dependency structures online and propagate lower-level dependency structures effectively without any information about these structures given in advance.

The chapter starts with an introduction to the XCS classifier system, including an overview of the most important theoretical achievements as well as a discussion of the problem of crossover disruption in XCS. Section 11.3 introduces several problems for which the disruptive effects of standard crossover operators lead to inefficient search in standard XCS. Section 11.4 provides an empirical study of the performance of XCS with and without the model building and sampling techniques of BOA on a variety of Boolean function problems. The chapter concludes with an outlook on the future impact of this line of research.

11.2 The XCS Classifier System

The creation of the accuracy-based classifier system XCS [40] can be considered a milestone in classifier system research. The XCS system solves the most severe challenges in LCSs. By deriving fitness values from an accuracy estimate of reward predictions, instead of from reward predictions themselves, the problem of strong overgenerals has overcome [27, 28]. Generalization is achieved by enforcing a niche-based reproduction combined with population-wide deletion. Thus, XCS is designed not only to evolve solutions that yield maximum feedback (or reward), but also to evolve a complete and accurate *payoff map* of all possible solutions for all possible problem instances.

This section introduces the XCS classifier system. For the interested reader, further details on the system including a precise algorithmic description can be found elsewhere [5, 14]. After the introduction, we proceed with a theoretic overview and crossover considerations.

11.2.1 System Introduction

The accuracy-based classifier system XCS is a learning classifier system (LCS) that learns distributed problem solutions online. A solution is represented by a population of classifiers that is evolved by a combination of two general learning principles: (1) a reinforcement learning-based gradient mechanism optimizes predictions and updates fitness estimates and (2) an evolutionary-based mechanism evolves classifier condition structures.

Problem Types

XCS may be applied to two basic problem types, classification problems and reinforcement learning problems. In this paper we focus on classification problems.

A classification problem is defined by a set of problem instances $s \in \mathcal{S}$. Each problem instance belongs to one class $a \in \mathcal{A}$ (traditionally termed an action in LCSs). In machine learning terms, s may be termed a feature vector and a a concept class. The mapping from $\mathcal{S}$ to $\mathcal{A}$ is represented by a *target concept* belonging to a set of concepts (i.e., the *concept space*). The goal is to learn the target concept. Most desirable properties of such a learning system are that the learner learns a maximally *accurate* problem solution, measured usually by the percentage of correct problem instance classifications, and a maximally *general* problem solution, which can be characterized as a solution that generalizes well to other (unseen) problem instances.

Since this chapter focuses on binary classification problems, we constrain our problem space to $\mathcal{S} = \{0, 1\}^l$ and use two classes $\mathcal{A} = \{0, 1\}$. Such binary classification problems with two classes are often referred to as *Boolean function problems.*

Knowledge Representation

The population of classifiers represents a problem solution. Each classifier can be regarded as an expert, endowed with a confidence measure, in the problem subspace its condition is satisfied.

Each classifier consists of five main components and several additional estimates: (1) *Condition part* C specifies when the classifier is applicable. (2) *Action part* $A \in \mathcal{A}$ specifies the proposed action (or classification). (3) *Reward Prediction* $R \in \Re$ estimates the average reward received when executing action A given condition C is satisfied. (4) *Reward prediction error* ε estimates the mean absolute deviation of R with respect to the actual reward. (5) *Fitness* F estimates the scaled, relative accuracy (scaled, inverse error) with respect to other, overlapping classifiers.

In the binary case, condition part C is coded by $C \in \{0, 1, \#\}^l$ where l is the number of attributes and # is a don't-care symbol that matches both 0 and 1; for example, a condition #10## matches an instance 01011 or any other 5-bit instance that contains a substring 10 starting in the second position. Each condition thus identifies a hyperrectangle in which the classifier is applicable, or *matches.*

Each classifier maintains several additional parameters. The *action set size estimate as* estimates the moving average of the action sets it is applied in. The *time stamp ts* specifies the time when the classifier participated in its last GA competition. The *experience* counter *exp* counts the number of parameter updates the classifier underwent so far. The numerosity *num* specifies the number of identical micro-classifiers, this (macro-)classifier actually represents.

Learning Interaction

XCS usually starts with an empty population. Initial classifiers are generated by a covering mechanism described below. Later, the genetic discovery component is in charge of generating new classifiers.

Given current problem instance $s \in \mathcal{S}$ at iteration time t, the *match set* $[M]$ is formed, containing all classifiers in $[P]$ whose conditions match s. If some action is not represented in $[M]$, a covering mechanism is applied. In this case, new classifiers are generated whose condition parts match the current input (each attribute in the condition part is set to a #-symbol with probability $P_\#$) and whose action parts specify the unrepresented actions.

Given a match set $[M]$, XCS estimates the payoff for each possible action by forming a *prediction array* $P(A)$, which reflects the fitness-weighted reward predictions for each possible action:

$$P(A) = \frac{\sum_{\text{cl}.A=A \wedge \text{cl} \in [M]} \text{cl}.R \cdot \text{cl}.F}{\sum_{\text{cl}.A=A \wedge \text{cl} \in [M]} \text{cl}.F}. \tag{11.1}$$

We use the dot notation to refer to classifier parameters in XCS. The prediction array is used to determine the appropriate classification. Several action selection policies (i.e., behavioral policies) may be applied. Usually, XCS chooses actions randomly during learning, and it chooses the best action $A_{\max} = \texttt{argmax}_A P(A)$ during testing. All classifiers in $[M]$ that specify the chosen action A comprise the action set $[A]$.

After the execution of the chosen action, feedback is received in the form of scalar reward $R \in \Re$, which is used to update classifier parameters. Next, the GA may be applied. Finally, the successive problem instance is received and the iteration time t is increased by one. The overall learning process is illustrated in Fig. 11.1.

Rule Evaluation

XCS iteratively updates its population of classifiers with respect to the successive problem instances received from the environment using gradient-decent techniques.

In a classification problem, parameters of each classifier in the current action set $[A]$ are updated with respect to the immediate feedback r. Reward prediction error ε of each classifier in $[A]$ is updated by

$$\varepsilon \leftarrow \varepsilon + \beta(|r - R| - \varepsilon), \tag{11.2}$$

where parameter $\beta \in [0, 1]$ denotes the learning rate influencing accuracy and adaptivity of the moving average reward prediction error. Next, reward prediction R of each classifier in $[A]$ is updated by

$$R \leftarrow R + \beta(r - R), \tag{11.3}$$

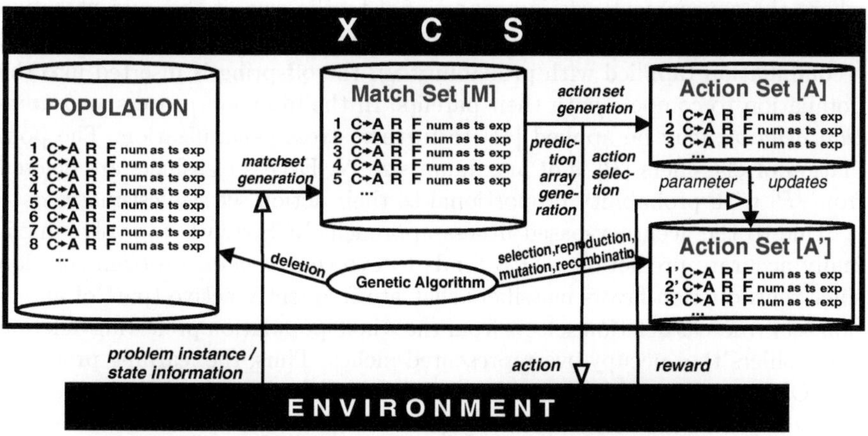

Fig. 11.1. XCS learning process

with the same notation as in the update of ε. The fitness value of each classifier in $[A]$ is updated with respect to its current scaled relative accuracy κ', which is derived from the current reward prediction error ε as follows:

$$\kappa = \begin{cases} 1 & \text{if } \varepsilon < \varepsilon_0, \\ \alpha \left(\frac{\varepsilon}{\varepsilon_0}\right)^{-\nu} & \text{otherwise,} \end{cases} \tag{11.4}$$

$$\kappa' = \frac{\kappa \cdot num}{\sum_{\text{cl} \in [A]} \text{cl}.\kappa \cdot \text{cl}.num}. \tag{11.5}$$

Parameter κ measures the current absolute accuracy of a classifier, which is scaled by a power function. The relative accuracy κ' reflects the relative accuracy with respect to the other classifiers in the current action set. The fitness estimate F is then updated with respect to κ' as follows:

$$F \leftarrow F + \beta(\kappa' - F). \tag{11.6}$$

Fitness loosely reflects the moving average, set-relative accuracy of a classifier. The action set size estimate as is updated similar to the reward prediction R but with respect to the current action set size $|[A]|$.

Each time the parameters of a classifier are updated, experience counter 'exp' is incremented. If genetic reproduction is applied to classifiers of the current action set, all time stamps ts are set to the current iteration t.

Rule Evolution

XCS applies a GA for rule evolution. Genetic reproduction is invoked in the current action set $[A]$ if the average time since the last GA application (stored in parameter ts) upon the classifiers in $[A]$ exceeds threshold θ_{GA}.

The GA selects two parental classifiers using tournament selection [12] where the tournament size is proportional to the size of the current action set $[A]$. After mutation (changing an attribute in C with probability μ) and recombination (applied with probability χ), the offspring is inserted into the population to compete with their parents. In the insertion process, *subsumption deletion* may be applied [41] to further stress generalization. The population of classifiers $[P]$ is of maximum size N. Excess classifiers are deleted from $[P]$ with probability proportional to their action set size estimates as.

Due to the accuracy-based fitness approach, XCS strives to evolve maximally accurate predictions. XCS tends to evolve a *general* problem solution since reproduction favors classifiers that are frequently active (part of an action set) whereas deletion selects from the whole population preferring deletion of classifiers that occupy overrepresented niches. Thus, the learning processes in XCS are designed to achieve one common goal: to evolve a *complete, maximally accurate, and maximally general* representation of the underlying payoff-map. This representation was previously termed the optimal solution representation $[O]$ [25, 26].

11.2.2 A Note on XCS Learning Complexity

Computational complexity theory on XCS has shown that XCS is able to PAC-learn k-DNF problems [3, 7]. To ensure structural growth, it was shown that XCS's population size needs to grow proportionally to l^k if there is no fitness guidance to the optimal solution [9], where l is the number of attributes in the problem and k is the maximum size of a clause in a k-DNF problem. The analysis is purely based on selection pressure and mutation influences. Crossover was not considered in the analysis. Neither potential *disruptive* effects of crossover were considered nor any types of *innovative* effects. These effects are investigated and analyzed in Sect. 11.3.

11.3 Structural Processing in XCS

While many studies of XCS have shown that the system is able to reliably learn challenging Boolean function problems [9, 41] as well as real-world data classification problems [1, 2], few results have focused on the crossover operator. Traditionally, simple crossover operators, such as one-point and uniform crossover, were applied with a certain fixed probability.

This section studies crossover mechanisms in simple *count ones problem* instances [9]. The problem is comparable to a one-max problem in GAs in that the initial specialization of each relevant attribute yields equal accuracy (and thus fitness) increase; it differs, however, in its final solution since many overlapping subsolutions need to be maintained.

The count ones problem is defined for binary problems of length l, in which $k \leq l$ bits are relevant. A problem instance is in class 1 if more than half of its k relevant bits are one. Thus, the more ones (or zeros) a classifier condition specifies, the more accurate the classifier will be. Table 11.1 lists some classifiers with expected reward prediction R and error ε values.

11.3.1 Effective Recombination in the Count Ones Problem

To illustrate the equally strong fitness pressure on all relevant attributes in the count ones problem, we ran XCS on a small count ones problem instance with $k = 5$ and $l = 10$ (10/5). Starting with a completely general population

Table 11.1. Expected reward prediction R and reward prediction error ε on the count ones problem ($l = 5$, $k = 5$) for classifiers with action part $A = 1$

C	R	ε	C	R	ε	C	R	ε	C	R	ε
#####	500.0	500.0	####0	312.5	429.7	##0#0	125.0	218.8	111##	1000.0	0.0
1####	687.5	429.7	11###	875.0	218.8	00###	125.0	218.8	11##1	1000.0	0.0
##1##	687.5	429.7	##11#	875.0	218.8	#0#01	250.0	375.8	0#0#0	0.0	0.0
0####	312.5	429.7	#1##1	875.0	218.8	110##	750.0	375.0	0##00	0.0	0.0

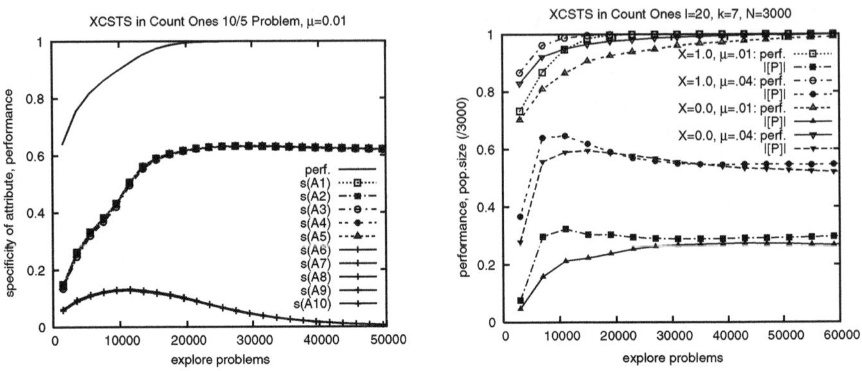

Fig. 11.2. The specificity curves on the 10/5 count ones problem (left-hand side) show that XCS exhibits strong specialization effects on the relevant condition attributes. The 20/7 count ones problem on the right-hand side shows how crossover helps to solve the problem effectively and reliably

($P_{\#} = 1.0$), Fig. 11.2 (left-hand side) shows how specificity, that is, the percentage of specialized attributes in classifier conditions, behaves. Soon, fitness pressure causes the reproduction of classifiers that specify more of the k relevant attributes. Specificity increases in all relevant attributes while it stays low on all irrelevant attributes.

Performance of XCS without and with crossover as well as different mutation rates in the 20/7 count ones problem are shown on the right-hand side of Fig. 11.2. While a larger mutation rate helps increasing specificity and thus performance initially, uniform crossover ensures effective recombination yielding maximally accurate problem solutions. When crossover is not applied, performance is more strongly dependent on mutation rate and a completely accurate performance is reached less reliably with the chosen population size ($N = 3{,}000$).

The results show that if the problem structure consists of BBs of order one, then uniform crossover yields effective recombination and the problem is solved more efficiently when crossover is applied. However, problems may consist of BBs of larger order as illustrated in the following section. On such problems, standard crossover and mutation operators yield poor performance.

11.3.2 Crossover Disruption in Hierarchical Problem Structures

We construct BB-hard classification problems by forming a two-level hierarchical problem structure. On the lower-level, small Boolean functions are evaluated which provide the input to the higher level. For example, we may use a parity-count ones combination in which the smaller lower-level blocks are evaluated by parity functions, whose results are used as input to the count-ones function, effectively counting the number of parity blocks that evaluate to one.

Table 11.2. Expected reward prediction R and reward prediction error ε for exemplar condition parts in the hierarchical 3-parity, 5-count ones problem with $l = 18$ (3 additional problem-irrelevant bits) and classifiers with action part $A = 1$

C	R	ε	C	R	ε
### ### ### ### ### ###	500.0	500.0	101 ### 111 ### ### ###	500.0	500.0
#1# ### #11 #1# #11 ###	500.0	500.0	### 000 ### ### 000 ###	125.0	218.8
111 ### ### ### ### ###	687.5	429.7	101 111 ### 100 ### ###	750.0	375.0
### ### 111 ### ### ###	687.5	429.7	100 010 111 ### ### ###	1000.0	0.0
### #1# ### 100 ##1 ###	687.5	429.7	100 ### 001 010 ### ###	1000.0	0.0
### 0## ### ### 000 ###	312.5	429.7	### 100 ### 111 010 ###	1000.0	0.0
### 111 ### 010 ### ###	875.0	218.8	### 000 ### 011 110 ###	0.0	0.0
##1 111 ##0 100 #0# ###	875.0	218.8	000 000 000 ### ### ###	0.0	0.0

For readability, the lower-level 3-parity blocks are tightly coded and separated by spaces. Parity blocks evaluate to one if there is an odd number of ones in the block

Note that we are not interested in creating a problem to force BB processing for its own sake. In fact, many indications in nature and engineering suggest that typical natural problems are hierarchical and nearly decomposable [20, 21, 37]. The introduced problems are intended to serve as the test-of-fire for machine learning systems that attempt to solve such hierarchical and decomposable problems.

How can XCS solve this problem? If a parity block is not completely specified, its probability of evaluating to one remains unchanged. Thus, to discover the structure of the higher level function, first, lower-order parity blocks need to be discovered and then recombined effectively. Table 11.2 shows exemplar conditions with expected reward prediction and error values for the hierarchical 3-parity, 5-count ones problem. The next section shows that XCS is able to solve this problem efficiently only if it recombines the parity blocks effectively without disrupting them.

Since entropy in the class distribution decreases only if a classifier condition specifies all parity-relevant attributes, the parity problem is a hard classification problem. Nonetheless, other lower-level functions possibly with different numbers of relevant attributes may be used in the introduced two-level problem design.

Performance of XCS on the hierarchical 3-parity, 5-count ones problem is shown in Fig. 11.3.[1] It can be seen that XCS is not able to solve the problem if uniform crossover is applied. One and two-point crossover are slightly less disruptive than uniform crossover but still yield poor performance. Since the attributes are randomly distributed over the input string, the potential

[1] All results in this chapter are averaged over ten experiments. Performance is assessed by test trials in which no learning takes place and the better classification is chosen. During learning, classifications are chosen at random. Parameters were set as follows: $N = 20{,}000$, $\beta = 0.2$, $\alpha = 1$, $\varepsilon_0 = 10$, $\nu = 5$, $\theta_{\mathrm{GA}} = 25$, $\chi = 1.0$, $\mu = 0.01$, $\theta_{\mathrm{del}} = 20$, $\delta = 0.1$, $\theta_{\mathrm{sub}} = 20$, and $P_{\#} = 0.6$

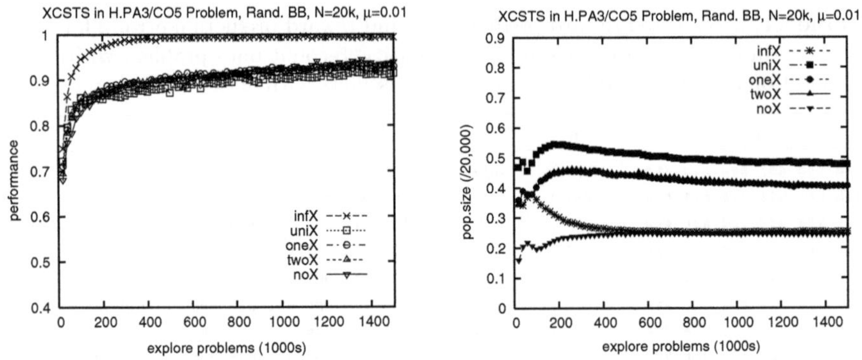

Fig. 11.3. Performance (a) and population size (b) of XCS ($N = 20k$) in the hierarchical 3-parity, 5-count ones problem. Efficient BB recombination strongly improves XCS's performance. One-point and two-point crossover are only beneficial if the BBs are tightly coded. Mutation alone gradually improves performance but is much less effective that BB-wise crossover

recombinatory benefits of one-point or two-point crossover are overshadowed by their disruptive effects, preventing efficient learning. Mutation alone slowly improves performance but does not reach significantly higher accuracy than the runs with crossover.

To investigate the potential of EDA operators in XCS, we also applied XCS with BB-wise uniform crossover to the same problem. BB-wise uniform crossover is given information about the important BBs in advance; then, it proceeds similarly as standard uniform crossover, but it exchanges entire BBs (parity blocks) instead of single bits. Consequently, BB-wise uniform crossover provides effective mixing without any disruption of BBs. XCS with BB-wise crossover solves the problem efficiently and reliably (see Fig. 11.3).

We can conclude that, to solve hierarchical problems efficiently, XCS necessitates a mechanism for identifying lower-level BBs. Once the BB identification is successful, effective BB processing and recombination can be applied.

11.3.3 Building Block Identification and Processing

To face the BB-challenge (linkage learning) in XCS, it is necessary to develop a mechanism that learns effective recombination online. Because of the success in the area of effective identification and processing of BBs using EDAs, advanced EDAs are a good candidate for providing such a mechanism.

However, the evolutionary component in XCS differs from the usual GA application in several respects. Due to XCS's niche reproduction in action sets and since action sets are generally rather small compared to the whole population, structure extraction is hard to apply successfully in an action set alone. On the other hand, extracting structural information from the whole classifier population makes it difficult to generate classifier offspring for the current problem niche.

Our investigations show that at least two structure identification mechanisms are suitable for competent BB processing in XCS: (1) *marginal product models* (MPMs), used in ECGA [22], and (2) *Bayesian networks with decision trees*, used in BOA [34, 35]. The former is easier to understand and to apply but is limited to the identification and processing of non-overlapping BBs only. The latter is more complicated but is able to model overlapping dependency structures as well.

In the following section, we show how to integrate either mechanism into the XCS framework. We show that both mechanisms are suitable to learn the *global* lower-level problem structure and can be used to generate or improve *local* classifier offspring. Since other chapters of this book give a concise introduction to ECGA and BOA, we focus primarily on discussing how the mechanisms of ECGA and BOA are integrated into XCS.

BB Identification with the ECGA Model Builder

ECGA [22] uses a marginal product model (MPM) to model and sample candidate solutions. MPM can encode non-overlapping subsets of problem features (BBs) that interact significantly. To measure model quality, ECGA uses a minimum description length (MDL) metric, which consists of the sum of the model complexity (MC) and the compressed population complexity (CPC). While MC biases the model building toward simpler models (with smaller BBs), CPC biases the model building toward more accurate models (with all dependencies covered).

The model building in ECGA starts with each attribute forming one block. Each iteration merges two blocks that yield the maximum gain in the sum of MC and CPC. The learning is terminated when no more model improvement can be achieved by merging two blocks. The MDL mechanisms used to learn MPMs in XCS were taken from the available ECGA implementation [29].

BB Identification with the BOA Model Builder

The Bayesian optimization algorithm (BOA) [34, 35] uses BNs [24, 32] to model and sample candidate solutions. The structure of a BN is defined by a directed acyclic graph where the nodes correspond to different attributes and the directed edges between the nodes encode conditional dependencies. BNs are a more general class of models than MPMs; while the MPM assumes structural independence between attribute subsets, BNs also allow the modeling of overlapping attribute subsets. In this work, we use BNs with decision trees [15, 18], where decision trees are used to store conditional probabilities for each attribute.

Also in BOA, a greedy algorithm is used to learn a BN that adds new edges iteratively based on the resulting improvement of model quality, starting with a network with no edges (dependencies). In this work, model quality is measured using a combination of the Bayesian Dirichlet metric with likelihood equivalence [16, 23] and the Bayesian information criterion [36]; this metric was previously used in BOA with decision graphs [33] and the hierarchical BOA (hBOA) [34], and is described in the chapter on the hierarchical BOA along with the learning and sampling algorithms.

11.3.4 Dependency Structures in XCS

Similarly to the BB-identification mechanisms in ECGA and BOA, which search for BBs in the selected population of individuals, it is possible to learn dependency structures from the current population in XCS. However, two aspects need to be considered: (1) Selection from the global population is not straightforward. (2) To use the available implementations of ECGA and BOA, classifiers need to be suitably transferred into binary representation.

To identify important BBs, the model needs to be built from the best individuals in the population of XCS. To select the best individuals, we use a filtering mechanism that extracts the most accurate classifiers out of the current population. The mechanism extracts those classifiers that have a minimum experience θ_{be} and a maximum error $\theta_{b\varepsilon}$. The parameters were set to $\theta_{be} = 20$ and $\theta_{b\varepsilon} = 400$ throughout the subsequent experiments, filtering out the young and high-error classifiers.

Given a filtered population, we need to transform the classifier set into a representation suitable to build the model. In order to use the available implementation of model building and sampling algorithms of BOA and ECGA, classifiers are encoded as binary string. Specifically, each condition attribute is encoded by two bits: The first bit encodes whether the condition attribute is general (i.e., don't care) or specific. The second bit encodes the value of the attribute. If the attribute is a don't care symbol, we choose zero or one uniformly randomly for the second bit. Finally, the classification (action) is coded by an additional bit. Table 11.3 shows a set of classifiers and the corresponding encoding that is used to learn the MPM or the Bayesian network with decision graphs. With a binary coded set of individuals at hand, we are able to learn the model in the form of either an MPM or a Bayesian network.

Table 11.3. Sample condition and action parts and their corresponding binary encoding for model learning

C	A	Binary encoding	C	A	Binary encoding
##11##	1	10 11 01 01 11 10 1	0#11##	0	00 11 01 01 11 11 0
##00##	1	11 11 00 00 10 11 1	001###	1	00 00 01 10 10 10 1
0#1###	0	00 11 01 11 11 10 0	10##0#	0	01 00 11 11 00 10 0
0#0###	0	00 10 00 11 10 10 0	000###	1	00 00 00 11 10 10 1
0#11##	1	00 11 01 01 11 10 1	01#1##	0	00 01 11 01 11 11 0

Spaces are added for clarity. If an attribute is a don't care symbol, the second bit in the corresponding binary code is chosen randomly

Since XCS applies a steady-state niche GA, the dependency structure does not need to be rebuilt every time step. We rebuild the network after a fixed number of time steps θ_{bs}, set to $N/2$ in our experiments. It seems appropriate to set θ_{bs} to half the population size since each iteration, two new classifiers are generated, so that after $N/2$ iterations, approximately N offspring classifiers will be generated.

11.3.5 Sampling From the Learned Dependency Structures

As shown above, recombination of parent classifiers using simple crossover operators may lead to disruptive effects, potentially destroying important BB structures. Once the dependency model is learned, XCS may use the model to recombine or directly generate offspring classifiers more effectively.

XCS generates offspring from parental classifiers selected from the current action set, which is selected from classifiers that match the current problem instance (match set). When using the globally learned probabilistic model to generate offspring, we consequently need to adjust the model to fit the local problem niche specified by the current problem instance. We investigate the following two options: (1) sample classifiers using the model with parameters updated to the local probability distribution of the current action set; and (2) update the selected classifier using the model with global or local probabilities using Markov–Chain Monte Carlo (MCMC) sampling [31].

Generating offspring by simply sampling from the global probabilistic model is not considered, because the probabilistic model created for the entire population of classifiers cannot be expected to accurately reflect the distribution in a specific niche. Similarly, updating parent classifiers using MCMC sampling with the probabilistic model with global probabilities is expected to be disruptive because of its bias toward the distribution of the entire population as opposed to the current problem niche. Both offspring generation methods are introduced next.

Sampling Using Local Probabilities

Reproducing classifiers in action sets yields offspring classifiers with average specificity that corresponds to the average specificity distribution in the action sets. Fitness may increase the average specificity due its pressure towards higher accuracy, which often leads to an implicit specialization pressure.

To sample offspring from the learned probabilistic model, the model parameters (marginal and conditional probabilities) should reflect the local specificity distribution in the current action set. To ensure this, we update the parameters of the model to reflect the best classifiers in the current action set, selecting classifiers from the set using the usual tournament selection mechanism. Consequently, the updated probabilistic model reflects the dependencies detected globally but also mimics the local probability distribution. The globally detected dependencies are thus combined with the local parameters, resulting in an offspring sampling mechanism that combines global and local problem knowledge. BBs that are important in the current problem niche will be recombined effectively as long as the global dependency structure or its parts apply in the local problem niche.

Structure Optimization

Another approach to sampling new classifiers is structure optimization, which starts by selecting a parent classifier from the action set using the usual tournament selection. The selected classifier is then updated using MCMC, which proceeds by perturbing the parent classifier and accepting each change with probability that depends on the likelihood of the instance before and after the change. Since sampling by perturbing an actual classifier from the local niche biases the sampling to this niche, in this scenario we can use both the global parameters as well as the local ones. However, without fitting the model parameters to the local niche, it can be expected that the more update steps we perform, the more likely we disrupt important BBs and overspecialize.

MCMC [31] was first introduced in the statistical physics literature in the 1950s in the so-called *Metropolis Algorithm*. As mentioned above, MCMC proceeds by making small changes to the current classifier and evaluating each potential change by computing the likelihoods before and after the change. Here each perturbation flips a random bit in the classifier. Each change is accepted with the probability equal to the ratio of the likelihood of the classifier before the change and the likelihood of the classifier after the change. The likelihood is determined directly from the probabilistic model by parsing the model for the particular classifier. To avoid zero likelihoods, all conditional are linearly normalized to values ranging from 0.05 to 0.95.

11.3.6 Experimental Evaluation

We evaluate XCS's performance on the aforementioned hierarchical problems and compare both offspring generation methods under several different settings. To learn the model structure, the population of XCS is filtered as described above. If the filtered population is empty, no model is learned. As long as no model is learned, XCS applies uniform crossover. Parameters are set as above if not stated differently.

The results in the hierarchical 3-parity, 5-count ones problem (Fig. 11.4) show that XCS/ECGA and XCS/BOA are able to successfully learn the problem. XCS/ECGA does not show any problems in solving the problem (Fig. 11.4a,b). All runs converge to the near-optimal solution nearly as fast as the informed BB-wise crossover runs. Even with a lower mutation rate (effectively decreasing supply and probability of randomly generating completely specified parity blocks), performance is hardly influenced.

Similarly, XCS/BOA successfully learns the problem (Fig. 11.4c,d). XCS/BOA is slightly slower than XCS/ECGA early in the run but then it reaches a

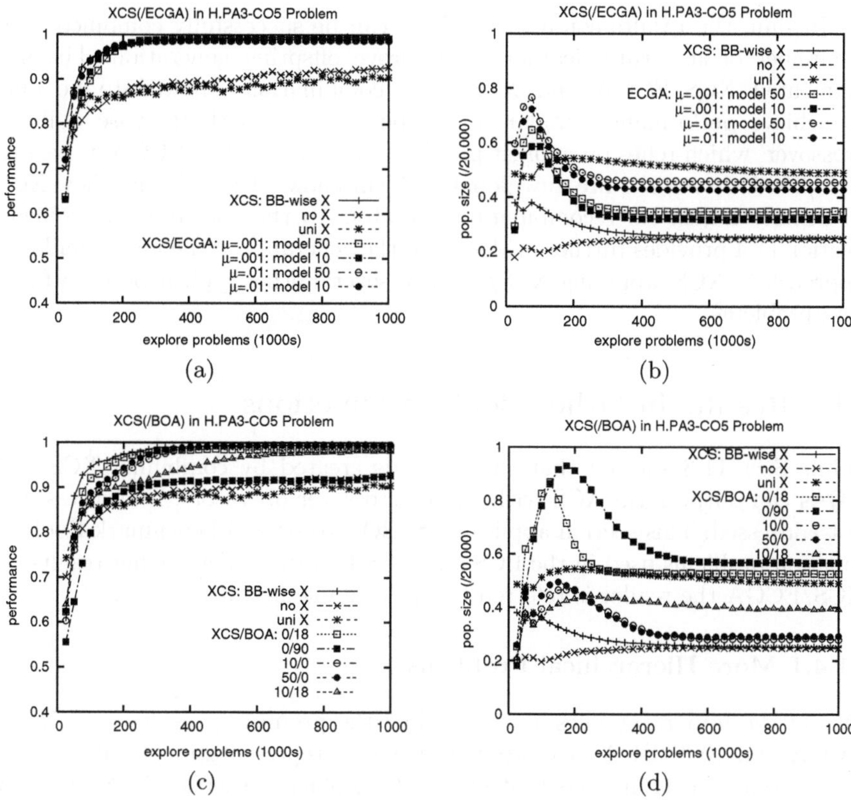

Fig. 11.4. The hierarchical 3-parity, 5-count ones problem can be efficiently solved with any of the tested model-based offspring generation methods. The ECGA combination appears slightly more robust in this case, indicating that the Bayesian network might include unnecessary, spurious dependencies that delay convergence. The A/B variations refer to the number A of selected classifiers used to set the probabilities to the local probability distribution and the number B of MCMC updates executed on the parent classifier (B=0 indicates that the model is sampled directly). The results confirm that if many MCMC updates are used with the global model with global parameters (0/90), learning becomes inefficient

slightly higher performance level. Apparently, BOA initially models spurious dependencies that may slow down the overall learning process. Since in this problem the propagation of all five BBs independently is nearly the most effective approach, the Bayesian learning algorithm appears to over-model and thus delay the learning early on. On the other hand, late in the run, BOA may allow a more effective prevention of BB disruption.

Performance of both methods clearly outperforms the runs without crossover as well as the runs with uniform crossover. Similar results were obtained on various parity-multiplexer problem instances including sizes 3–6 and 2–11 [5].

In sum, the results confirm that XCS can be successfully combined with a number of structural learners to improve offspring generation. The implemented XCS/ECGA and XCS/BOA combinations showed to be able to achieve performance similar to the performance with BB-wise uniform crossover, which relies on explicit problem knowledge. XCS/ECGA as well as XCS/BOA do not require any global problem knowledge and thus allow XCS to flexibly adjust its recombination operators to the encountered problem. Section 11.4 provides further evidence for the generality of the model-building approach in XCS, applying XCS/BOA to several other typical Boolean function problems.

11.4 Results in Other Boolean Functions

While Sect. 11.3 showed that the methods created by combining XCS and EDAs can outperform XCS with simple crossover operators, robustness was not addressed. This section applies XCS/BOA to several benchmark Boolean function problems used in the LCS and XCS literature. For further results on XCS/ECGA the reader is referred to the available literature [3, 5].

11.4.1 More Hierarchical Problems

One of the important questions regarding the promising results with XCS/ECGA and XCS/BOA is whether the binary representation by itself does not introduce a strong bias that makes the problem easier for XCS. In other words, the performance might have improved simply due to the conversion to the binary encoding and there was no need for advanced variation operators of ECGA and BOA.

Figure 11.5 shows the performance of several variants of XCS on the hierarchical 3-parity, 3-multiplexer problem ($N = 2k, \mu = 0.01, P_{\#} = 0.8$). XCS/BOA shows similar performance as in the previous test problems; for most settings, XCS/BOA finds the problem solution efficiently, but the algorithm overspecializes if too many MCMC steps with the global parameters are used. Even more importantly, the figure shows the performance of XCS with the same binary encoding as XCS/BOA but with a univariate probabilistic model that does not encode any interactions between attributes. The performance comparison shows that the coding by itself is insufficient for good XCS performance and that it is necessary to find an appropriate model of dependencies between the attributes. Additionally, due to the higher population size in the univariate probabilistic model, it can be seen that the coding induces specialization and diversification effects in the population.

Another important question is whether XCS/BOA is able to handle a large number of irrelevant attributes and focus its search on relevant attributes. To investigate this issue, we applied XCS/BOA to the hierarchical 3-parity, 3-multiplexer problem with a large number of irrelevant attributes ($l = 100$, that is, 201 bits in the binary coding). Figure 11.6 ($P_{\#} = 1.0$) shows that

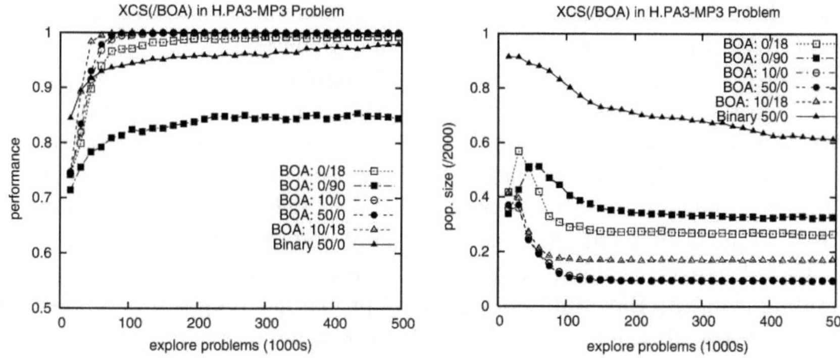

Fig. 11.5. Performance comparison of XCS with BNs and the univariate probabilistic model (with no interactions) confirms that (1) the binary coding alone is insufficient for efficiently solving the problem and (2) the binary coding induces an implicit specialization and diversification effect

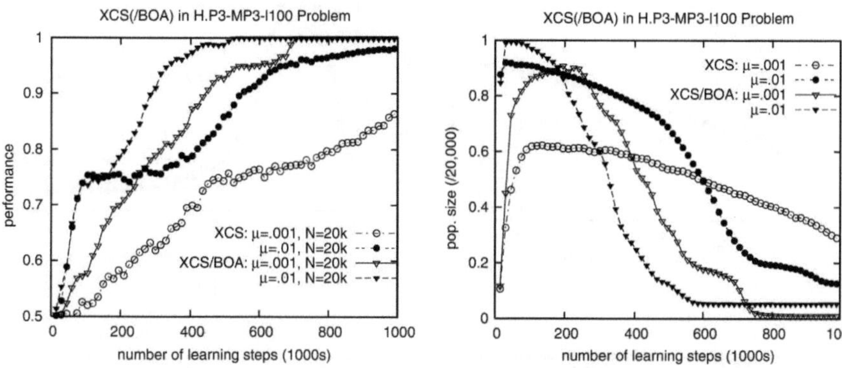

Fig. 11.6. XCS/BOA yields reliable performance on the 3-parity, 3-multiplexer problem with 91 additional irrelevant bits ($l = 100$)

XCS/BOA (setting with model sampling using 50 locally selected classifiers to update model probabilities) actually outperforms XCS with uniform crossover in the problem. While mutation also has a significant impact on the performance, both XCS/BOA runs reach 100% performance while neither simple XCS run does. The population sizes point out that there is more specialization and diversification in the XCS/BOA runs. However, the overall performance confirms that these effects are not random but targeted towards the more promising search subspaces.

11.4.2 The Multiplexer Problem

The multiplexer problem has been investigated throughout the LCS literature [11, 17, 39–41]. The problem is defined for instances encoded by binary strings of length $l = k + 2^k$ where k address bits encode the location of the correct classification located in one of the remaining 2^k bits.

Figure 11.7 shows the performance and population size curves for the 20- and 37-multiplexer problems ($P_{\#} = 0.8$). XCS/BOA exhibits slightly slower learning performance, which is most likely caused by the lower mutation rate ($\mu = 0.001$). In the 20-multiplexer case, we see yet another confirmation of the overspecialization effect when the global model with global probabilities is used to optimize local offspring with many update steps (setting 0/90). In the 37-multiplexer, though, we see that performance becomes unreliable when sampling from the Bayesian model. Model probabilities appear too noisy and additional specialization effects due to the chosen binary coding appear to stall

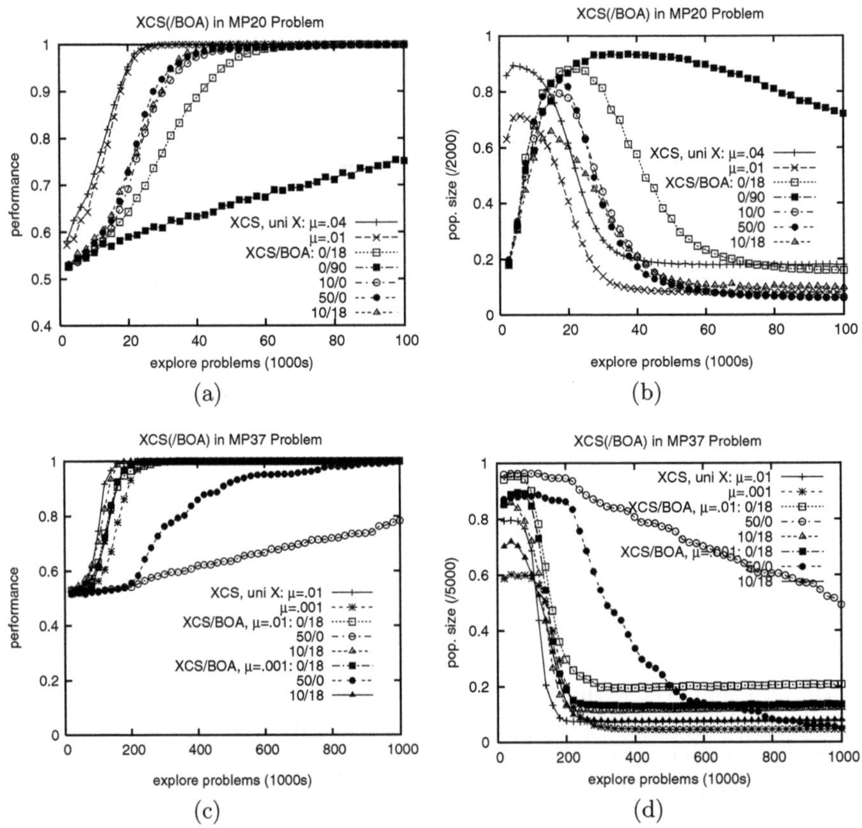

Fig. 11.7. On the 37-multiplexer, the BOA-based model learning and sampling slightly delays learning in XCS. Most likely, the reason for slower learning is that the specialization effect of mutation becomes too strong. In all experiments shown in this figure, XCS/BOA uses mutation rate $\mu = 0.001$

learning, yielding large population sizes and poor learning. The more cautious MCMC-based classifier optimization methods, on the other hand, learn the solution reliably and robustly. The most robust setting appears to optimize classifiers with the model adjusted to the local probability distribution (setting 10/18). This is certainly the most cautious setting causing least disruption but still providing effective recombination and structure optimization.

Since optimizing the classifier locally yielded most robust performance, we also ran experiments on the 70-multiplexer with this setting. The large Boolean function was recently solved successfully with the XCS system [11, 12]. Figure 11.8 (left-hand side) shows the performance of XCS in the 70-multiplexer problem with population sizes $N = 20\,k$, $N = 30\,k$, $N = 40\,k$, a mutation rate of $\mu = 0.01$, and an initially completely general population ($P_{\#} = 1.0$). The curves are averaged over 25 experiments. The graphs show that XCS with a population size of $40\,k$ solves the problem within $1,700\,k$ learning iterations. Decreasing the population size to $20\,k$ results in a much harder problem. All runs except one converged after $4,500\,k$ problems. The last run took more than $5{,}000\,k$ problems to find the optimal solution. Due to the small size of the population, XCS struggles to allocate reproductive opportunities to more accurate classifiers, often loosing the detected higher accurate classifiers.

Interestingly, XCS/BOA yields more robust performance in the 70 multiplexer. Figure 11.8 (right-hand side) shows the performance of XCS/BOA (setting 10/18) on the 70-multiplexer. Due to the model-based classifier optimization, performance becomes more reliable so that 100% accuracy is reached in all runs even when $N = 20\,k$. The comparison shows that XCS/BOA with the local offspring optimization mechanism yields a highly reliably search mechanism in XCS.

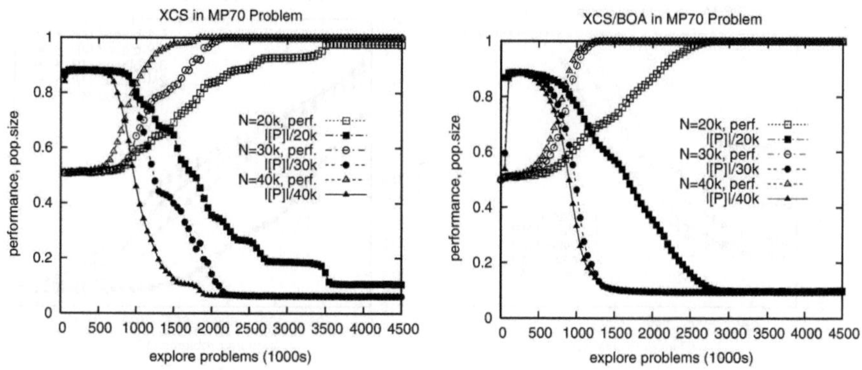

Fig. 11.8. XCSTS reliably solves the very large 70-multiplexer problem. A smaller population size delays the learning progress. Substituting uniform crossover with the Bayesian network-based recombination results in more effective search and thus faster and more reliable learning

11.4.3 The *xy*-Biased Multiplexer

The xy-biased multiplexer problem was introduced elsewhere [10, 11] to investigate fitness guidance. The problem combines the difficulty of the multiplexer problem iteratively. A first multiplexer function with x address bits chooses the y (biased-) multiplexer that decides on the current class of the problem. The xy-biased multiplexer is biased because the y multiplexer is slightly modified in that if all address bits are zero (or one) the result is a zero (one) regardless of the value bits, depending on whether the biased multiplexer is zero (or one) biased, respectively. This biases the problem in that the specialization of an address bit can increase accuracy slightly.

Figure 11.9 shows the performance of several variants of XCS on various large instances of the xy-biased multiplexer problem. Curves are averaged

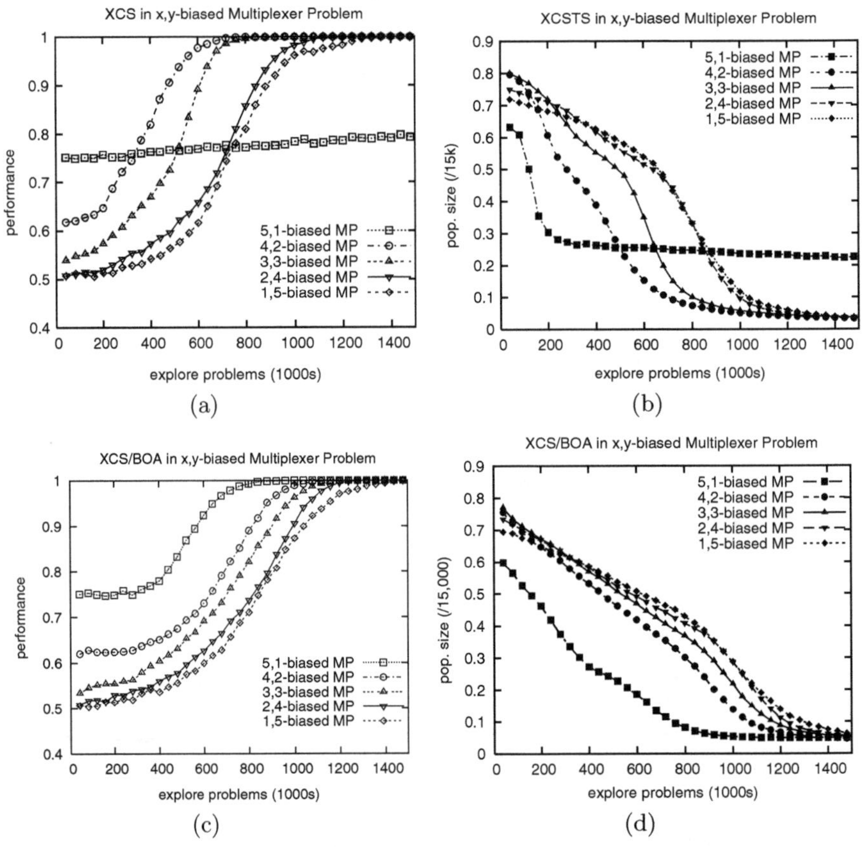

Fig. 11.9. Larger biased multiplexer problem instances are particularly challenging because the minimal order of problem difficulty increases. XCS/BOA can detect and propagate lower-level dependency structures more effectively than XCS with standard variation operators

over 20 runs and population size is set to $N = 15\,k$ ($P_{\#} = 0.8$, $\mu = 0.001$). To get an idea of how complex the final solutions are, we use the function $||[O]||$ that specifies the optimal classifier population [25, 26]. The measure defines the complexity of the problem as the size of the minimal, accurate, non-overlapping population that covers all environmental niches accurately. We note that $||[O]||(5,1) = 192(l = 69)$, $||[O]||(4,2) = 224(l = 84)$, $||[O]||(3,3) = 244(l = 83)$, $||[O]||(2,4) = 248(l = 78)$, and $||[O]||(1,5) = 252(l = 63)$. The $[O]$-measure suggests that $(5,1)$ is the simplest problem, while $(1,5)$ is the most difficult one. However, the plots in Fig. 11.9a show that the $(5,1)$ setting is very hard to solve for XCS with simple crossover and the solution is not learned even after $1,500k$ learning steps. In the XCS/BOA (10/18) setting (Fig. 11.9c), however, $(5,1)$ is solved the fastest, which indicates disruptive effects of uniform crossover.

The results indicate that on the xy-multiplexer, problem structure and fitness guidance are the main factors for a fast and reliable development of an accurate problem solution. For example, on the $(5,1)$ problem, the performance reaches the 75% level very fast but has a hard time to evolve a maximally accurate solution. The challenge in the $(5,1)$ problem is that the minimal order of difficulty is larger than one since the first bit of the five address bits is easily detected, but a specialization of any or even all of the other address bits does not increase accuracy until at least one value bit is correctly set.

On the simpler problem instances, XCS/BOA is outperformed by standard XCS recombination. The problem is simpler with respect to evolutionary search but becomes harder with respect to initial accuracy. The extreme initial gain in accuracy in the $(5,1)$ problem can cause disruption in the later learning progress. XCS/BOA detects relevant dependencies much more effectively and alleviates the disruptive effects of simple, uniform crossover.

11.5 Summary and Conclusions

This chapter focused on XCS, which is an accuracy-based learning classifier system. The chapter first argued that since many complex real-world systems are hierarchical and nearly decomposable [20, 37], solving difficult hierarchical classification problems represents an important challenge. It was shown that, analogously to GAs, also in LCSs simple crossover operators may be more disruptive than innovative. Especially in hierarchical problems in which subsets of attributes that interact on the lower-level need to be processed and recombined effectively, simple crossover operators can prevent effective learning.

The chapter then integrated the model building and sampling techniques from the extended compact genetic algorithm (ECGA) and the Bayesian optimization algorithm (BOA) into XCS, creating two competent LCSs: XCS/ECGA and XCS/BOA. XCS/ECGA and XCS/BOA were shown to provide efficient and reliable solutions for all difficult hierarchical problems that are intractable with XCS with standard variation operators.

Additionally to the hierarchical problems, we tested XCS and XCS/BOA on several benchmark binary classification problems used throughout the LCS literature and showed that XCS/BOA yields efficient and reliable performance in a variety of problem settings. XCS/BOA was able to solve the 70-multiplexer problem more reliably than simple XCS. In the xy-biased multiplexer problems, XCS/BOA was able to solve a large problem instance, which simple XCS was not able to solve. Finally, XCS/BOA was shown to be robust with respect to additional, problem-irrelevant bits, yielding more reliable performance than simple XCS.

In conclusion, XCS/ECGA and XCS/BOA are competent LCSs that combine advanced LCSs and EDAs to scalably solve difficult hierarchical problems. Results confirmed that XCS/BOA and XCS/ECGA are able to solve challenging binary problems efficiently, accurately and reliably. It is expected that this capability extends to all decomposable classification problems of bounded order. Further studies providing extended experimental evidence on this conjecture are in preparation.

There are several important directions for future research in this area. Both XCS/ECGA and XCS/BOA are currently applicable to discrete problems and one important area of future research is to extend these algorithms to handle real-valued conditions [4, 42]. Moreover, XCS is not restricted to the domain of classification and XCS/ECGA and XCS/BOA should be extensively tested on other types of problems suitable for XCS. Finally, XCS may be applied to more challenging predictive tasks, such as the prediction of actual changes in the world, similarly to the ACS system [8] or the modular MACS system [19], and an interesting area of future research is to analyze how incorporating the model building and sampling techniques from advanced EDAs would affect the performance of XCS on such predictive tasks.

References

[1] Bernadó, E., Llorà, X., and Garrell, J. M. (2002). XCS and GALE: A comparative study of two learning classifier systems and six other learning algorithms on classification tasks. In Lanzi, P. L., Stolzmann, W., and Wilson, S. W., (Eds.), *Advances in Learning Classifier Systems (LNAI 2321)*, pp. 115–132. Springer, Berlin Heidelberg New York

[2] Bernadó-Mansilla, E. and Garrell-Guiu, J. M. (2003). Accuracy-based learning classifier systems: Models, analysis, and applications to classification tasks. *Evolutionary Computation*, 11:209–238

[3] Butz, M. V. (2004). *Rule-based evolutionary online learning systems: Learning bounds, classification, and prediction.* PhD thesis, University of Illinois at Urbana-Champaign, Urbana, IL

[4] Butz, M. V. (2005a). Kernel-based, ellipsoidal conditions in the real-valued XCS classifier system. *GECCO 2005: Genetic and Evolutionary Computation Conference: Vol. 2*, pp. 1835–1842

[5] Butz, M. V. (2005b). Rule-based evolutionary online learning systems: A principled approach to LCS analysis and design. *Studies in Fuzziness and Soft Computing*. Springer, Berlin Heidelberg New York

[6] Butz, M. V. and Goldberg, D. E. (2004). Hierarchical classification problems demand effective building block identification and processing in LCSs. IlliGAL report 2004017, Illinois Genetic Algorithms Laboratory, University of Illinois at Urbana-Champaign

[7] Butz, M. V., Goldberg, D. E., and Lanzi, P. L. (2005). *Foundations of Learning Classifier Systems, Computational Complexity of the XCS Classifier System*, pp. 91–126. Studies in Fuzziness and Soft Computing. Springer, Berlin Heidelberg New York

[8] Butz, M. V., Goldberg, D. E., and Stolzmann, W. (2002). The anticipatory classifier system and genetic generalization. *Natural Computing*, 1:427–467

[9] Butz, M. V., Goldberg, D. E., and Tharakunnel, K. (2003a). Analysis and improvement of fitness exploitation in XCS: Bounding models, tournament selection, and bilateral accuracy. *Evolutionary Computation*, 11:239–277

[10] Butz, M. V., Kovacs, T., Lanzi, P. L., and Wilson, S. W. (2001). How XCS evolves accurate classifiers. *Proceedings of the Third Genetic and Evolutionary Computation Conference (GECCO-2001)*, pp. 927–934

[11] Butz, M. V., Kovacs, T., Lanzi, P. L., and Wilson, S. W. (2004a). Toward a theory of generalization and learning in XCS. *IEEE Transactions on Evolutionary Computation*, 8:28–46

[12] Butz, M. V., Sastry, K., and Goldberg, D. E. (2003b). Tournament selection in XCS. *Proceedings of the Fifth Genetic and Evolutionary Computation Conference (GECCO-2003)*, pp. 1857–1869

[13] Butz, M. V., Sastry, K., and Goldberg, D. E. (2004b). Strong, stable, and reliable fitness pressure in XCS due to tournament selection. *Genetic Programming and Evolvable Machines*, 6:53–77

[14] Butz, M. V. and Wilson, S. W. (2001). An algorithmic description of XCS. In Lanzi, P. L., Stolzmann, W., and Wilson, S. W., (Eds.), *Advances in Learning Classifier Systems: Third International Workshop, IWLCS 2000 (LNAI 1996)*, pp. 253–272. Springer, Berlin Heidelberg New York

[15] Chickering, D. M., Heckerman, D., and Meek, C. (1997). A Bayesian approach to learning Bayesian networks with local structure. Technical Report MSR-TR-97-07, Microsoft Research, Redmond, WA

[16] Cooper, G. F. and Herskovits, E. H. (1992). A Bayesian method for the induction of probabilistic networks from data. *Machine Learning*, 9: 309–347

[17] De Jong, K. A. and Spears, W. M. (1991). Learning concept classification rules using genetic algorithms. *IJCAI-91 Proceedings of the Twelfth International Conference on Artificial Intelligence*, pp. 651–656

[18] Friedman, N. and Goldszmidt, M. (1999). Learning Bayesian networks with local structure. In Jordan, M. I., (Ed.), *Graphical models*, pp. 421–459. MIT, Cambridge, MA

[19] Gérard, P. and Sigaud, O. (2003). Designing efficient exploration with MACS: Modules and function approximation. *Proceedings of the Fifth Genetic and Evolutionary Computation Conference (GECCO-2003)*, pp. 1882–1893

[20] Gibson, J. J. (1979). *The Ecological Approach to Visual Perception.* Lawrence Erlbaum Associates, Mahwah, NJ

[21] Goldberg, D. E. (2002). *The Design of Innovation: Lessons from and for Competent Genetic Algorithms.* Kluwer Academic, Boston, MA

[22] Harik, G. (1999). Linkage learning via probabilistic modeling in the ECGA. IlliGAL report 99010, Illinois Genetic Algorithms Laboratory, University of Illinois at Urbana-Champaign

[23] Heckerman, D., Geiger, D., and Chickering, D. M. (1994). Learning Bayesian networks: The combination of knowledge and statistical data. Technical Report MSR-TR-94-09, Microsoft Research, Redmond, WA

[24] Howard, R. A. and Matheson, J. E. (1981). Influence diagrams. In Howard, R. A. and Matheson, J. E., (Eds.), *Readings on the Principles and Applications of Decision Analysis*, Vol. 2, pp. 721–762. Strategic Decisions Group, Menlo Park, CA

[25] Kovacs, T. (1996). Evolving optimal populations with XCS classifier systems. Master's thesis, School of Computer Science, University of Birmingham, Birmingham, UK

[26] Kovacs, T. (1997). XCS classifier system reliably evolves accurate, complete, and minimal representations for boolean functions. In Roy, Chawdhry, and Pant, (Eds.), *Soft Computing in Engineering Design and Manufacturing*, pp. 59–68. Springer, London Berlin Heidelberg New York

[27] Kovacs, T. (2000). Strength or Accuracy? Fitness calculation in learning classifier systems. In Lanzi, P. L., Stolzmann, W., and Wilson, S. W., (Eds.), *Learning Classifier Systems: From Foundations to Applications (LNAI 1813)*, pp. 143–160. Springer, Berlin Heidelberg New York

[28] Kovacs, T. (2001). Towards a theory of strong overgeneral classifiers. *Foundations of Genetic Algorithms 6*, pp. 165–184

[29] Lobo, F. and Harik, G. (1999). Extended compact genetic algorithm in C++. IlliGAL report 99016, Illinois Genetic Algorithms Laboratory, University of Illinois at Urbana-Champaign

[30] Mühlenbein, H. (1992). How genetic algorithms really work: I.Mutation and Hillclimbing. In Männer, R. and Manderick, B., (Eds.), *Parallel Problem Solving from Nature*, pp. 15–25, Elsevier, Amsterdam Netherlands

[31] Neal, R. M. (1993). Probabilistic inference using Markov chain Monte Carlo methods. Technical Report CRG-TR-93-1, Department of Computer Science, University of Toronto

[32] Pearl, J. (1988). *Probabilistic Reasoning in Intelligent Systems: Networks of Plausible Inference.* Morgan Kaufmann, San Mateo, CA

[33] Pelikan, M. (2001). Bayesian optimization algorithm, decision graphs, and Occam's razor. *Proceedings of the Third Genetic and Evolutionary Computation Conference (GECCO-2001)*, pp. 511–518

[34] Pelikan, M. (2002). *Bayesian optimization algorithm: From single level to hierarchy.* PhD thesis, University of Illinois at Urbana-Champaign, Urbana, IL. Also IlliGAL Report No. 2002023

[35] Pelikan, M., Goldberg, D. E., and Cantu-Paz, E. (1999). BOA: The Bayesian optimization algorithm. *Proceedings of the Genetic and Evolutionary Computation Conference (GECCO-99)*, pp. 525–532

[36] Schwarz, G. (1978). Estimating the dimension of a model. *The Annals of Statistics*, 6:461–464

[37] Simon, H. A. (1969). *Sciences of the Artificial.* MIT, Cambridge, MA

[38] Thierens, D. and Goldberg, D. E. (1993). Mixing in genetic algorithms. *Proceedings of the Fifth International Conference on Genetic Algorithms*, pp. 38–45

[39] Wilson, S. W. (1994). ZCS: A zeroth-level classifier system. *Evolutionary Computation*, 2:1–18

[40] Wilson, S. W. (1995). Classifier fitness based on accuracy. *Evolutionary Computation*, 3(2):149–175

[41] Wilson, S. W. (1998). Generalization in the XCS classifier system. *Genetic Programming 1998: Proceedings of the Third Annual Conference*, pp. 665–674

[42] Wilson, S. W. (2001). Mining oblique data with XCS. In Lanzi, P. L., Stolzmann, W., and Wilson, S. W., (Eds.), *Advances in Learning Classifier Systems: Third International Workshop, IWLCS 2000 (LNAI 1996)*, pp. 158–174. Springer, Berlin Heidelberg New York

12

Military Antenna Design Using a Simple Genetic Algorithm and hBOA

Tian-Li Yu, Scott Santarelli, and David E. Goldberg

Summary. This chapter describes the optimization of an antenna design problem. A simple genetic algorithm (SGA) and the hierarchical Bayesian optimization algorithm (hBOA) were applied to this problem. Three objective functions were designed in an effort to find a solution that meets the system requirements/specifications. Empirical results indicate that the SGA and hBOA perform comparably when the objective function is "easy" (i.e., traditional mask). When the objective function more accurately reflects the true objective of the problem (i.e., "difficult"), however, hBOA consistently outperforms the SGA both computationally and electromagnetically.

Key words: Antenna design, Simple genetic algorithm, Hierarchical Bayesian optimization algorithm, Estimation of distribution algorithm, Competent genetic algorithm, Real-world application

12.1 Introduction

Simple genetic algorithms (SGAs) [3] have been applied to a wide variety of antenna applications over past decades [13]. Over roughly the same period, researchers in the GA field have devoted themselves to the design of *competent* genetic algorithms (GAs), such as estimation of distribution algorithms (EDAs), that solve hard problems *quickly*, *reliably*, and *accurately* [4]. Research in EDAs have shown that EDAs outperform SGAs in many artificial and real-world problems [1, 6, 8, 11].

In this chapter, we apply the hierarchical Bayesian optimization algorithm (hBOA) [8, 9] – one of the most successful EDAs – to the optimization of a constrained feed network for a linear antenna array. We designed three objective functions, from simple to difficult, to adequately solve the problem, and to demonstrate the capability of hBOA. The performance of hBOA is compared and contrasted with that of an SGA using the same number of function evaluations.

T.-L. Yu et al.: *Military Antenna Design Using a Simple Genetic Algorithm and hBOA*, Studies in Computational Intelligence (SCI) **33**, 275–289 (2006)
www.springerlink.com

This chapter is organized as follows. Section 12.2 describes the constrained feed network for a linear antenna array. The three manually-designed objective functions are described in Sect. 12.3. Details about the encoding, parameter settings, and implementation for both the SGA and hBOA can be found in Sect. 12.4. Following this, Sect. 12.5 compares and contrasts the results for the SGA and hBOA. Finally, in Sect. 12.6, we summarize our results and draw some interesting conclusions concerning the fundamental differences between simple and competent GAs.

12.2 Problem Statement

This section describes an antenna system designed for space-based and airborne radar applications. The goal of this system is to produce a far-field radiation pattern having at least 30 dB sidelobes over a 20% bandwidth. This is accomplished by implementing an optimized, constrained-feed network. The following overview is intended to provide the reader with enough background information to understand the details of the system optimization. For further information about the system design and implementation, the reader is referred to [7].

The following discussion assumes an ideal system. A single section of the system is shown in Fig. 12.1. At the front end is an N-element, linear array. It can be shown that when the array is illuminated by a plane wave, the element excitations can be computed via the following [5]:

$$a_n(\theta) = \mathrm{e}^{\mathrm{j}\frac{2\pi}{\lambda} nd \sin\theta}, \tag{12.1}$$

where n is the element index, λ is the wavelength of the incoming plane wave, d is the inter-element spacing, θ is the angle of incidence with respect to the normal, and j is the square root of -1.

Each element is connected to a single input port of an N by M Rotman lens. Thus, the element excitations become the inputs to the Rotman lens. It can be shown that the Rotman lens output signals are described by the following:

$$I_i(\theta) = \sum_{n=1}^{N} \mathrm{e}^{\mathrm{j}\frac{2\pi}{\lambda} nd\left(\sin\theta - \frac{i}{N}\right)}, \tag{12.2}$$

where i is the Rotman lens output index. Each output signal is then multiplied by a complex weight, w_i.

Next, these signals are input to an M by M Butler matrix, the outputs of which are computed from:

$$J_m(\theta) = \sum_{i=1}^{M} I_i(\theta) w_i \mathrm{e}^{\mathrm{j}2\pi i\left(\frac{m}{M}\right)\left(\frac{\lambda_0}{\lambda}\right)}, \tag{12.3}$$

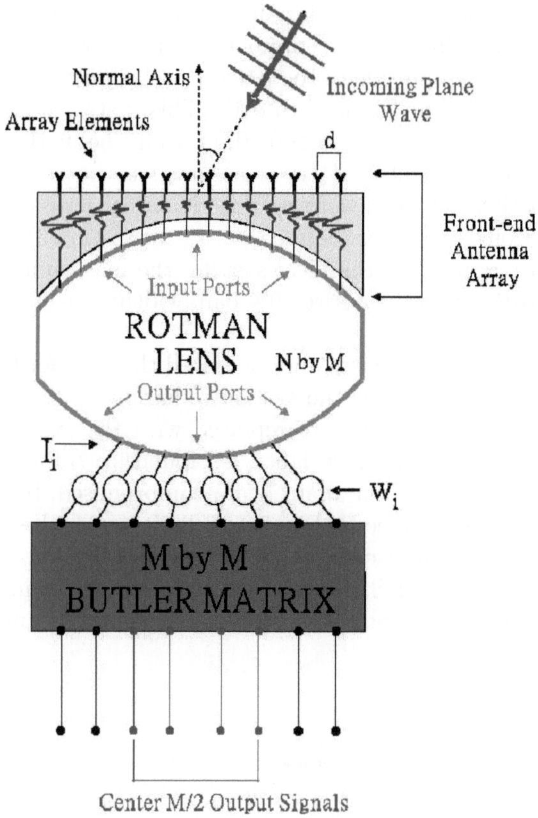

Fig. 12.1. Single section of antenna system, including front-end array, Rotman lens, and Butler matrix

where m is the Butler-matrix output index, and λ_0 is the center-frequency wavelength of the system. The center $M/2$ output signals from each of P sections are time-delayed, weighted (e.g., fixed weights like a Taylor distribution, etc.), and combined to compute the final radiation pattern of the system.

In order to optimize the system, the set of complex weights, w_i, must be determined for each of P sections, such that the final radiation pattern exhibits -30 dB sidelobes over a 20% bandwidth. For our particular system, the specifications and parameter values are as follows:

- Frequency band of operation: 9.0–11.0 GHz.
- Center frequency $f_0 = 10.0$ GHz (where $\lambda_0 = c/f_0$, and c is the velocity of electromagnetic waves in free space, roughly $3 \times 10^8\,\mathrm{m\,s^{-1}}$).
  $N = 64$.
  $d = 0.5\lambda_0$.
  $M = 8$.
  $P = 3$.

Figure 12.2 shows far-field radiation patterns for an ideal system at 9.0, 10.0, and 11.0 GHz. The x-axis represents U-space (where $U = \sin\theta$), and the y-axis shows the amplitude of the pattern measured in decibels (dB). The peak of each pattern is normalized to 0 dB. This system was optimized for a beam-steering angle of 45° (i.e., $u \simeq 0.7071$) using the method of alternating projections [7]. The weights, w_i, were assumed to be real-valued, and the weights for a particular index i were assumed identical across the three sections (i.e., w_5 for section 1 = w_5 for section 2 = w_5 for section 3). Note that when these weights are applied to the ideal system, the maximum sidelobe level is well below −30 dB across the entire 20% bandwidth, thus meeting the system requirements.

We built three Rotman lenses and measured the transfer function for each of them. Figure 12.3 illustrates the system radiation patterns when the ideal Rotman lens transfer functions are replaced with the experimental transfer functions. We applied the same weights, w_i, found in [7] for the ideal system. Not only is the integrity of the main beam compromised, but also the maximum sidelobe level well exceeds the −30 dB limit across the entire frequency band. In other words, the weights that we successfully implemented for the ideal system cause the experimental system to break down. Thus, we need to re-optimize the system now that the experimental Rotman lens data is incorporated.

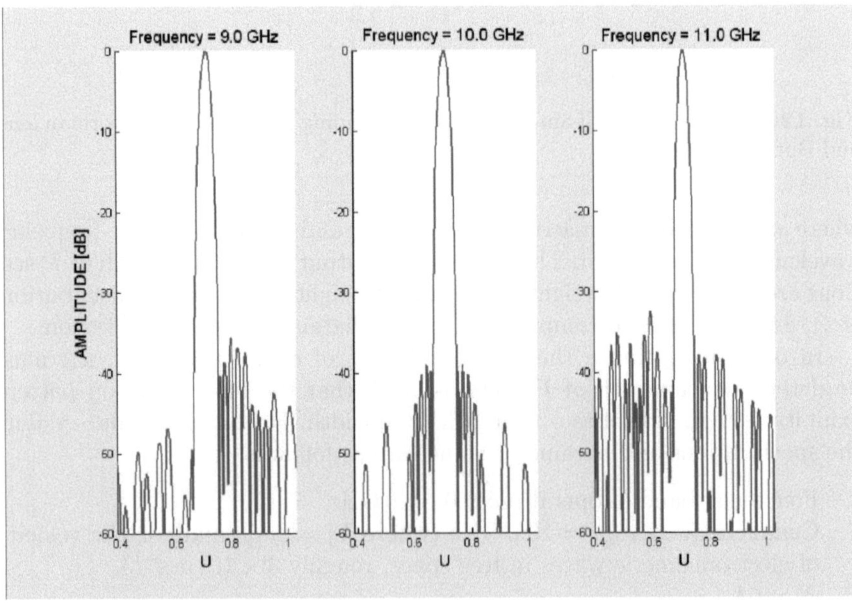

Fig. 12.2. Far-field radiation patterns as a function of frequency for the ideal system. The ideal system is optimized by applying the method of alternating projections [7] to 12.3

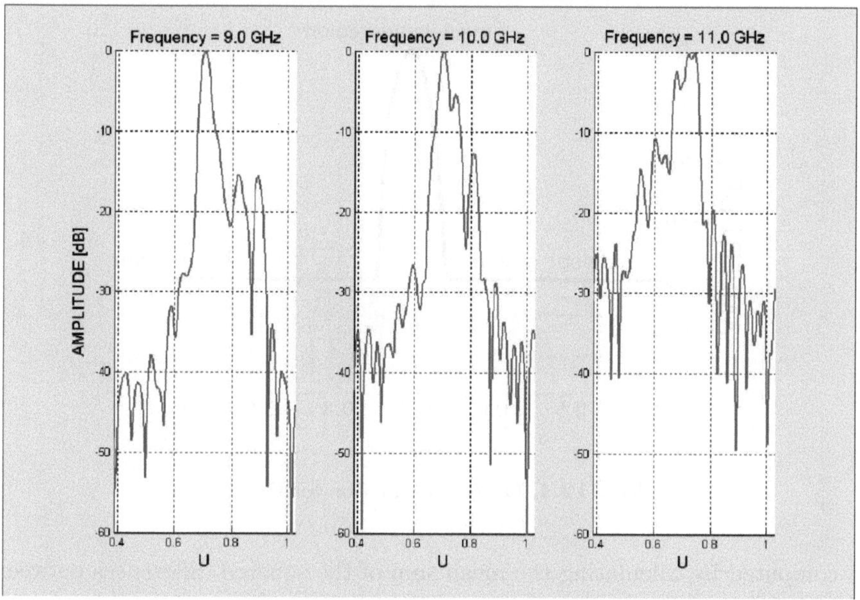

Fig. 12.3. Far-field radiation patterns as a function of frequency when the experimental Rotman lens data is incorporated into the system model and the ideal system weights from [7] are applied. The figure indicates that the weights obtained from the ideal system mathematically cannot be directly applied to the real system. The weights need to be tuned again based on the real system

12.3 Objective Function

The objective function is essentially a subroutine written in MATLAB, which was used by both the SGA and hBOA to evaluate potential solutions to the problem (i.e., chromosomes). When the user inputs a set of 24 complex weights, the subroutine computes the corresponding far-field radiation patterns for five discrete frequencies (9.0, 9.48, 10.0, 10.52, and 11.0 GHz). This experiment employed three variations of the objective function, which are described below as Cases 1, 2, and 3. (Note that for this experiment, we defined fitness such that lower values correspond to higher quality solutions. Traditionally, fitness is defined such that higher values correspond to higher quality solutions).

Figure 12.4 shows the objective function for Case 1. The gray curve is a typical far-field radiation pattern produced by the system for a given frequency and set of complex weights. The x-axis represents u-space (i.e., $\sin\theta$), and the y-axis measures the normalized amplitude of the pattern in decibels. The mask represents the objective function, showing a main-beam and sidelobe region. For this case, we perform a point-by-point subtraction of the mask from the pattern. For a given frequency and set of complex weights, an error value E_k

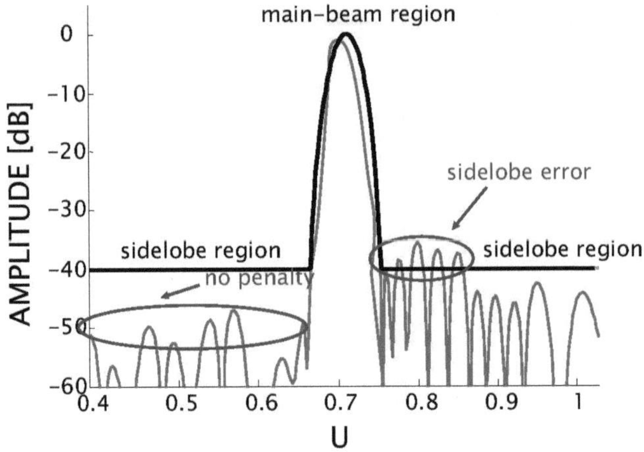

Fig. 12.4. Objective function for Case 1

is computed by calculating the mean sum of the squared differences between the pattern and mask:

$$E_k(w, f_k) = \frac{1}{U}\left[\sum_{i \in \text{main-beam region}} (\text{pattern}_i - \text{mask}_i)^2 + \sum_{i \in \text{sidelobe region and pattern}_i > \text{mask}_i} (\text{pattern}_i - \text{mask}_i)^2\right], \quad (12.4)$$

where w represents the vector of complex weights, f_k is the kth discrete frequency, and U represents the total number of points in the radiation pattern. Note that no penalty is administered when the pattern lies below the mask in the sidelobe region (i.e., if the difference between the pattern and mask is negative, it is not used in the computation). In essence, we are trying to force the pattern to conform to the mask in the main-beam region while forcing the pattern to lie below the mask in the sidelobe region. Also note that we are "overshooting" by trying to force the algorithm to find a solution with −40 dB sidelobes in hopes that it will at least be able to obtain −30 dB sidelobes. This lack of efficiency is an inherent weakness of this approach. The overall Case 1 fitness value, $F_1(w)$, is the average of the error across the entire frequency band:

$$F_1(w) = \frac{1}{K}\sum_{k=1}^{K} E_k, \quad (12.5)$$

where K is the total number of discrete frequencies (and is equal to 5 for our experiment).

Figure 12.5 shows the general objective function used for both Cases 2 and 3. For Case 2, for a given frequency and set of complex weights, the error has

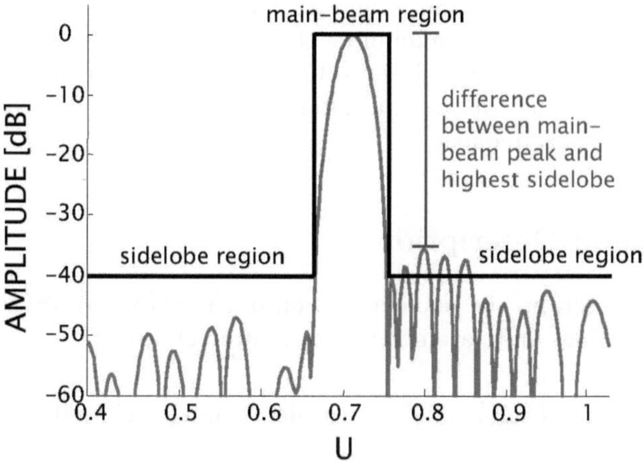

Fig. 12.5. Objective function for Cases 2 and 3

two components, the first of which is

$$E_{k,1}(w, f_k) = [1 - \text{pattern}(u_0)]^2 , \tag{12.6}$$

where u_0 is the desired steering angle of the pattern peak ($u_0 = 0.7071$, corresponding to $\theta = 45°$, for this experiment). In essence we need to ensure that the peak of the normalized pattern in the main-beam region coincides with the desired steering angle u_0. The second error component is as follows:

$$E_{u,2}(w, f_k) = \text{MSL}^2, \tag{12.7}$$

where MSL refers to the "maximum sidelobe level" (i.e., the maximum level of the radiation pattern in the sidelobe region). In other words, we are trying to maximize the difference between the normalized pattern peak and the maximum sidelobe level as illustrated in Fig. 12.5. The overall Case 2 fitness value, $F_2(w)$, is the mean summation of the error components across the entire frequency band:

$$F_2(w) = \frac{1}{K} \sum_{k=1} K \left(E_{k,1} + E_{k,2}\right). \tag{12.8}$$

It is clear that the objective function for Case 2 involves only two subtractions, rather than a point-by-point comparison of the pattern to the mask – this property renders Case 2 more computationally efficient than Case 1. Similar to Case 1, however, the overall fitness value for a given set of complex weights is the average of the error across the entire frequency band.

Case 3 is identical to Case 2, except the overall fitness value, $F_3(w)$, is equal to the maximum error across frequency:

$$F_3(w) = \max_k \left(E_{k,1} + E_{k,2}\right). \tag{12.9}$$

In other words, Cases 1 and 2 aim at minimizing the mean error across frequency, whereas Case 3 minimizes the maximum error across frequency. Among the three objective functions, Case 3 is the most relevant to our particular problem, since we are ultimately trying to minimize the maximum sidelobe level across frequency.

12.4 Method Description

This section describes the implementation of the SGA and hBOA. In particular, the encodings and parameter settings are detailed.

12.4.1 Implementation of the Simple Genetic Algorithm

Here we describe our first optimization approach, which involves the implementation of an SGA. Recall that our antenna system consists of three sections, each containing eight complex weights (i.e., amplitude and phase). Our goal then is to obtain a set of 24 complex weights that will allow us to meet the system requirements outlined in Sect. 12.2.

We used the chromosome representation shown in Fig. 12.6. We chose a binary encoding scheme and chose to represent each complex weight with 16 bits (i.e., 8 bits amplitude and phase). Therefore, the length of our chromosome was: 24 weights $\times$ 2 components/weight $\times$ 8 bits/component $=$ 384 bits. We arbitrarily chose to encode the complex-weight amplitudes along the first half of the chromosome and the phases along the latter half. Both the amplitudes and phases are numbered sequentially along their respective halves of the chromosome. We restricted the amplitudes to lie in the interval [0, 1] and the phases to lie in the interval $[0\ 2\pi]$. It has been shown that Gray coding are mutation friendly and can be beneficial for real-world problems [2, 14]. Therefore, we used an 8-bit gray code for both the amplitude and phase encoding schemes.

At the start of the algorithm, we formed a random population of 200 chromosomes (parents). The population size is so set via a series of experiments

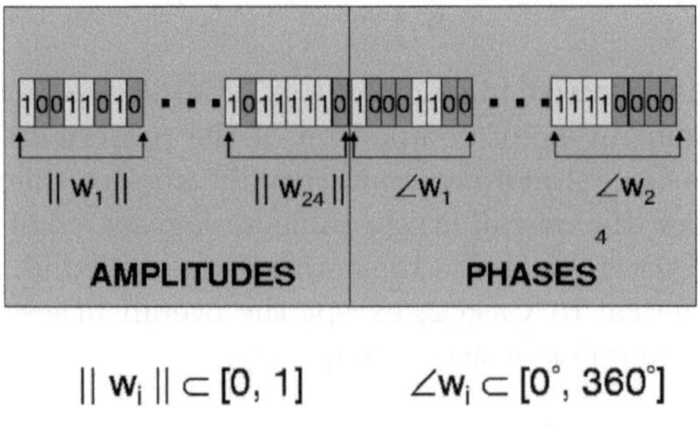

Fig. 12.6. Chromosomal encoding scheme

given the limitation of a one-million function evaluations. Each member of the population was evaluated and ranked (the details of the objective function is described in Sect. 12.3). Then, we formed a mating pool of 200 individuals via binary tournament selection [3]. Next, two individuals from the mating pool (i.e., parents) were chosen randomly to create a child via two-point crossover. This process was repeated until 200 children had been generated. Each child was passed to an operator having a constant mutation rate of 0.005. The children were evaluated, ranked, and proceeded to become the parents of the next generation. For this experiment, we always ran the SGA for 5,000 iterations with a constant population size of 200 for a total of one-million objective-function evaluations.

12.4.2 Implementation of the Hierarchical Bayesian Optimization Algorithm

This section describes the implementation and parameter settings of our second optimization approach – hBOA. Details about hBOA can be found elsewhere [8–10, 12]. Here we only focus on the encoding and parameter settings.

The encoding scheme is exactly the same as that used for the SGA. Each candidate solution was encoded into a 384-bit binary string. Also, gray coding was used. Given the same number of function evaluations (one-million), a population size of 5,000 was found to be suitable. and hence the maximum number of generations was set to 200 for a total of one-million function evaluations (i.e., same as SGA case).

Restricted tournament selection was adopted for the purpose of preserving good candidate solutions [8]. The window size was set to the problem size (384) to perform niching globally; the tournament size was set to 12 based on empirical observations. Bit-wise mutation was used, with a mutation probability of 0.005. A series of experiments showed that the maximum number of incoming edges for a single node in the Bayesian network should be limited to 4 so as to avoid unnecessary linkage complexity. Elitism was adopted. Every generation, the worst half of the parents is replaced by the newly generated candidate solutions. In each generation, parental candidate solutions were evaluated, and the bottom half were replaced by newly generated offspring.

12.5 Empirical Results: SGA versus hBOA

This section analyzes the results for Cases 1, 2, and 3 by considering the computational performance of SGA versus hBOA. We also compare SGA and hBOA on the basis of sidelobe attenuation.

Each case was run three times for both the SGA and hBOA, the results of which are tabulated in Tables 12.1–12.3. For convenience, the runs are sorted according to fitness from best to worst. (Note that for this experiment, we defined fitness such that *lower* values correspond to higher quality solutions.

Table 12.1. The results of the SGA and hBOA for CASE 1

			SGA			
F (GHz)	9.00	9.48	10.0	10.52	11.00	F_1 ($\times 10^{-3}$)
Run 1 error ($\times 10^{-3}$)	0.2009	0.0743	0.1307	0.0711	0.1212	0.1196
Run 2 error ($\times 10^{-3}$)	0.2310	0.1471	0.2196	0.1225	0.1629	0.1766
Run 3 error ($\times 10^{-3}$)	0.2920	0.1117	0.2456	0.0889	0.1793	0.1835
Mean error ($\times 10^{-3}$)	0.2413	0.1110	0.1986	0.0942	0.1545	0.1599
Standard deviation ($\times 10^{-3}$)	0.0464	0.0364	0.0603	0.0261	0.0300	0.0351
			hBOA			
F (GHz)	9.00	9.48	10.0	10.52	11.00	F_1 ($\times 10^{-3}$)
Run 1 error ($\times 10^{-3}$)	0.1912	0.0800	0.1407	0.0619	0.1232	0.1194
Run 2 error ($\times 10^{-3}$)	0.2034	0.0760	0.1433	0.0588	0.1167	0.1196
Run 3 error ($\times 10^{-3}$)	0.1927	0.0785	0.1447	0.0608	0.1225	0.1198
Mean error ($\times 10^{-3}$)	0.1958	0.0782	0.1429	0.0605	0.1208	0.1196
Standard deviation ($\times 10^{-3}$)	0.0067	0.0020	0.0020	0.0016	0.0036	0.0002

Although the SGA and hBOA give solutions with comparable mean errors, hBOA's performance is more stable and consistent

Table 12.2. The results of the SGA and hBOA for CASE 2

			SGA			
F (GHz)	9.00	9.48	10.0	10.52	11.00	F_1 ($\times 10^{-3}$)
Run 1 error ($\times 10^{-3}$)	3.0140	2.0519	2.2108	1.3882	1.7584	2.0847
Run 2 error ($\times 10^{-3}$)	2.8931	1.9392	2.5851	2.0724	1.8038	2.2587
Run 3 error ($\times 10^{-3}$)	4.1890	1.6209	2.0234	1.4106	2.8395	2.4167
Mean error ($\times 10^{-3}$)	3.3654	1.8707	2.2731	1.6237	2.1339	2.2534
Standard deviation ($\times 10^{-3}$)	0.7158	0.2235	0.2860	0.3887	0.6115	0.1661
			hBOA			
F (GHz)	9.00	9.48	10.0	10.52	11.00	F_1 ($\times 10^{-3}$)
Run 1 error ($\times 10^{-3}$)	2.3269	1.1383	1.5282	1.5482	1.6063	1.6296
Run 2 error ($\times 10^{-3}$)	2.1459	1.2727	1.5955	1.5421	1.9166	1.6946
Run 3 error ($\times 10^{-3}$)	2.1133	1.3130	1.6334	1.5957	1.8963	1.7103
Mean error ($\times 10^{-3}$)	2.1954	1.2413	1.5857	1.5620	1.8064	1.6782
Standard deviation ($\times 10^{-3}$)	0.1151	0.0915	0.0533	0.0293	0.1736	0.0428

The solutions produced by hBOA have smaller mean errors and a smaller variance

Traditionally, fitness is defined such that *higher* values correspond to higher quality solutions). Table 12.1 contains the results for Case 1. If we compare the performance of the best run for each algorithm (i.e., Run 1), we see that the overall fitness value F_1 is practically identical for the SGA and hBOA. In addition, Figure 12.7 shows that there is virtually no difference between the maximum sidelobe levels of the resulting radiation patterns at 9.0 GHz. Thus, one may be tempted to assume that both algorithms perform equally well for this objective function. However, it is also evident from the table that the mean fitness across runs is 33% higher for the SGA compared to hBOA, and

Table 12.3. The results of the SGA and hBOA for CASE 3

SGA						
F (GHz)	9.00	9.48	10.0	10.52	11.00	F_1 ($\times 10^{-3}$)
Run 1 error ($\times 10^{-3}$)	0.9338	0.4556	0.8370	0.1015	0.0007	0.9338
Run 2 error ($\times 10^{-3}$)	1.0124	0.3198	0.8681	0.0953	0.0026	1.0124
Run 3 error ($\times 10^{-3}$)	1.4588	0.3618	0.1016	0.0240	0.0006	1.4588
Mean error ($\times 10^{-3}$)	1.1350	0.3791	0.6022	0.0736	0.0013	1.1350
Standard deviation ($\times 10^{-3}$)	0.2832	0.0695	0.4338	0.0431	0.0011	0.2832
hBOA						
F (GHz)	9.00	9.48	10.0	10.52	11.00	F_1 ($\times 10^{-3}$)
Run 1 error ($\times 10^{-3}$)	0.0019	0.0019	0.0019	0.0019	0.0019	0.0019
Run 2 error ($\times 10^{-3}$)	0.0019	0.0019	0.0019	0.0019	0.0019	0.0019
Run 3 error ($\times 10^{-3}$)	0.0019	0.0019	0.0019	0.0019	0.0019	0.0019
Mean error ($\times 10^{-3}$)	0.0019	0.0019	0.0019	0.0019	0.0019	0.0019
Standard deviation ($\times 10^{-3}$)	0.0000	0.0000	0.0000	0.0000	0.0000	0.0000

hBOA consistently outperformed the SGA in all runs, and the solution quality of hBOA is extremely stable

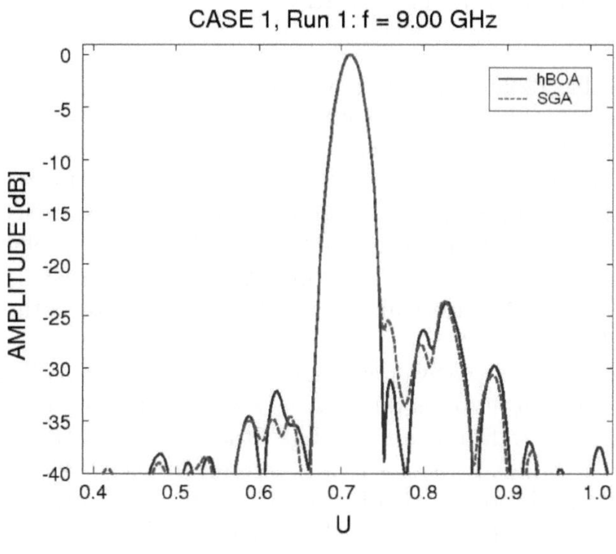

Fig. 12.7. SGA versus hBOA performance, CASE 1, Run 1. *Solid*: hBOA, *dotted*: SGA. The performances were comparable

the standard deviation of the fitness across runs for the SGA exceeds that for hBOA by a factor of 175! These results imply that, given enough runs, the best-quality SGA solution may be comparable to the best-quality hBOA solution; however, hBOA seems to be a much more consistent and reliable search mechanism.

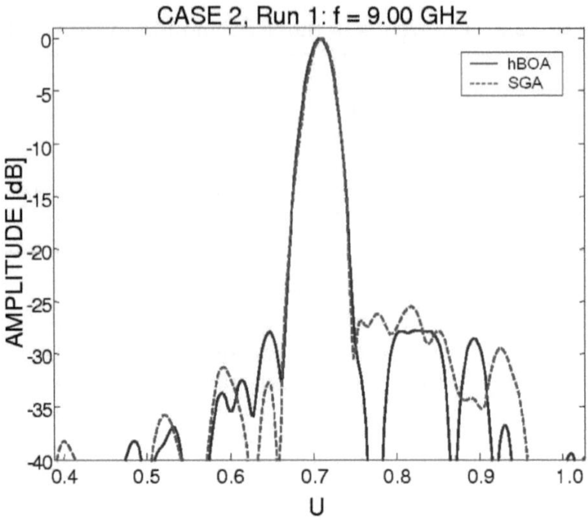

Fig. 12.8. SGA versus hBOA performance, CASE 2, Run 1. *Solid*: hBOA, *dotted*: SGA. hBOA slightly outperformed the SGA

Case 2 (Table 12.2 and Fig. 12.8) exhibits the same general trends as seen for Case 1, namely the mean of the fitness F2 across runs is 34% greater for the SGA versus hBOA, and the standard deviation of F2 across runs for the SGA exceeds that for hBOA by almost a factor of 4. However, for this case, the best-quality hBOA solution is 28% better than the best-quality SGA solution. This is apparent in the figure, where we see that the highest sidelobe resulting from the SGA solution exceeds that of hBOA by approximately 2.5 dB. Thus, for this case, the SGA performance trails that of hBOA for both best fitness and average fitness.

The performance of the SGA is by far the worst for Case 3 (Table 12.3 and Fig. 12.9). hBOA outperforms the SGA by several orders of magnitude when comparing both best fitness and average fitness. In addition, the standard deviation of the fitness F_3 across runs is zero for hBOA! The sidelobe attenuation of hBOA solution is -27 dB, which is quite close to the desired value of -30 dB. By contrast, in the best SGA solution, it is difficult to identify a mainlobe and worst-case sidelobes exist at -6 dB.

Overall, these results are not surprising. The different objective functions represent drastically different solution spaces. Case 1 involves forcing a function to a mask, which is considered a GA-easy problem because taking average makes the fitness function landscape smooth. Thus, we see that in this case the SGA performs comparably to hBOA. Case 2 is somewhat more difficult because the objective is to minimize the averaged maximal errors over the bandwidth. Case 3 is the most difficult problem for GAs, because the min/max nature of the objectives give rise to a solution space that contains many local minima. The SGA, therefore, easily fell into some local minimum

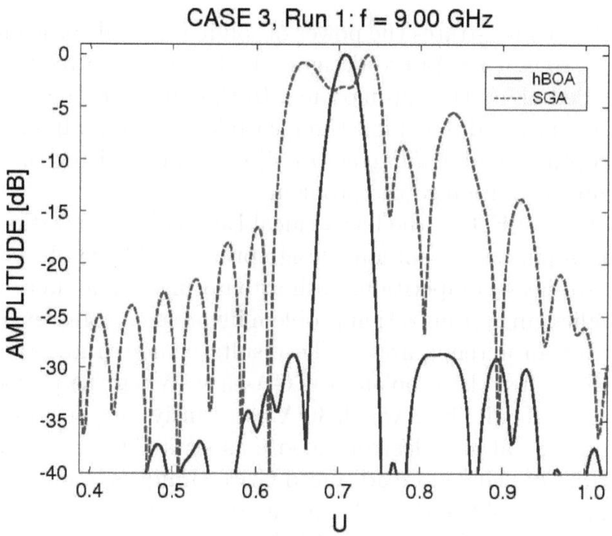

Fig. 12.9. SGA versus hBOA performance, CASE 3, Run 1. *Solid*: hBOA, *dotted*: SGA. The SGA totally failed on finding solutions with an acceptable quality, while hBOA was able to give a solution with the sidelobe attenuation of only −27 dB

and was not capable of exploring the landscape globally. hBOA, on the other hand, was able to better identify the linkage of the problem, which allowed it to recombine salient pieces of information without disrupting good building blocks. The solutions obtained from hBOA for Case 3 correspond to -27 dB sidelobes across the entire frequency band of operation. Thus, hBOA allowed us to come within 3 dB of our goal (i.e., −30 dB across frequency). This result is quite satisfactory, considering the slightly degraded state of the experimental Rotman lenses as discussed in Sect. 12.2.

12.6 Summary and Conclusions

This chapter investigated the application of both a simple GA and hBOA to the optimization of a constrained feed network for an antenna linear array. Three objective functions were designed to compare the performance of the SGA and hBOA. The first objective function was designed to be GA-easy but did not completely reflect the desired objective of the problem. The third objective function was GA-difficult but was much more relevant to the problem objectives. For the GA-easy objective function, the performance of the SGA and hBOA were comparable. When the objective functions became more complicated (i.e., GA-difficult), the competent GA technique demonstrated significant improvement over the SGA. In all three cases, the quality of the solutions that hBOA gave is more reliable than that the SGA gave.

This chapter demonstrates the power of solving difficult problems of a new breed of GA – EDAs. EDAs are designed to learn the correlations between variables encoded along the chromosome. In this manner, the recombination operator can be designed to exploit these correlations and minimize the probability of disrupting good *building blocks* [4] while maximizing the probability of finding a global solution to the problem.

A specific type of EDA – the hierarchical Bayesian optimization algorithm (hBOA) – was applied to an antenna design problem. The SGA and hBOA engaged in head-to-head competition, both attempting to find an acceptable solution to a challenging optimization problem involving a complex, constrained feed network for an antenna array. The results demonstrated that the SGA competes hBOA when the problem was GA-easy. When the problem became more and more difficult, however, hBOA constantly outperformed the SGA in both computational and electromagnetic aspects. This case study demonstrates the utility of using more advanced GA techniques to obtain acceptable solution quality as problem difficulty increases.

To conclude, for simple problems, SGAs are preferred since they are computationally inexpensive and the solution quality is comparable to that of competent GAs. However, for difficult problems, competent GA techniques should be adopted, because based on our observations, competent GA techniques are able to achieve higher-quality solutions than SGAs. For most of the real-world applications, there is no easy way to determine the problem difficulty for GAs. In this case, it is always beneficial to adopt competent GAs since competent GAs perform well on both simple and difficult problems.

References

[1] Ahn, C. W. (2005). *Theory, Design, and Application of Efficient Genetic and Evolutionary Algorithms.* Doctoral dissertation, Gwangju Institude of Science and Technology

[2] Caruana, R., and Schaffer, J. (1988). Representation and hidden bias: Gray vs. binary coding for genetic algorithms. *Proceedings of the Fifth International Conference on Machine Learning*, pp. 153–161

[3] Goldberg, D. E. (1989). *Genetic algorithms in search, optimization, and machine learning.* Addison-Wesley, Reading, MA

[4] Goldberg, D. E. (2002). *The design of innovation: Lessons from and for competent genetic algorithms.* Kluwer Academic, Boston, MA

[5] Kraus, J. D. (1988). *Antennas*, 2nd ed. McGraw-Hill, NY, USA

[6] Larrañaga, P., and Lozano, J. A., (Eds.) (2002). *Estimation of distribution algorithms.* Kluwer Academic, Boston, MA

[7] Mailloux, R. J. (2001). A low-sidelobe partially overlapped constrained feed network for time-delayed subarrays. *IEEE Transactions on Antennas and Propagation*, 49(2):280–291

[8] Pelikan, M. (2002). *Bayesian optimization algorithm: From single level to hierarchy.* Ph.d. thesis, University of Illinois at Urbana-Champaign, Urbana, IL (also IlliGAL Report No. 2002023)

[9] Pelikan, M., and Goldberg, D. E. (2000). Hierarchical problem solving by the Bayesian optimization algorithm. *Genetic and evolutionary computation conference (GECCO-2000)*, pp. 267–274

[10] Pelikan, M., and Goldberg, D. E. (2001). Escaping hierarchical traps with competent genetic algorithms. *Genetic and evolutionary computation conference (GECCO-2001)*, pp. 511–518

[11] Pelikan, M., and Goldberg, D. E. (2003). Hierarchical BOA solves Ising spin glasses and MAXSAT. *Genetic and Evolutionary Computation Conference (GECCO-2003)*, pp. 1271–1282 (also IlliGAL Report No. 2003001)

[12] Pelikan, M., Goldberg, D. E., and Sastry, K. (2001). Bayesian optimization algorithm, decision graphs, and Bayesian networks. *Genetic and evolutionary computation conference (GECCO-2001)*, pp. 519–526

[13] Rahmat-Samii, Y., and Michielssen, E., (Eds.) (1999). *Electromagnetic optimization by genetic algorithms.* Wiley, NY, USA

[14] Rowe, J., Whiteley, D., Barbluescu, L., and Watson, J. P. (2004). Properties of gray and binary representations. *Evolutionary Computation*, 12(1):47–76

13

Feature Subset Selection with Hybrids of Filters and Evolutionary Algorithms

Erick Cantú-Paz

Summary. The performance of classification algorithms is affected by the features used to describe the labeled examples presented to the inducers. Therefore, the problem of feature subset selection has received considerable attention. Approaches to this problem based on evolutionary algorithms (EAs) typically use the wrapper method, treating the inducer as a black box that is used to evaluate candidate feature subsets. However, the evaluations might take a considerable time and the wrapper approach might be impractical for large data sets. Alternative filter methods use heuristics to select feature subsets from the data and are usually considered more scalable than wrappers to the dimensionality and volume of the data. This chapter describes hybrids of evolutionary algorithms (EAs) and filter methods applied to the selection of feature subsets for classification problems. The proposed hybrids were compared against each of their components, two feature selection wrappers that are in wide use, and another filter-wrapper hybrid. The objective of this chapter is to determine if the proposed evolutionary hybrids present advantages over the other methods in terms of accuracy or speed. The experiments used are decision tree and naive Bayes (NB) classifiers on public-domain and artificial data sets. The experimental results suggest that the evolutionary hybrids usually find compact feature subsets that result in the most accurate classifiers, while beating the execution time of the other wrappers.

13.1 Introduction

The problem of classification in machine learning consists of using labeled examples to induce a model that classifies objects into a set of known classes. The objects are described by a vector of features, some of which may be irrelevant or redundant and may have a negative effect on the accuracy of the classifier. There are two basic approaches to feature subset selection: wrapper and filter methods [1]. Wrappers treat the induction algorithm as a black box that is used by the search algorithm to evaluate each candidate feature subset. While giving good results in terms of the accuracy of the final classifier, wrappers are computationally expensive and may be impractical for large

E. Cantú-Paz: *Feature Subset Selection with Hybrids of Filters and Evolutionary Algorithms*, Studies in Computational Intelligence (SCI) **33**, 291–314 (2006)
www.springerlink.com

data sets. Filter methods are independent of the classifier and select features based on properties that good feature sets are presumed to have, such as class separability or high correlation with the target. Although filter methods are much faster than wrappers, filters may produce disappointing results, because they completely ignore the induction algorithm.

This chapter presents experiments with hybrid algorithms that combine the strengths of filters and wrappers and attempt to avoid their weaknesses. The hybrids consist of an evolutionary algorithm (EA) used in its traditional role as a wrapper, but initialized with the output of a filter. The objective of this study is to determine if the EA-filter hybrids present advantages over EAs and conventional feature selection algorithms in terms of accuracy or speed. The experiments described in this chapter use public-domain and artificial data sets and consider decision tree and naive Bayes (NB)classifiers, which can be induced quickly and have been shown to have good accuracy in many problems [2].

I proposed earlier [3] the combination of EAs and filters studied in this chapter. My previous paper reported promising experiments initializing a simple genetic algorithm (GA) with the output of a filter and using NB as the induction algorithm. The present chapter extends my previous work [3] in several directions: (1) It examines the scalability of the hybrids by experimenting with additional data sets that are larger and of higher dimensionality than in the original study; (2) since the performance of wrapper algorithms depends on the classifier, this chapter presents experiments with decision trees in addition to the NB considered previously; (3) the experiments consider a simple distribution estimation algorithm in addition to the simple GA of my previous work; (4) experiments examine the effect of the single additional parameter introduced in the filter-EA hybrids; and (5) the proposed filter-EA hybrids are compared with a recently proposed combination of filters and wrappers.

The goal of this study was to maximize the accuracy of classification. The experiments demonstrate that, in most cases, the hybrids find subsets that result in the best accuracy (or in an accuracy not significantly different from the best), while finding compact feature subsets, and performing faster than popular wrapper methods.

Section 13.2 reviews previous applications of EAs to feature subset selection. Section 13.3 describes the class separability filter and the proposed hybridization with EAs. Section 13.4 describes the algorithms, data sets, and the fitness evaluation method used in the experiments reported in Sect. 13.5. Finally, Sect. 13.6 concludes this chapter with a summary and a discussion of future research directions.

13.2 Feature Selection

Reducing the dimensionality of the vectors of features that describe each object presents several advantages. As mentioned above, irrelevant or redundant features may affect negatively the accuracy of classification algorithms. In addition, reducing the number of features may help decrease the cost of acquiring data and might make the classification models easier to understand.

There are numerous techniques for dimensionality reduction [4]. Some common methods seek transformations of the original variables to lower dimensional spaces. For example, principal components analysis (PCA) reduces the dimensions of the data by finding orthogonal linear combinations with the largest variance. In the mean square error sense, PCA yields the optimal linear reduction of dimensionality. However, the principal components that capture most of the variance are not necessarily useful to discriminate among objects of different classes. Moreover, the linear combinations of variables obscure the effect of the original variables on class discrimination. For these reasons, this chapter focuses on techniques that select subsets of the original variables.

Among feature subset selection algorithms, wrapper methods have received considerable attention [1, 2, 4]. Wrappers are attractive because they seek to optimize the accuracy of a classifier, tailoring their solutions to a specific inducer and a domain. Wrappers search for a good feature subset using the induction algorithm to evaluate the merit of candidate subsets, and numerous search algorithms have been used to search for feature subsets [5]. In some small problems, it is possible to use exhaustive search, but it quickly becomes impractical as the dimensionality d of the data grows, as there are 2^d possible feature subsets. Popular wrappers use greedy search strategies in forward selection and backward elimination modes. Forward selection methods start from the empty set and iteratively add features, while backward elimination starts from the full set of features and iteratively eliminates what it considers the least important ones. There are variations of these greedy algorithms that try to alleviate the problem of getting stuck in local optima by using different strategies to add or remove features. This chapter compares the results of the proposed algorithms against popular greedy forward selection and backward elimination algorithms that add or delete a single feature in each iteration and which are described later.

The remainder of this section focuses on EAs as feature selectors. For comprehensive reviews of feature selection methods, the reader can refer to the survey articles by Kohavi and John [2], Blum and Langley [6], and Guyon and Elisseeff [4].

Applying EAs to the feature selection problem is straightforward: the chromosomes of the individuals contain one bit for each feature, and the value of the bit determines whether the feature will be used in the classification. Using the wrapper approach, the individuals are evaluated by training the classifiers using the feature subset indicated by the chromosome and using the resulting accuracy to calculate the fitness.

Siedlecki and Sklansky [7] were the first to describe the application of GAs in this way. GAs have been used to search for feature subsets in conjunction with several classification methods such as neural networks [8, 9], decision trees [10], k-nearest neighbors [11–14], rules [15], and naive Bayes [16, 17]. GAs usually deliver good results as wrappers, but exceptions have been reported where simpler (and faster) algorithms result in higher accuracies on particular data sets [5, 16].

Besides selecting feature subsets, GAs can extract new features by searching for vector of numeric coefficients that is used to transform linearly the original features [11, 12]. In this case, a value of zero in the transformation vector is equivalent to avoiding the feature. Raymer et al. [13, 18] combined the linear transformation with explicit feature selection flags in the chromosomes, and reported an advantage over the pure transformation method.

Inza et al. [16] pioneered the use of distribution estimation algorithms (DEAs) to search for optimal feature subsets. Similarly to traditional EAs, DEAs select the most promising solutions to create the next generation, but DEAs do not use the traditional crossover or selection operators. Instead, DEAs create a probabilistic model of the selected individuals and use the model to generate new solutions. Sophisticated DEAs explicitly model the relationships among the variables of a problem to avoid the disruption of groups of related variables that might prevent a simple GA from reaching the global optimum. However, multiple experimental studies suggest that – in terms of accuracy – DEAs do not significantly outperform simple GAs when searching for feature subsets [16, 17, 19, 20]. Despite this, DEAs seem like a good fit for the problem of feature selection and the present work investigates, for the first time, DEAs initialized with the output of a filter.

The wrappers' evaluation of candidate feature subsets can be computationally expensive on large data sets. Filter methods are computationally efficient and offer an alternative to wrappers. GAs have been used as filters in regression problems to optimize a cost function derived from the correlation matrix between the features and the target value [21]. Lanzi used GAs as a filter in classification problems by minimizing the inconsistencies present in subsets of the features [22]. An inconsistency between two examples occurs if the examples match with respect to the feature subset considered, but their class labels disagree. Lanzi demonstrated that this filter method efficiently identifies feature subsets that were as predictive as the original set of features (the results were never significantly worse). However, the accuracy using the reduced subsets was not much different (better or worse) than with all the features. The present study shows that the proposed methods can reduce the dimensionality of the data and increase the predictive accuracy considerably.

13.3 Class Separability

The idea of using a measure of class separability to select features has been used in machine learning and computer vision [4, 23]. The filter described in this section calculates the class separability of each feature using the Kullback–Leibler (KL) divergence between histograms of feature values. For each feature, there is one histogram for each class. Numeric features are discretized using $\sqrt{|D|}/2$ equally-spaced bins, where $|D|$ is the number of examples in the training data. The histograms are normalized dividing each bin count by the total number of elements to estimate the probability that the jth feature takes a value in the ith bin of the histogram given a class k, $p_j(i|k)$. For each feature j, the filter calculates the class separability as

$$\Delta_j = \sum_{m=1}^{c} \sum_{n=1}^{c} \delta_j(m,n), \tag{13.1}$$

where c is the number of classes and $\delta_j(m,n)$ is the KL divergence between histograms corresponding to classes m and n:

$$\delta_j(m,n) = \sum_{i=1}^{b} p_j(i|m) \log\left(\frac{p_j(i|m)}{p_j(i|n)}\right), \tag{13.2}$$

where b is the number of bins in the histograms. Of course, other metrics could be used instead of KL divergence.

The features are then sorted in descending order of the divergence Δ_j (larger divergence mean better separability). Heuristically, the filter considers that two features are redundant if their divergences differ by less than 1%, and if there is a decrease of 50% or more in the divergences of consecutive features, the filter considers the remainder features as irrelevant.

These heuristics may fail if the thresholds chosen are not adequate for a particular classification problem. The heuristics were calibrated using artificial data sets that are described later. However, perhaps the major disadvantage of this and most filter methods is that they ignore pairwise (or higher) interactions among variables. It is possible that features that appear irrelevant (not discriminative) when considered alone are relevant when considered in conjunction with other variables. For example, consider the two-class data displayed in Fig. 13.1. Each of the features alone does not have discriminative power, but taken together the two features perfectly discriminate the two classes.

The initial motivation for the present work was to investigate methods to alter the feature subset output by the filter to examine feature combinations that the filter cannot consider. A simple way of doing this is to represent the feature subset as a vector of ones and zeroes that indicate which features were selected by the filter, and then randomly flip elements of this vector to produce several new alternative feature subsets. It is possible that some of these feature subsets perform better than the subset output by the filter. This simple idea can be improved in two ways. The first improvement is to recognize that the filter does not output only a binary decision for including each feature, but it also gives an indication of their relative importance that can be used to bias the inclusion of features in the new feature subsets. A feature that is deemed important by the filter should have a higher chance of being included in new feature subsets than a feature considered unimportant. Second, we can select

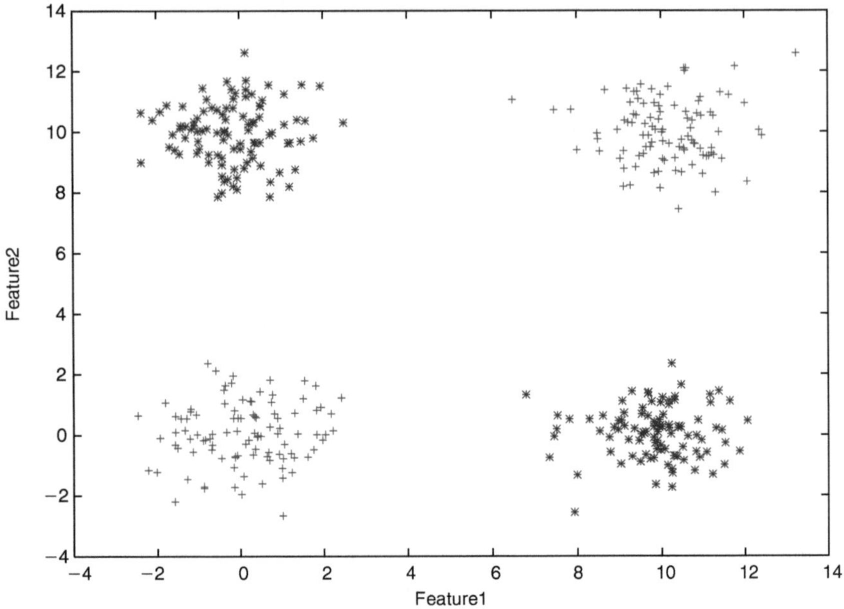

Fig. 13.1. Example of a data set where each feature considered alone does not discriminate between the two classes, but the two features taken together discriminate the data perfectly

the best subsets and examine how often each feature appears in them and bias the generation of new subsets to prefer features that appear often in the selected solutions. Iterating this procedure essentially results in a distribution estimation algorithm. As discussed in the previous section, EAs and DEAs have been used extensively for feature subset selection, but never in the role of trying to improve the output of a filter.

After executing the filter algorithm, we have some knowledge about the relative importance of each feature considered individually. This knowledge can be incorporated into the EAs by using the KL distances to initialize the EA. The distances Δ_j are linearly normalized between $\epsilon = 0.1$ and $1-\epsilon = 0.9$ to obtain the probability p_j that the jth bit in the chromosomes is initialized to 1 (and thus that the corresponding feature is selected). By making the lower and upper limits of p_j different from 0 and 1, the EA can explore combinations that include features that the filter had eliminated. It also allows a chance to delete features that the filter identified as important. The parameter ϵ regulates how much importance is given to the output of the filter. Setting $\epsilon = 0$ results in the most biased initialization, while setting $\epsilon = 0.5$ corresponds to the usual unbiased initialization of EAs. Section 13.5.3 will examine the effect of the parameter ϵ in detail.

After the EA is initialized with the output of the filter, the EA runs as a wrapper feature selection algorithm. The EA manipulates a population of candidate feature subsets using the conventional randomized EA operators. Each candidate solution is evaluated using an estimate of the accuracy of a classifier on the feature subset indicated in the chromosome and the best solution is reported to the user.

13.4 Methods

This section describes the data sets and algorithms used in this study, the method used to evaluate the fitness, and the method used to compare the algorithms.

13.4.1 Data Sets and Algorithms

The data sets used in the experiments are described in Table 13.1. With the exception of Random21 and Redundant21, the data sets are available in the UCI repository [24]. Random21 and Redundant21 are two artificial data sets with 21 features each and were proposed originally by Inza et al. [16]. The target concept of these two data sets is whether the first nine features are closer to (0,0,. . . ,0) or (9,9,. . . ,9) in Euclidean distance. The features were generated uniformly at random in the range [3,6]. All the features in Random21 are random, and the first, fifth, and ninth features are repeated four times each in Redundant21.

The classifiers used in the experiments were a naive Bayes (NB) and a decision tree (DT). These classifiers were chosen for their speed and simplicity, and they have been considered in numerous studies of feature selection. Of course, the proposed hybrid method can be used with any other supervised classifiers. In the NB, the probabilities for nominal features were estimated

Table 13.1. Description of the data used in the experiments

			Features		
Domain	**Instances**	**Classes**	**Numeric**	**Nominal**	**Missing**
Anneal	898	6	9	29	Y
Arrhythmia	452	16	206	73	Y
Credit-A	690	2	6	9	Y
Euthyroid	3,163	2	7	18	Y
Ionosphere	351	2	34	–	N
Isolet-5	1,559	26	617	–	N
Pendigits	10,992	10	16	–	N
Pima	768	2	8	–	N
Segmentation	2,310	7	19	–	N
Soybean	683	19	–	35	Y
Random21	2,500	2	21	–	N
Redundant21	2,500	2	21	–	N

from the data using maximum likelihood estimation (their observed frequencies in the data) and applying the Laplace correction. Numeric features were assumed to have a normal distribution. The DT split the data using the Gini criterion and the trees were pruned using Quinlan's pessimistic error pruning [25]. Missing values in the data were ignored by the algorithms.

The GA used uniform crossover with probability 1.0, and mutation with probability $1/l$, where l was the length of the chromosomes and corresponds to the total number of features in each problem. The population size was set to $\lfloor 3\sqrt{l} \rfloor$, following the gambler's ruin model for population sizing that asserts that the population size required to reach a solution of a particular quality is $O(\sqrt{l})$ [26]. Promising solutions were selected with pairwise binary tournaments without replacement. The algorithms were terminated after observing no improvement of the best individual over consecutive generations. Inza et al. [16] and Cantú-Paz [17] used similar algorithms and termination criterion.

In addition to the GA, the experiments consider a simple distribution estimation algorithm that assumes that the variables (bits) that represent the problem are independent of each other, and therefore models the population as a product of Bernoulli distributions. The algorithm maintains a vector p of length equal to the problem's length, l. Each element of p contains the probability that a sample will take the value 1. If the Bernoulli trial is not successful the sample will be 0. Usually, all positions of p are initialized to 0.5 to simulate the usual uniform random initialization of simple GAs, but we initialize p with the normalized outputs of the filter as described in the previous section. New individuals are obtained by sampling consecutively from each position of p and concatenating the values obtained. We used the same number of individuals as the simple GA ($\lfloor 3\sqrt{l} \rfloor$) in each generation. The probabilities vector is updated once every generation by selecting the top 20% of offspring and computing the proportion of selected individuals with a 1 in each position. The algorithm iterates until all positions in p contain either zero or one.

The compact GA [27], PBIL [28], and the UMDA [29] are other DEAs that use univariate models. They differ from the algorithm used here in the method to update the probabilities vector. Preliminary experiments with a compact GA yielded results that were not significantly different from those obtained with the simple algorithm described above.

We compare the results of the class separability filter and the EAs with two greedy feature selection wrappers. Greedy feature selection algorithms that add or delete a single feature from the candidate feature subset are common. There are two basic variants: sequential forward selection (SFS) and sequential backward elimination (SBE). Forward selection starts with an empty set of features. In the first iteration, the algorithm considers all feature subsets with only one feature. The feature subset with the highest accuracy is used as the basis for the next iteration. In each iteration, the algorithm tentatively adds to the basis each feature not previously selected and retains the feature subset that results in the highest estimated performance. The search terminates after the accuracy of the current subset cannot be improved by adding any other feature. Backward elimination works in an analogous way, starting from the full set of features and tentatively deleting each feature not deleted previously.

We also compare the results of the evolutionary filter-wrapper hybrids with a filter-wrapper hybrid recently proposed by Das [30]. Das' algorithm starts with an empty feature set and in each iteration greedily adds one feature, so it is similar to SFS. The differences lie in the way the feature is selected and in how the algorithm terminates.

In each iteration, using a user-supplied filter, Das' algorithm ranks the features that have not been selected so far and adds the highest-ranking feature to the feature subset. For this ranking, the algorithm uses the class separability filter described in Sect. 13.3. Then, a "re-weighting" classifier is trained using the current subset and is used to classify the instances in the training set. The weights of the instances are updated using the standard AdaBoost procedure [31] (giving more weight to instances misclassified by the classifier) and the algorithm iterates.

We follow Das and train the re-weighting classifier ignoring the instances' weights. The weights are only used by the filter to rank the unselected features in each iteration. In this way, the filter is asked to identify the feature that best discriminates the instances that are hard to classify using the features selected previously.

The algorithm is stopped when the accuracy of a "stopping" classifier trained with all the selected features does not improve from the previous iteration. Das argued that using the accuracy on the training set was adequate for stopping the algorithm. We performed experiments using cross-validation estimates of the accuracy, but confirmed that the results were not different. In the experiments reported in the next section the "re-weighting" and "stopping" classifiers were of the same type (either a DT or NB, as indicated in the experiments).

The algorithms were developed in C++ and compiled with g++ version 2.96 using -O2 optimizations. The experiments were executed on a single processor of a Linux (Red Hat 7.3) workstation with dual 2.4 GHz Intel Xeon processors and 1024 MB of memory. A Mersenne Twister random number generator [32] was used in the EAs and the data partitioning. The random number generator was initialized with 32-bit numbers obtained from `www.random.org`.

13.4.2 Measuring Fitness

Since we are interested in classifiers that generalize well, the fitness calculations must include an estimate of the generalization of the classifiers using the candidate subsets. We estimate the generalization of the classifiers using cross-validation. In k-fold cross-validation, the data D is partitioned randomly into k non-overlapping sets, $D_1, \ldots, D_k$. At each iteration i (from 1 to k), the classifier is trained with $D \backslash D_i$ and tested on D_i. Since the data are partitioned randomly, it is likely that repeated cross-validation experiments return different results. Although there are well-known methods to deal with "noisy" fitness evaluations in EAs [33], we chose to limit the uncertainty in the accuracy estimate by repeating tenfold cross-validation experiments until the standard deviation of the accuracy estimate drops below 1% (or a maximum of five repetitions). This heuristic was proposed by Kohavi and John [2] in their study of wrapper methods for feature selection, and was adopted by Inza et al. [16]. We use the accuracy estimate as the fitness function.

Even though cross-validation can be expensive computationally, the cost was not prohibitive in our case, since the data sets were relatively small and the classifiers are very efficient. If larger data sets or other inducers were used, we would deal with the uncertainty in the evaluation by other means, such as increasing slightly the population size (to compensate for the noise in the evaluation) or by sampling the training data. We defer a discussion of possible performance improvements until the final section.

The fitness measure does not include any term to bias the search toward small feature subsets. However, the algorithms found small subsets, and in some cases the algorithms consistently found the smallest subsets that describe the target concepts. This suggests that the data sets contained irrelevant or redundant features that decreased the accuracy of the classifiers. Extending the fitness evaluation to explicitly penalize large feature subsets is a possibility for future work.

13.4.3 Comparing the Algorithms

To evaluate the generalization accuracy of the feature selection methods, we used five iterations of twofold cross-validation (5×2 cv). In each iteration, the data were randomly divided in halves. One half was input to the feature selection algorithms. The final feature subset found in each experiment was used to train a final classifier (using the entire training data), which was then tested on the other half of the data that has not been shown to the algorithm during the search. The accuracy results presented in Tables 13.2 and 13.6 are the means of the ten tests.

Note that the data input to the feature selection method is further divided during the search to estimate the accuracy using (possibly multiple) tenfold cross-validations as described in the previous section. Using an outer 5×2 cross-validation allows us to use a proper statistical test to compare the results [34, 35] and to avoid over-fitting in the comparison of feature selection algorithms, an issue that has received attention recently [36, 37].

To determine if the differences among the algorithms were statistically significant, we used a combined F test proposed by Alpaydin [35]. Let $p_{i,j}$ denote the difference in the accuracy rates of two classifiers in fold j of the ith iteration of 5×2 cv, $\bar{p} = (p_{i,1} + p_{i,2})/2$ denote the mean, and $s_i^2 = (p_{i,1}\text{-}\bar{p})^2 + (p_{i,2}\text{-}\bar{p})^2$ the variance, then

$$f = \frac{\sum_{i=1}^{5} \sum_{j=1}^{2} (p_{i,j})^2}{2\sum_{i=1}^{5} s_i^2} \tag{13.3}$$

Table 13.2. Means of the accuracies found in the 5×2 cv experiments using the naive Bayes classifier

Domain	All	Filter	FiltGA	sGA	FiltDEA	DEA	SFS	SBE	Boost
Anneal	89.93	**93.43**	**93.07**	**92.47**	**94.36**	93.49	90.36	**93.47**	**93.30**
Arrhythmia	56.95	62.08	**64.16**	59.78	**66.59**	**64.77**	58.67	59.73	59.29
Credit-A	71.30	72.52	71.97	71.42	71.65	71.44	**81.25**	77.28	**81.57**
Euthyroid	87.33	89.06	**94.20**	**94.92**	**94.40**	**94.03**	**94.57**	**94.48**	**94.56**
Ionosphere	83.02	**89.57**	**90.54**	**88.95**	**91.56**	**90.02**	85.23	**89.17**	87.29
Isolet	**80.24**	80.24	84.00	81.08	**86.32**	**85.43**	51.28	61.13	15.65
Pendigits	**85.72**	**85.82**	**84.63**	**84.77**	**86.77**	**85.66**	**86.16**	**85.99**	77.21
Pima	**74.87**	**74.45**	**75.49**	**75.29**	**75.62**	**75.65**	**73.46**	**74.45**	**75.39**
Random21	93.89	82.24	**95.41**	92.45	**95.34**	**94.11**	82.12	80.61	82.02
Redundant	77.12	80.29	83.68	**86.70**	**87.44**	**90.34**	79.74	80.32	80.79
Segment	79.92	85.40	**87.97**	84.73	**88.29**	86.19	**90.85**	**91.28**	80.03
Soybean	84.28	86.01	81.23	81.79	83.81	**94.93**	78.63	86.27	80.56
Mean	80.38	81.76	83.86	82.86	85.18	85.51	79.36	81.18	75.64
Best	3	4	8	6	10	9	5	6	4

The best result and those not significantly different from the best are displayed in **bold**, and the number of times each method found these results is reported in the last line

is approximately F distributed with 10 and 5 degrees of freedom. We rejected the null hypothesis that the two algorithms have the same error rate at a 0.05 significance level if $f > 4.74$ [35]. Care was taken to ensure that all the algorithms used the same training and testing data in the two folds of the five cross-validation experiments.

13.5 Experiments

The next two subsections presents experiments with NB and DTs. Each subsection presents experiments that compare the proposed filter-GA and filter-DEA hybrids against (1) their filter and EA components separately, (2) the popular SFS and backward elimination wrappers, (3) a recently proposed filter-wrapper hybrid, and (4) classifiers built on all the features. Section 13.5.3 presents experiments that evaluate the effect of the parameter ϵ on the filter-EA hybrids.

13.5.1 Experiments with Naive Bayes

Table 13.2 has the mean accuracies obtained with each feature selection method. The best observed result in the table is highlighted in **bold** type as well as those results that according to the combined F test are not significantly different from the best at a 0.05 significance level. There are two immediate observations that we can make from the results. First, the feature selection algorithms result in an improvement of accuracy over using a NB with all the features. However, this difference is not always significant (Isolet, Pendigits, Pima). Second, the proposed filter-DEA hybrid reaches the highest accuracy or accuracies that are not significantly different from the highest more often than the other algorithms (in 10 out of 12 data sets). In terms of accuracy, there seems to be a small difference between the filter-DEA hybrid and the DEA that was initialized randomly (which found best accuracies in nine of the data sets). The filter-GA hybrid also did slightly better than the GA initialized randomly, finding solutions indistinguishable from the best for eight data sets. The other wrappers (SFS and SBE) and Das' hybrid did noticeably worse than the evolutionary wrappers in these data sets.

In terms of the size of the final feature subsets (Table 13.3), forward sequential selection consistently found the smallest subsets. This was expected, since this algorithm is heavily biased toward small subsets (because it starts from an empty set and adds features only when they show improvements in accuracy). However, as noted above, in most cases SFS resulted in significantly worse accuracies than the proposed EA hybrids. The proposed hybrids usually found significantly – and sometimes substantially – smaller feature subsets than the filter alone or the EAs that were initialized randomly. This is interesting because the smaller subsets did not result in accuracy degradations.

Table 13.4 shows the mean number of feature subsets examined by each algorithm. In most cases, the GAs examine fewer subsets than SFS and SBE, and the Filter-GA examined much fewer subsets than the GA initialized at random. This suggests that the search of the Filter-GA was highly biased towards good solutions. In contrast, while the Filter-DEA examined fewer subsets than the simple DEA, the difference is not as marked as with the GAs.

Table 13.3. Means of the sizes of final feature subsets in experiments using the naive Bayes classifier

Domain	Original	Filter	FiltGA	sGA	FiltDEA	DEA	SFS	SBE	Boost
Anneal	38	23.8	12.8	22.1	12.6	23.1	**5.4**	16.4	29.20
Arrhythmia	279	212.5	86.2	138.9	79.5	137.8	**3.9**	261.1	11.10
Credit-A	15	**1**	4.7	8.40	4.3	7.3	3.1	10.1	3.0
Euthyroid	25	**1**	6.3	13.7	6.7	13.5	**1.3**	1.2	10.2
Ionosphere	34	31.0	11.2	16.0	10.3	14.4	**4.4**	30.9	**4.6**
Isolet	617	617	224.2	317.6	218.3	296.6	12.1	221.5	**4.8**
Pendigits	16	16.00	9.90	13.30	9.8	10.4	10.60	9.60	**7.20**
Pima	8	4.3	**2.9**	4.9	**2.8**	3.7	**1.6**	5.3	4.2
Random21	21	**10.2**	**10.3**	13.6	**10**	12.9	**9.3**	12.6	10.6
Redundant	21	8.8	**8.1**	10.6	**8.1**	10.4	**8.6**	9.1	9.0
Segmentation	19	11.0	9.9	9.6	8.6	8.8	**4.0**	7.7	**3.2**
Soybean	35	32.9	19.50	21.7	22.2	**13.5**	**10.6**	30.7	24.6
Mean	94.00	80.79	33.83	49.20	32.77	46.03	6.24	51.35	10.14

The best result and those not significantly different from the best are in **bold**

Table 13.4. Means of the number of feature subsets examined by each algorithm using the naive Bayes classifier

Domain	FiltGA	sGA	FiltDEA	DEA	SFS	SBE
Anneal	**38.84**	48.08	252	293.4	225.50	569.20
Arrhythmia	**105.23**	120.26	3,100	3,835	1,356.0	4,706.9
Credit-A	**25.56**	**27.89**	93.6	98.4	53.50	71.20
Euthyroid	**36.00**	**37.50**	204.8	257.6	55.8	324.8
Ionosphere	**38.48**	**41.98**	288	257.4	170.5	131.5
Isolet	63.60	**36.00**	113.70	99.00	105.4	287.1
Pendigits	28.80	48.00	67.2	70.8	124.00	94.30
Pima	**12.73**	20.36	25.6	32.8	**18.5**	24.1
Random21	**35.74**	64.61	105	126	168.0	147.9
Redundant21	**32.99**	42.62	67.2	70	159.9	193.9
Segmentation	**37.92**	**30.08**	65.8	84	84.8	160.3
Soybean	**42.60**	**42.60**	241.2	257.6	342.5	171.5
Mean	41.54	46.67	385.34	456.83	238.70	573.56

The best result and those not significantly different from the best are in **bold**

The number of examined subsets can be used as a coarse surrogate for the execution time, but the actual times depend on the number of features present in each candidate subset and may vary considerably from what we might expect. The execution times (user time in CPU seconds) for the entire 5×2 cv experiments are reported in Table 13.5. For the filter method, the time reported includes the time to compute and sort class separabilities and the time to evaluate the NB on the feature subset found by the filter method. The proposed filter method is by far the fastest algorithm, beating its closest competitor by two orders of magnitude. However, the filter found significantly less accurate results for nine of the 12 datasets. Among the wrapper methods, SFS and the hybrid of the filter and the GA are the fastest. The filter-EA hybrids were faster than the EAs initialized randomly, but the differences are more pronounced in the GA than in the DEA.

13.5.2 Experiments with Decision Trees

The mean accuracies found by feature selection methods using DTs are reported in Table 13.6. The relative performance of different algorithms is more pronounced with DTs than with NB. The filter-DEA hybrid always finds feature subsets that result in the best accuracy or accuracies that are not significantly different from the best. In contrast, SFS and Boosting always find results that are significantly lower from the best. In terms of accuracy, the results suggest that initializing the EAs with the filter output is advantageous.

The size of the resulting feature subsets is reported in Table 13.7. As in the experiments with the NB, SFS clearly dominates the other algorithms. The second smallest feature subsets were usually found by the Boosting filter.

Table 13.5. Execution time (in CPU seconds) of the 5×2cv experiments with each feature selection algorithm using the naive Bayes classifier

Domain	Filter	FiltGA	sGA	FiltDEA	DEA	SFS	SBE	Boost
Anneal	**0.28**	44.2	66.4	203.1	325.4	26.1	190	12.56
Arrhythmia	**4.37**	926.0	1,322.9	19,091	31,837	775	32,497	14.75
Credit-A	**0.05**	5.5	7.6	15.7	21.9	3.2	9.9	0.28
Euthyroid	**0.31**	62.4	91.9	257.7	434.1	21.2	290.3	4.63
Ionosphere	**0.12**	9.9	12.8	54.9	57.9	10.4	22.1	0.26
Isolet	**115**	19,689	24,287	17,775	21,987	46,757	2,88,980	107
Pendigits	**4.94**	368	626	629	835	504	594	29
Pima	**0.03**	2.1	2.8	3.1	4.6	0.9	2.3	0.26
Random21	**0.46**	44.8	80.6	112.1	150.2	71.9	119.6	3.60
Redundant21	**0.45**	44.0	54.6	65.9	78.3	67.1	148.6	3.11
Segmentation	**0.64**	77.3	65.5	105.1	136.6	31.6	138.6	1.78
Soybean	**1.81**	94.5	99.7	482.9	503.7	137.2	293.4	41.08
Mean	10.71	1,780	2,226	3,232	4,697	4,033	26,940	18.19
Median	0.46	53.60	73.50	157.60	237.80	49.35	169.30	4.12

The Filter method is always the fastest algorithm (denoted with **bold** type)

Table 13.6. Means of the accuracies found in the 5×2cv experiments using the decision tree

Domain	All	Filter	FiltGA	sGA	FiltDEA	DEA	SFS	SBE	Boost
Anneal	**87.54**	**85.79**	85.95	**86.17**	**85.69**	**86.14**	83.96	83.96	35.26
Arrhythmia	54.20	57.96	57.08	66.59	**67.30**	**69.12**	64.25	59.12	58.23
Credit-A	81.91	**85.51**	**84.43**	82.23	**85.87**	84.71	79.78	**84.78**	78.32
Euthyroid	94.12	90.71	**96.50**	**97.33**	**96.36**	**97.22**	96.13	**97.04**	95.61
Ionosphere	88.15	88.15	**88.71**	87.97	**89.12**	**88.03**	80.68	**88.32**	80.97
Isolet	9.52	9.52	38.03	28.46	**42.01**	35.54	21.73	–	21.26
Pendigits	**89.49**	**89.49**	85.90	**88.86**	**88.73**	**89.03**	71.13	78.65	66.84
Pima	**73.70**	**72.14**	**73.23**	**72.97**	**72.66**	**72.08**	66.35	**73.23**	66.77
Random21	72.90	**74.43**	**74.05**	72.83	**74.18**	72.79	71.56	72.48	63.87
Redundant	72.92	**74.23**	**74.27**	**73.45**	**73.80**	**73.74**	72.94	**73.46**	63.98
Segment	**93.88**	**94.29**	**93.99**	**93.96**	**94.48**	**94.15**	85.88	88.51	70.15
Soybean	89.76	90.11	**95.12**	91.97	**95.54**	92.03	88.15	91.41	84.41
Mean	75.67	76.03	78.94	78.57	80.48	79.55	73.55	81.00	65.47
Best	4	7	8	6	12	8	0	5	0

The best result and those not significantly different from the best are displayed in **bold**

However, the accuracies reached using the feature subsets found by SFS and Boosting were always lower than the best results. Notably, the filter-DEA hybrid that always found the most accurate results found feature subsets that are approximately one fourth of the size of the original feature set. The filter-GA hybrid also successfully found compact subsets that were on average approximately one third of the original set. As in the case with NB, the EAs

Table 13.7. Means of the sizes of final feature subsets in experiments using the decision tree

Domain	Original	Filter	FiltGA	sGA	FiltDEA	DEA	SFS	SBE	Boost
Anneal	38	23.8	9.7	19.7	9.8	19.6	1.6	8.6	1.3
Arrhythmia	279	212.5	86.8	139.4	84.5	129.9	**4.4**	191.2	**6.9**
Credit-A	15	**1**	3.5	7.4	3.4	7.2	3.4	12.8	3.1
Euthyroid	25	**1**	5.7	13.6	6.6	13.6	2.6	5.5	4.3
Ionosphere	34	31	11.2	16.4	11.1	16.5	**3.1**	31.8	30.2
Isolet	617	617	213.9	343.1	98.1	190.1	**8.7**	–	12.5
Pendigits	16	16	9.3	11.6	10.7	12.5	7.0	14.30	**6.3**
Pima	8	4.3	**3.1**	4.2	3.0	4.0	**2.1**	6.6	8.00
Random21	21	10.2	9.9	12.7	9.7	13.1	**6.8**	17.2	**6.7**
Redundant	21	8.8	10.7	11.8	9.7	12.0	6.7	16.7	**3.4**
Segment	19	11	10.7	10.7	9.4	10.2	**4.1**	14.6	4.7
Soybean	35	32.9	**19.4**	24.3	**19.7**	**20.1**	22.7	32.8	**17.9**
Mean	94.00	80.79	32.83	51.24	22.98	37.40	6.10	31.37	8.78

The best result and those not significantly different from the best are in **bold**

initialized randomly found noticeably larger feature subsets than the EAs initialized with the filter's output.

The number of subsets examined is reported in Table 13.8 and the execution times of each algorithm is in Table 13.9. There is a noticeable increase from the execution times with NB and the differences between feature selection algorithms are more pronounced with DTs than with naive Bayes. The distance filter and the boosting filter are again the fastest algorithms, but as mentioned above the accuracies of DTs trained on the features identified with these methods were always significantly different from the best result. Among the wrapper algorithms, the filter-GA hybrid is clearly the fastest and SBE the slowest.

To better appreciate the relative performance of the algorithms we tried, Fig. 13.2 contains plots of the average accuracy versus average execution time. Ideal algorithms would be close to the bottom right corner of the plots, which represents minimum time and maximum accuracy. The plots clearly show that the filter is the fastest algorithm on average and that SBE is the slowest. When the EAs are hybridized with the filter, they produce more accurate results and converge faster than without the hybridization.

13.5.3 Effect of Normalizing the Filter Output

Recall that in the previous experiments the outputs of the filter are linearly normalized between $\epsilon = 0.1$ and $1 - \epsilon = 0.9$ and used as the probabilities of including features in the initial populations for the EAs. This section examines the impact of ϵ on the results with experiments on the segmentation and the two artificial data sets used previously. For the experiments in this section, the value of ϵ varies between 0, which is the setting that gives the most weight

Table 13.8. Means of the number of feature subsets examined by each algorithm using the decision tree

Domain	FiltGA	sGA	FiltDEA	DEA	SFS	SBE
Anneal	36.99	**22.19**	315.00	343.80	96.60	195.6
Arrhythmia	**100.22**	**107.42**	3,695.00	3,840.00	1,494.30	20,754.30
Credit-A	**27.48**	**32.60**	54.65	59.90	63.40	89.90
Euthyroid	**33.00**	**39.00**	214.40	201.60	84.80	304.60
Ionosphere	**34.99**	**26.24**	295.20	356.40	131.80	104.90
Isolet	**46.79**	**49.15**	**66.64**	**65.23**	–	197.21
Pendigits	**37.20**	**42.00**	80.40	87.60	98.90	174.2
Pima	**16.12**	**11.03**	20.00	28.80	21.00	17.40
Random21	**28.87**	**30.24**	110.60	148.40	135.90	90.60
Redundant	**31.62**	**32.99**	109.20	128.80	135.60	97.90
Segment	**18.31**	27.46	102.20	130.20	86.2	86.8
Soybean	**32.90**	**43.91**	87.15	90.13	290.1	343.1
Mean	37.04	38.09	429.20	456.74	239.87	1,871.38

The best result and those not significantly different from the best are in **bold**

Table 13.9. Execution time (in CPU seconds) of the 5×2 cv experiments with each feature selection algorithm using the decision tree

Domain	Filter	FiltGA	sGA	FiltDEA	DEA	SFS	SBE	Boost
Anneal	**0.54**	19	19	80	126	114	384	1.20
Arrhythmia	**8.8**	2,984	2,444	62,353	70,600	2,993	1,34,880	13
Credit-A	**0.1**	690	1,090	998	1,320	2,560	12,520	12
Euthyroid	**1.5**	225	454	1,522	1,326	299	1,433	9.8
Ionosphere	**0.6**	44.4	58.3	249.4	475.5	124.2	2,790	70.8
Isolet	**46.1**	31,089	34,583	96,843	1,32,780	1,12,322	–	111.7
Pendigits	**17.7**	3,003	4,163	5,822	7,759	22,533	87,994	532
Pima	**1.3**	16.1	15.5	19.2	53.0	22.1	123.5	7.7
Random21	**0.9**	551	665	993	1,884	635	2,457	57.6
Redundant	**2.1**	554	696	1,014	1,339	367	4,559	13.3
Segment	**3.3**	251	295	735	969	2,033	12,634	66.4
Soybean	**2.56**	101	111	427	435	243	890	16
Mean	7.13	3,293.96	3,716.15	14,254	18,255	12,020	23,696	75.88
Median	1.80	401.00	559.50	995.50	1,323.00	501.00	2,790.00	14.65

The Filter method is always the fastest algorithm (denoted with **bold** type)

to the filter outputs, to 0.5 where the filter outputs are ignored and the GA is initialized without bias.

The results of the experiments with the Random21, Redundant, and Segmentation data are in Figs. 13.3–13.5. Each panel in the figures shows results with the NB and DT classifiers. The top row in each figure corresponds to results with the GA-Filter and the bottom row to experiments with the Filter-DEA. The figures show the average accuracy of 5×2 cv experiments,

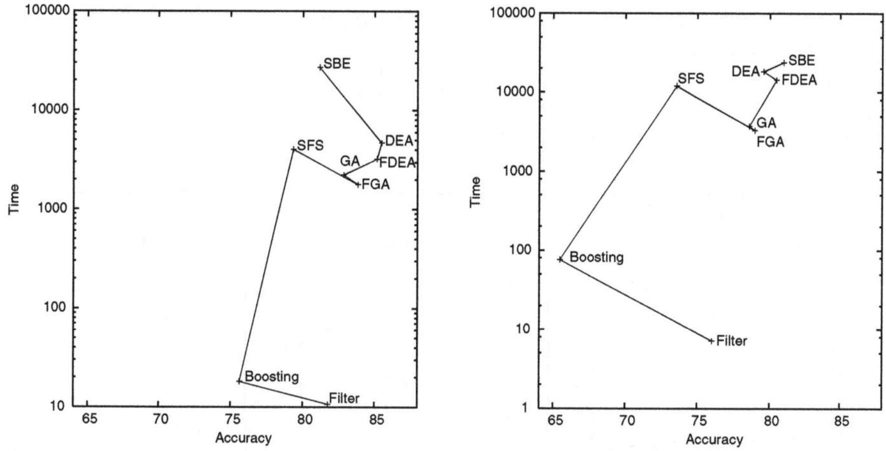

(a) Naive Bayes (b) Decision Trees

Fig. 13.2. Plot of mean accuracies versus mean execution time for experiments with the naive Bayes and decision trees

the average number of features in the final subsets, and the execution time in CPU seconds.

Examining the results on Random21 in Fig. 13.3, it is clear that as ϵ increases the accuracy tends to decrease, more features are included in the feature subsets, and the execution time is longer. Qualitatively, the results with the GA and the DEA are very similar.

The filter correctly identifies that only nine features are relevant in this data set, that is, the KL divergence for the relevant features is much higher than for the irrelevant ones. Making $\epsilon = 0$ gives the most weight to the outputs of the filter and is the best choice for these data, because the EA is initialized with very small probabilities of considering the irrelevant features and very high probabilities of considering the relevant ones. Increasing ϵ diminishes the influence of the filter on the initialization of the EAs and there are greater chances that the irrelevant features are considered.

The accuracy of the DT is noticeably lower than the NB. This data set is difficult for axis-parallel DTs for two reasons. The most important reason is that the data are separated into classes by a hyperplane that is not parallel to any of the axes. The DT is forced to approximate this "oblique" hyperplane by a series of axis-parallel partitions that resemble a staircase. The other reason is that the tree building algorithm consistently selects irrelevant features that accidentally appear relevant in lower levels of the tree after the data have been partitioned several times.

For the Redundant21 data, Fig. 13.4 shows that as the value of ϵ increases, more features are included in the final subsets, and the execution time is

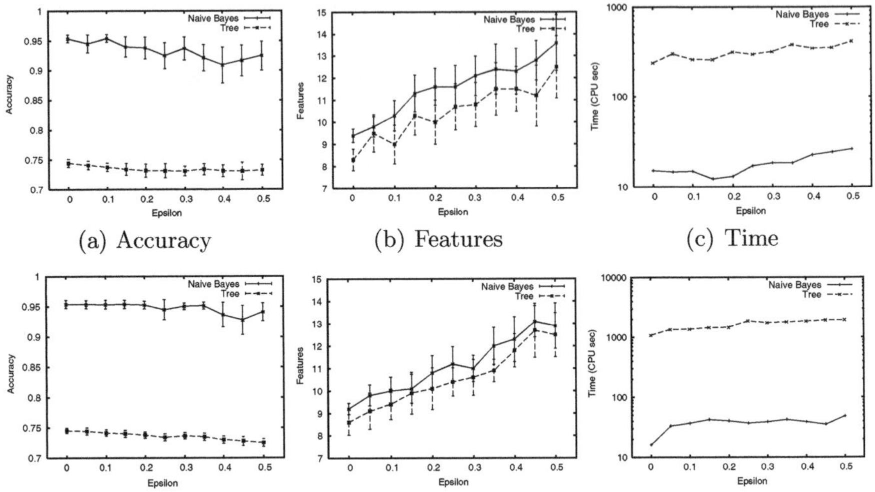

(a) Accuracy (b) Features (c) Time

(d) Accuracy (e) Features (f) Time

Fig. 13.3. Experimental results on Random21 varying the value of ϵ and using a GA (*top row*) and an DEA (*bottom row*) as the wrapper. The left panels present the classification accuracy, the middle panels present the number of features, and the right panels present the execution time in CPU seconds. The error bars represent 95% confidence intervals

longer. The Redundant21 data is difficult for DTs for the same reasons that Noisy is difficult, and again DTs have noticeably lower accuracies than NB.

In these data, the first nine features define the class of each instance and the first, fifth, and ninth features are copied four times each, resulting in a total of 21 features. The KL divergence for the repeated features is, of course, the same, and since all the features contribute the same to define the class, all KL divergences should be equal, and there should be no effect of ϵ. However, these data were generated randomly, and the ninth feature appears to be the less relevant because it has the lowest KL divergence. After scaling the KL divergences, the probability that the ninth feature is included in the initial population is only ϵ. The figure shows that the accuracy is significantly lower than the best accuracy only for low settings of ϵ (0 for the GA, and 0 and 0.05 for the DEA). It is expected that as ϵ increases, the ninth feature (or any of its copies) has a higher chance of being considered.

The results on the Segmentation data in Fig. 13.5 show that the accuracy with the DT is mostly unaffected by the choice of ϵ but the accuracy with the NB benefits from smaller values of ϵ (i.e., with larger influence of the biased initialization). With all settings of ϵ, the GA and DEA reduced the number of features from 26 to between 8 and 10.5 on average. The execution time of the GA seems insensitive to the setting of ϵ, but the DEA has the same increasing trend in execution time that we observed with the other data sets.

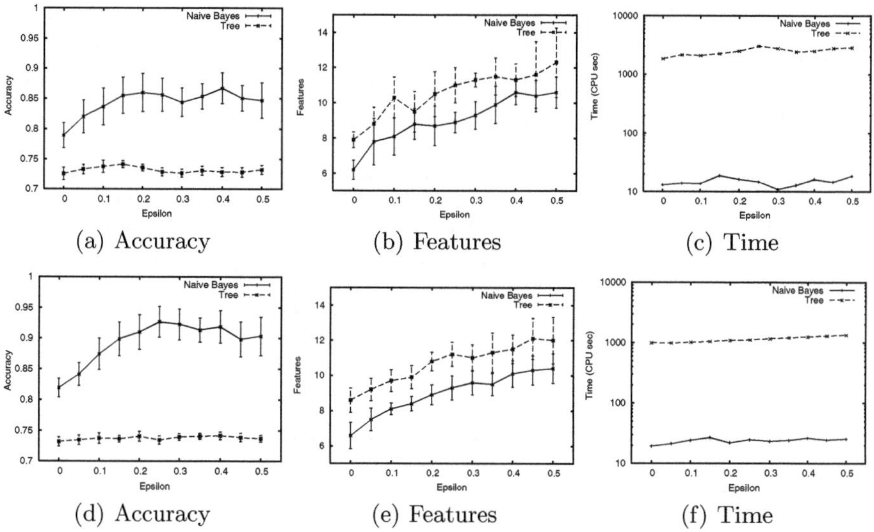

Fig. 13.4. Experimental results on Redundant21 varying the value of ϵ and using a GA (*top row*) and an DEA (*bottom row*) as the wrapper. The left panels present the classification accuracy, the middle panels present the number of features, and the right panels present the execution time in CPU seconds. The error bars represent 95% confidence intervals

The experiments in this subsection suggest that the optimal choice of ϵ depends on the data set. The setting of $\epsilon = 0.1$ used in the previous subsections never resulted in a significantly lower accuracy than with the optimal setting on these three data sets. The previous subsections showed that $\epsilon = 0.1$ often resulted in improvements over the unbiased initialization, and therefore 0.1 seems like a good initial setting for experimenting with this parameter.

13.6 Conclusions

The results of experiments with the proposed filter-EA hybrids suggest that these algorithms find feature subsets that result in accurate classifiers more often than the other feature selection algorithms examined. Together with the increased accuracy, the filter-EA hybrids identify compact feature subsets that reduce the dimensionality to one fourth or one third of the original data, on average. Other methods, notably SFS and the Boosting filter-wrapper hybrid identify much smaller feature subsets, but these subsets result in substantially lower accuracy.

In terms of execution time, the simple class-separability filter and the boosting filter-wrapper hybrid are clearly much faster than the wrapper algorithms. Notably, the filter sometimes finds results that are more accurate than

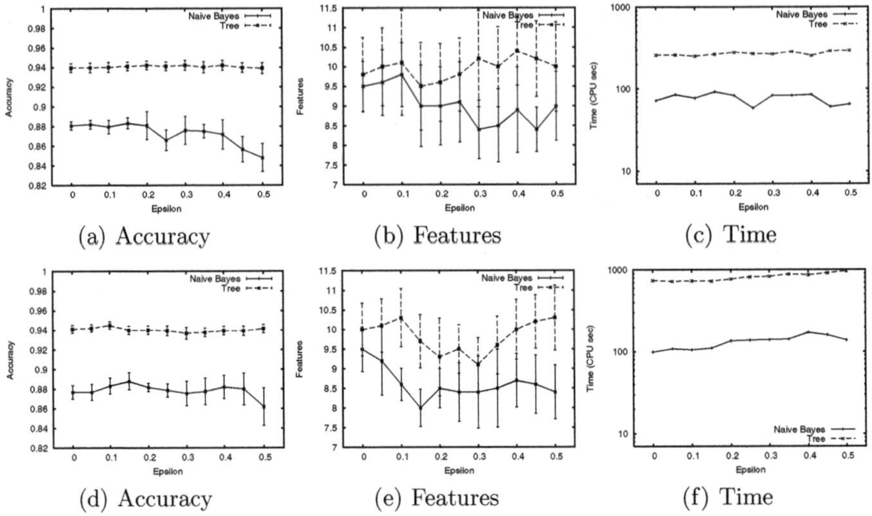

Fig. 13.5. Experimental results on Segmentation varying the value of ϵ and using a GA (*top row*) and an DEA (*bottom row*) as the wrapper. The left panels present the classification accuracy, the middle panels present the number of features, and the right panels present the execution time in CPU seconds. The error bars represent 95% confidence intervals

the wrappers that are slower by orders of magnitude. These results confirm that it is a good idea to begin the analysis with a simple and fast filter algorithm. On average, however, the EAs and the EA-filter hybrids found more accurate feature subsets than the filter.

In machine learning applications, EAs are frequently assessed as being competitive or superior than traditional methods in terms of accuracy, but usually require much more computational resources [38]. In contrast, the experiments reported here suggest that EAs – and especially the EA-filter hybrids – can find more accurate results and in shorter times than traditional wrapper algorithms.

Comparing the two EAs, the results indicate that the DEA and the filter-DEA hybrid find slightly more accurate solutions and more compact feature subsets than the GA and the filter-GA, but at a considerably higher cost. The experiments also indicate that biasing the EAs with the output of the filter is beneficial in terms of accuracy as well as the compactness of the selected feature subsets and execution time.

This work can be extended with experiments with other EAs, classification methods, additional data sets, and other filter algorithms. There are numerous opportunities to improve the computational efficiency of the algorithms to deal with much larger data sets. In particular, sub-sampling the training sets and parallelizing the fitness evaluations seem like promising alternatives. Note that SFS and SBE are inherently serial methods and cannot benefit from parallelism as much as EAs. Additional future work should explore efficient methods to deal with the accuracy estimates, instead of the relatively expensive multiple cross-validations used in this chapter.

Acknowledgments

I would like to thank Martin Pelikan and Chandrika Kamath for useful discussions on this topic. I am also indebted to the anonymous reviewers who provided valuable comments that improved the presentation of the chapter.

References

[1] G. John, R. Kohavi, and K. Phleger, "Irrelevant features and the feature subset problem," in *Proceedings of the 11th International Conference on Machine Learning*, pp. 121–129, Morgan Kaufmann, San Francisco, CA 1994

[2] R. Kohavi and G. John, "Wrappers for feature subset selection," *Artificial Intelligence*, vol. 97, no. 1–2, pp. 273–324, 1997

[3] E. Cantú-Paz, "Feature subset selection, class separability, and genetic algorithms," in *Genetic and Evolutionary Computation Conference – GECCO-2004*, K. Deb et al., (Eds.), Springer, Berlin Heidelberg New York, 2004

[4] I. Guyon and A. Elisseeff, "An introduction to variable and feature selection," *Journal of Machine Learning Research*, vol. 3, pp. 1157–1182, 2003

[5] A. Jain and D. Zongker, "Feature selection: Evaluation, application and small sample performance," *IEEE Transactions on Pattern Analysis and Machine Intelligence*, vol. 19, no. 2, pp. 153–158, 1997

[6] A. Blum and P. Langley, "Selection of relevant features and examples in machine learning," *Artificial Intelligence*, vol. 97, no. 1–2, pp. 245–271, 1997

[7] W. Siedlecki and J. Sklansky, "A note on genetic algorithms for large-scale feature selection," *Pattern Recognition Letters*, vol. 10, pp. 335–347, 1989

[8] F. Z. Brill, D. E. Brown, and W. N. Martin, "Genetic algorithms for feature selection for counterpropagation networks," Tech. Rep. No. IPC-TR-90-004, University of Virginia, Institute of Parallel Computation, Charlottesville, 1990

[9] T. W. Brotherton and P. K. Simpson, "Dynamic feature set training of neural nets for classification," in *Evolutionary Programming IV*, J. R. McDonnell, R. G. Reynolds, and D. B. Fogel, (Eds.), pp. 83–94, MIT Cambridge, MA, 1995

[10] J. Bala, K. De Jong, J. Huang, H. Vafaie, and H. Wechsler, "Using learning to facilitate the evolution of features for recognizing visual concepts," *Evolutionary Computation*, vol. 4, no. 3, pp. 297–311, 1996

[11] J. D. Kelly and L. Davis, "Hybridizing the genetic algorithm and the K nearest neighbors classification algorithm," in *Proceedings of the Fourth International Conference on Genetic Algorithms*, R. K. Belew and L. B. Booker, (Eds.), pp. 377–383, Morgan Kaufmann, San Mateo, CA, 1991

[12] W. F. Punch, E. D. Goodman, M. Pei, L. Chia-Shun, P. Hovland, and R. Enbody, "Further research on feature selection and classification using

genetic algorithms," in *Proceedings of the Fifth International Conference on Genetic Algorithms*, S. Forrest, (Ed.), pp. 557–564, Morgan Kaufmann, San Mateo, CA, 1993

[13] M. L. Raymer, W. F. Punch, E. D. Goodman, P. C. Sanschagrin, and L. A. Kuhn, "Simultaneous feature scaling and selection using a genetic algorithm," in *Proceedings of the Seventh International Conference on Genetic Algorithms*, T. Bäck, (Ed.), pp. 561–567, Morgan Kaufmann, San Francisco, CA, 1997

[14] M. Kudo and K. Sklansky, "Comparison of algorithms that select features for pattern classifiers," *Pattern Recognition*, vol. 33, no. 1, pp. 25–41, 2000

[15] H. Vafaie and K. A. De Jong, "Robust feature selection algorithms," in *Proceedings of the International Conference on Tools with Artificial Intelligence.* pp. 356–364, IEEE Computer Society, USA 1993

[16] I. Inza, P. Larrañaga, R. Etxeberria, and B. Sierra, "Feature subset selection by Bayesian networks based optimization," *Artificial Intelligence*, vol. 123, no. 1–2, pp. 157–184, 1999

[17] Erick Cantú-Paz, "Feature subset selection by estimation of distribution algorithms," in *GECCO 2002: Proceedings of the Genetic and Evolutionary Computation Conference*, W. B. Langdon, E. Cantú-Paz, K. Mathias, R. Roy, D. Davis, R. Poli, K. Balakrishnan, V. Honavar, G. Rudolph, J. Wegener, L. Bull, M. A. Potter, A. C. Schultz, J. F. Miller, E. Burke, and N. Jonoska, (Eds.), pp. 303–310, Morgan Kaufmann, San Francisco, CA, 2002

[18] M. L. Raymer, W. F. Punch, E. D. Goodman, L. A. Kuhn, and A. K. Jain, "Dimensionality reduction using genetic algorithms," *IEEE Transactions on Evolutionary Computation*, vol. 4, no. 2, pp. 164–171, 2000

[19] I. Inza, P. Larrañaga, and B. Sierra, "Feature subset selection by Bayesian networks: A comparison with genetic and sequential algorithms," *International Journal of Approximate Reasoning*, vol. 27, no. 2, pp. 143–164, 2001

[20] I. Inza, P. Larrañaga, and B. Sierra, "Feature subset selection by estimation of distribution algorithms," in *Estimation of Distribution Algorithms: A new tool for Evolutionary Computation*, P. Larrañaga and J. A. Lozano, (Eds.), Kluwer Academic, Dordrecht Hingham, MA 2001

[21] M. Ozdemir, M. J. Embrechts, F. Arciniegas, C. M. Breneman, L. Lockwood, and K. P. Bennett, "Feature selection for in-silico drug design using genetic algorithms and neural networks," in *IEEE Mountain Workshop on Soft Computing in Industrial Applications.* pp. 53–57, IEEE, USA 2001

[22] P.L. Lanzi, "Fast feature selection with genetic algorithms: A wrapper approach," in *IEEE International Conference on Evolutionary Computation.* pp. 537–540, IEEE, USA 1997

[23] I.-S. Oh, J.-S. Lee, and C. Suen, "Analysis of class separation and combination of class-dependent features for handwritting recognition,"

IEEE Transactions on Pattern Analysis and Machine Intelligence, vol. 21, no. 10, pp. 1089–1094, 1999
[24] C.L. Blake and C.J. Merz, *"UCI repository of machine learning databases,"* 1998
[25] J. R. Quinlan, "Induction of decision trees," *Machine Learning*, vol. 1, pp. 81–106, 1986
[26] G. Harik, E. Cantú-Paz, D. E. Goldberg, and B. L. Miller, "The gambler's ruin problem, genetic algorithms, and the sizing of populations," *Evolutionary Computation*, vol. 7, no. 3, pp. 231–253, 1999
[27] G. R. Harik, F. G. Lobo, and D. E. Goldberg, "The compact genetic algorithm," in *Proceedings of 1998 IEEE Iternational Conference on Evolutionary Computation*, Institute of Electrical and Electronics Engineers, pp. 523–528, IEEE Service Center, Piscataway, NJ, 1998
[28] S. Baluja, "Population-based incremental learning: A method for integrating genetic search based function optimization and competitive learning," Tech. Rep. No. CMU-CS-94-163, Carnegie Mellon University, Pittsburgh, PA, 1994
[29] H. Mühlenbein, "The equation for the response to selection and its use for prediction," *Evolutionary Computation*, vol. 5, no. 3, pp. 303–346, 1998
[30] Sanmay Das, "Filters, wrappers and a boosting-based hybrid for feature selection," in *Proceedings of the 18th International Conference on Machine Learning*, Carla Brodley and Andrea Danyluk, (Eds.), pp. 74–81, Morgan Kaufmann, San Francisco, CA, 2001
[31] Y. Freund and R. E. Schapire, "Experiments with a new boosting algorithm," in *Proceedings of the Thirteenth International Conference on Machine Learning*, L. Saitta, (Ed.), pp. 148–156, Morgan Kaufmann, San Mateo, CA, 1996
[32] M. Matsumoto and T. Nishimura, "Mersenne twister: A 623-dimensionally equidistributed uniform pseudorandom number generator," *ACM Transactions on Modeling and Computer Simulation*, vol. 8, no. 1, pp. 3–30, 1998
[33] B. L. Miller and D. E. Goldberg, "Genetic algorithms, selection schemes, and the varying effects of noise," *Evolutionary Computation*, vol. 4, no. 2, pp. 113–131, 1996
[34] T. G. Dietterich, "Approximate statistical tests for comparing supervised classification learning algorithms," *Neural Computation*, vol. 10, no. 7, pp. 1895–1924, 1998
[35] E. Alpaydin, "Combined 5×2 cv F test for comparing supervised classification algorithms," *Neural Computation*, vol. 11, pp. 1885–1892, 1999
[36] J. Reunanen, "Overfitting in making comparisons between variable selection methods," *Journal of Machine Learning Research*, vol. 3, pp. 1371–1382, 2003

[37] C. Ambroise and G. J. McLachlan, "Selection bias in gene extraction on the basis of microarray gene-expression data," *Proceedings of the National Academy of Sciences*, vol. 99, no. 10, pp. 6562–6566, 2002

[38] E. Cantu-Paz and C. Kamath, "On the use of evolutionary algorithms in data mining," in *Data Mining: A Heuristic Approach*, H. Abbass, R. Sarker, and C. Newton, (Eds.), pp. 48–71. IDEA Group, Hershey, PA, 2002

14

BOA for Nurse Scheduling

Jingpeng Li and Uwe Aickelin

Summary. Our research has shown that schedules can be built mimicking a human scheduler by using a set of rules that involve domain knowledge. This chapter presents a Bayesian Optimization Algorithm (BOA) for the nurse scheduling problem that chooses such suitable scheduling rules from a set for each nurse's assignment. Based on the idea of using probabilistic models, the BOA builds a Bayesian network for the set of promising solutions and samples these networks to generate new candidate solutions. Computational results from 52 real data instances demonstrate the success of this approach. It is also suggested that the learning mechanism in the proposed algorithm may be suitable for other scheduling problems.

Key words: Nurse Scheduling, Bayesian Optimization Algorithm, Probabilistic Modeling

14.1 Introduction

Personnel scheduling problems have been studied extensively over the past three decades (see survey papers by Baker, Tien and Kamiyama, Bradley and Martin, Bechtold et al., Ernst et al. [4, 5, 7, 14, 27]). Personnel scheduling is the problem of assigning employees to shifts or duties over a scheduling period so that certain constraints (organizational and personal) are satisfied. It involves constructing a schedule for each employee within an organization in order for a set of tasks to be fulfilled. In the domain of healthcare, this is particularly challenging because of the presence of a range of different staff requirements on different days and shifts. In addition, unlike many other organizations, healthcare institutions work 24 hours a day for every single day of the year. In this chapter, we focus on the development of new approaches for nurse scheduling.

Most nurse scheduling problems are extremely difficult and complex. Tien and Kamiyama [27], for example, say nurse scheduling is more complex than the traveling salesman problem. A general overview of various approaches for nurse scheduling can be found in [9, 10, 25]. Early research [18, 28, 29]

J. Li and U. Aickelin: *BOA for Nurse Scheduling*, Studies in Computational Intelligence (SCI) **33**, 315–332 (2006)
www.springerlink.com

concentrated on the development of mathematical programming models. To reduce computational complexity, researchers had to restrict the problem dimensions and consider a smaller size of constraints in their models, resulting in solutions that are too simple to be applied in real hospital situations.

The above observations have led to other attempts, trying to solve the real nurse scheduling problems within reasonable time. Besides heuristic approaches (e.g., [6, 26]), artificial intelligence approaches such as constraint programming [20], expert systems [11] and knowledge based systems [24] were investigated with some success. Since the 1990s, most papers have tackled the problem with metaheuristic approaches, including simulated annealing (Isken and Hancock 1991), tabu search [12] and genetic algorithms [2].

This chapter solves a nurse scheduling problem arising at a major UK hospital [1, 13]. In this problem, the number of nurses is fixed (up to 30) and the target is to create a weekly schedule by assigning each nurse one out of up to 411 predefined shift patterns in the most efficient way. Due to the limitation of the traditional mathematical programming approach, a number of metaheuristic approaches have been explored for this problem. For example, in [1–3] various approaches based on Genetic Algorithms (GAs) are presented, in [17] an approach based on a learning classifier system is investigated, in [8] a tabu search hyperheuristic is introduced, and in [16] a Bayesian Optimization Algorithm (BOA) is described. In this chapter, we demonstrate a novel BOA approach.

In our proposed BOA, we try to solve the problem by applying a suitable rule, from a set containing a number of available rules, for each nurse's assignment. A schedule can be then created from a rule string (or a rule sequence) corresponding to nurses from the first to the last. To evolve the rule strings Bayesian networks [21] are used. The Bayesian network in our case is a directed acyclic graph with each node corresponding to a nurse/rule pair, by which a schedule will be constructed step by step. The causal relationship between two nodes is represented by a directed edge between the two nodes.

Based on the idea of using probabilistic models, the BOA builds a Bayesian network for the set of promising solutions and generates new candidate solutions by sampling these networks [22, 23]. In our proposed BOA for nurse scheduling, the conditional probabilities of all nurse/rule pairs in the network are first computed according to an initial set of rule strings. Subsequently, new instances of each node are generated by using conditional probabilities to obtain new rule strings. A new set of rule strings is thus generated, some of which will replace previous strings based on roulette-wheel fitness selection. If the stopping criterion is not reached, the conditional probabilities for all nodes in the network are updated again using the current set of rule strings.

14.2 The Nurse Scheduling Problem

The problem described in this chapter is that of creating weekly schedules for wards containing up to 30 nurses at a major UK hospital. These schedules must respect working contracts and meet the demand for a given number of nurses of different grades on each shift, while being perceived to be fair by the nurse themselves. The day is partitioned into three shifts: two day shifts called "earlies" and "lates", and a longer night shift. Until the final scheduling stage, "earlies" and "lates" are merged into day shifts. Due to hospital policy, a nurse would normally work either days or nights in a given week, and due to the difference in shift length, a full week's work normally includes more days than nights. For example, a full-time nurse works 5 days or 4 nights, whereas a part-time nurse works 4 days or 3 nights, 3 days or 3 nights, or 3 days or 2 nights. However, exceptions are possible and some nurses specifically must work both day- and night- shifts in one week.

As described in [13], the problem can be decomposed into three independent stages. The first stage uses a knapsack model to check if there are enough nurses to meet demand. If not, additional nurses are allocated to the ward, so that the second stage will always admit a feasible solution. The second stage is the most difficult and is concerned with the actual allocations of days or nights to be worked by each nurse. The third stage then uses a network flow model to assign those on days to "earlies" and "lates". The first and the third stages, and in this chapter we only concern with the highly constrained second stage.

The numbers of days or nights to be worked by each nurse defines the set of feasible weekly work patterns for that nurse. These will be referred to as shift patterns in the following. For example, (1111100 0000000) would be a pattern where the nurse works the first 5 days and no nights. Depending on the working hours of a nurse, there are a limited number of shift patterns available to her/him. Typically, there will be between 20 and 30 nurses per ward, 3 grade-bands, 9 part time options and 411 different shift patterns. Depending on the nurses' preferences, the recent history of patterns worked, and the overall attractiveness of the pattern, a penalty cost is allocated to each nurse-shift pattern pair. These values were set in close consultation with the hospital and range from 0 (perfect) to 100 (unacceptable), with a bias to lower values. Further details can be found in [12].

This second stage problem is described as follows. Each possible weekly shift pattern for a given nurse can be represented as a 0–1 vector with 14 elements, where the first 7 elements represent the 7 days of the week and the last 7 elements the corresponding 7 nights of the week. A "1"/"0" in the vector denotes a scheduled day or night "on"/"off". For each nurse there are a limited number of shift patterns available to her/him. For instance, a full-time nurse working either 5 days or 4 nights has a total of 21 (i.e., C_7^5) feasible day shift patterns and 35 (i.e., C_7^4) feasible night shift patterns. Typically, there are between 20 and 30 nurses per ward, 3 grade-bands, 9 part time options and 411 different shift patterns. Depending on the nurses' preferences, the recent history of patterns worked, and the overall attractiveness of the pattern, a penalty cost is allocated to each nurse-shift pattern pair. These values were set in close consultation with the hospital and range from 0 (perfect) to 100 (unacceptable), with a bias to lower values. Further details can be found in [12].

This problem can be formulated as follows. The decision variable x_{ij} assumes 1 if nurse i works shift pattern j and 0 otherwise. Let parameters m, n, p be the number of shift patterns, nurses and grades, respectively. Parameter a_{jk} is 1 if shift pattern j covers day/night k, 0 otherwise. q_{is} is 1 if nurse i is of grade s or higher, 0 otherwise. Furthermore, p_{ij} is the preference cost of nurse i working shift pattern j, and $F(i)$ is the set of feasible shift patterns for nurse i. Then we can use the following integer programming formulation from [12]:

$$\text{Min} \quad \sum_{i=1}^{n} \sum_{j \in F(i)}^{m} p_{ij} x_{ij}, \tag{14.1}$$

$$\text{s.t.} \quad \sum_{j \in F(i)} x_{ij} = 1, \quad \forall i = \{1, \ldots, n\}, \tag{14.2}$$

$$\sum_{j \in F(i)} \sum_{i=1}^{n} q_{is} a_{jk} x_{ij} \geq R_{ks}, \quad \forall k = \{1, \ldots, 14\},\ s = \{1, \ldots, p\},$$

$$x_{ij} \in \{0, 1\}, \quad \forall i, j. \tag{14.3}$$

The objective function (14.1) is to minimize total preference cost of all nurses. Constraint (14.2) ensures that every nurse works exactly one shift pattern from his/her feasible set, and constrain (14.3) ensures that the demand for nurses is fulfilled for every grade on every day and night. This problem can be regarded as a multiple-choice set-covering problem. The sets are given by the shift pattern vectors and the objective is to minimize the cost of the sets needed to provide sufficient cover for each shift at each grade. The constraints described in (14.2) enforce the choice of exactly one pattern (set) from the alternatives available for each nurse.

14.3 A BOA for Nurse Scheduling

In the nurse scheduling problem we are tackling, the number of nurses is fixed (approximately 30 depending on data instance), and the goal is to create weekly schedules by assigning each nurse one shift pattern in the most efficient way. The problem can be solved by using a suitable rule, from a rule set that contains a number of available rules, for each nurse's assignment. Thus, a potential solution is represented as a rule string, or a rule sequence connecting all nurses.

We chose this rule-base building approach, as the longer-term aim of our research is to model the explicit learning of a human scheduler. Human schedulers can provide high quality solutions, but the task is tedious and often requires a large amount of time. Typically, they construct schedules based on rules learnt during scheduling. Due to human limitations, these rules are typically simple. Hence, our rules will be relatively simple, too. Nevertheless, human generated schedules are of high quality due to the ability of the scheduler to switch between the rules, based on the state of the current solution. We envisage the proposed BOA to perform this role.

14.3.1 The Construction of a Bayesian Network

Bayesian networks are also called directed graphical models, in which each node corresponds to one variable, and each variable corresponds to one position in the strings representing the solutions. The relationship between two variables is represented by a directed edge between the two corresponding nodes.

Bayesian networks are often used to model multinomial data with both discrete and continuous variables by encoding the relationship between the variables contained in the modeled data, which represents the structure of a problem. Furthermore, they are used to generate new data instances or variables instances with similar properties as those of given data.

Figure 14.1 is the Bayesian network constructed for our nurse scheduling problem, which is a hierarchical and acyclic directed graph representing the solution structure of the problem. The node $N_{ij}(i \in \{1,2,\dots,m\};\ j \in \{1,2,\dots,n\})$ in the network is a nurse/rule pair which denotes that nurse i is assigned by rule j, where m is the number of nurses and n is the number

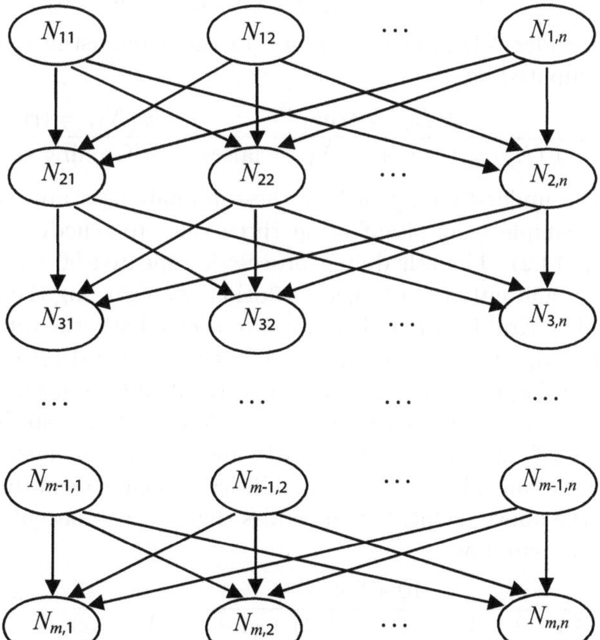

Fig. 14.1. A Bayesian network for nurse scheduling

of available rules. The directed edge from node N_{ij} to node $N'_{i+1,j}$ denotes a causal relationship of "N_{ij} causing $N'_{i+1,}$", i.e., if nurse i is scheduled by rule j then the next nurse will be scheduled by rule j'. In this network, a potential solution is shown as a directed path from nurse 1 to nurse m connecting m nodes.

14.3.2 Learning Based on the Bayesian Network

In BOAs, the structure of the Bayesian network can be either fixed [23] or variable [19]. In our proposed nurse scheduling model, we use a fixed network structure because all variables are fully observed. In essence, our Bayesian network is a fixed nurse-size vector of rules and the goal of learning is to find the variable values of all nodes N_{ij} that maximize the likelihood of the training data containing a number of independent cases.

Thus, the learning in our case amounts to "counting" based on a multinomial distribution and we use the symbol "#" meaning "the number of" in the following equations. It calculates the conditional probabilities of each possible value for each node given all possible values of its parent nodes. For example, for node $N'_{i+1,j}$ with a parent node N_{ij}, its conditional probability is

$$\begin{aligned} P(N_{i+1,j'}|N_{ij}) &= \frac{P(N_{i+1,j'},\ N_{ij})}{P(N_{ij})} \\ &= \frac{\#(N_{i+1,j'} = \text{true},\ N_{ij} = \text{true})}{\#(N_{i+1,j'} = \text{true},\ N_{ij} = \text{true}) + \#(N_{i+1,j'} = \text{false}\ N_{ij} = \text{true}}. \end{aligned} \quad (14.4)$$

Note that nodes N1j have no parents. In this circumstance, their probabilities are computed as

$$P(N_{1j}) = \frac{\#(N_{1j} = \text{true})}{\#(N_{1j} = \text{true}) + \#(N_{1j} = \text{false})} = \frac{\#(N_{1j} = \text{true})}{\#\text{Training sets}}. \quad (14.5)$$

To help the understanding of how these probabilities are computed, let us consider a simple example of using three rules to schedule three nurses (shown in Fig. 14.2). The scheduling process is repeated 50 times, and each time, rules are randomly used to get a solution, disregarding the feasibility of the solution. The Arabic numeral adjacent to each edge is the total number of times that this edge has been used in the 50 runs. For instance, if a solution is obtained by using rule 2 to schedule nurse 1, rule 3 to nurse 2 and rule 1 to nurse 3, then there exists a path of "N12 → N23 → N31", and the count of edge "N12 → N23" and edge "N23 → N31" are increased by one, respectively.

Therefore, we can calculate the probabilities of each node at different states according to the above count. For the nodes that have no parents, their probabilities are computed as:

$$P(N_{11}) = \frac{10+2+3}{(10+2+3)+(5+11+4)+(7+5+3)} = \frac{15}{50}, \quad P(N_{12}) = \frac{20}{50},$$
$$P(N_{13}) = \frac{15}{50}.$$

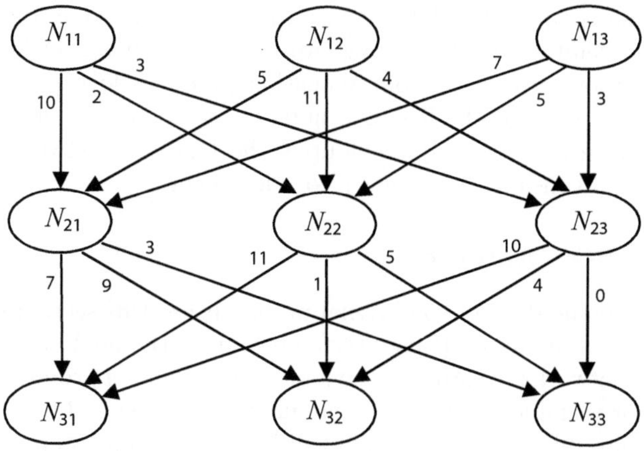

Fig. 14.2. A dummy problem with three nurses and three rules

For all other nodes (with parents), the conditional probabilities are:

$$P(N_{21}|N_{11}) = \frac{10}{10+2+3} = \frac{10}{15}, \quad P(N_{22}|N_{11}) = \frac{2}{15}, \quad P(N_{23}|N_{11}) = \frac{3}{15},$$
$$P(N_{21}|N_{12}) = \frac{5}{5+11+4} = \frac{5}{20}, \quad P(N_{22}|N_{12}) = \frac{10}{20}, \quad P(N_{23}|N_{12}) = \frac{4}{20},$$
$$P(N_{21}|N_{13}) = \frac{7}{7+5+3} = \frac{7}{15}, \quad P(N_{22}|N_{13}) = \frac{5}{15}, \quad P(N_{23}|N_{13}) = \frac{3}{15},$$
$$P(N_{31}|N_{21}) = \frac{7}{7+9+3} = \frac{7}{19}, \quad P(N_{32}|N_{21}) = \frac{9}{19}, \quad P(N_{33}|N_{21}) = \frac{3}{19},$$
$$P(N_{31}|N_{22}) = \frac{11}{11+1+5} = \frac{11}{17}, \quad P(N_{32}|N_{22}) = \frac{1}{17}, \quad P(N_{33}|N_{22}) = \frac{5}{17},$$
$$P(N_{31}|N_{23}) = \frac{10}{10+4+0} = \frac{10}{14}, \quad P(N_{32}|N_{23}) = \frac{4}{14}, \quad P(N_{33}|N_{23}) = \frac{0}{14}.$$

These probability values can be used to generate new rule strings, or new solutions. Since the first rule in a solution has no parents, it will be chosen from nodes N_{1j} according to their probabilities. The next rule will be chosen from nodes N_{ij} according to the probabilities conditioned on the previous nodes. This building process is repeated until the last node N_{mj} has been chosen, where m is number of the nurses. A path from nurse 1 to nurse m is thus formed, representing a new potential solution. Since all probability values for each nurse are normalized, we suggest the roulette-wheel method as a suitable strategy for rule selection [15].

For further clarity, consider the following example of scheduling five nurses with two rules (1: random allocation, 2: allocate nurse to low-cost shifts). In the beginning of the search, the probabilities of choosing rule 1 or rule 2 for each nurse is equal, i.e., 50%. After a few iterations, due to the selection pressure and reinforcement learning, we experience two solution pathways: Because pure low-cost or random allocation produces low quality solutions, either rule 1 is used for the first 2–3 nurses and rule 2 on remainder or vice versa. Therefore for this simple example, the Bayesian network learns "use rule 2 after 2–3 times of using rule 1" or vice versa.

14.3.3 Our BOA Approach

Based on the estimation of conditional probabilities, this section introduces a BOA for nurse scheduling. It uses techniques from the field of modeling data by Bayesian networks to estimate the joint distribution of promising solutions. The nodes, or variables, in the Bayesian network correspond to the individual nurse/rule pairs by which a schedule will be built step by step.

The conditional probability of each variable in the Bayesian network is computed according to an initial set of promising solutions. Subsequently, each new instance for each variable is generated by using the corresponding conditional probabilities, until all variables have been generated. Hence, in our case, a new rule string has been obtained. Another set of rule strings will be generated in this way, some of which will replace previous strings based on fitness selection. If the stopping criterion not reached, the conditional probabilities for all nodes in the Bayesian network are updated again using the current set of promising rule strings. In more detail, the steps of our BOA for nurse scheduling are described as follows:

1. Set $t = 0$, and generate an initial population $P(0)$ at random;
2. Use roulette-wheel to select a set of promising rule strings $S(t)$ from $P(t)$;
3. Compute the conditional probabilities for the values of each node according to this set of promising solutions;
4. For each nurse's assignment, use the roulette-wheel method to select one rule according to the conditional probabilities of all available nodes for this nurse, thus obtaining a new rule string. A set of new rule strings $O(t)$ will be generated in this way;
5. Create a new population $P(t+1)$ by replacing some rule strings from $P(t)$ with $O(t)$, and set $t = t + 1$;
6. If the termination conditions not met, go to step 2.

14.3.4 Four Heuristic Rules for Solution Building

Using our domain knowledge of nurse scheduling, the BOA can choose from the following four rules. The first two rules are quite simple, while the rest twos are a bit complex. Note that there are many more heuristic rules that could be used to build schedules.

(1) *"Random" Rule.*

This "Random" rule is used to assign shift patterns to individual nurses in a totally random manner. Its purpose is to introduce randomness into the search thus enlarging the search space, and most importantly to ensure that the proposed algorithm has the ability to escape from local optima. This rule mirrors much of a scheduler's creativeness to come up with different solutions if required.

(2) *"k-Cheapest" Rule.*

The second "*k-Cheapest*" rule is purely towards the solution cost, and disregards the feasibility of the solution. For each nurse's assignment, it randomly selects one candidate shift pattern from a set containing the cheapest k shift patterns available to that nurse, in an effort to reduce the solution cost of a schedule as much as possible.

(3) *"Overall Cover" Rule.*

The "Overall Cover" rule is designed to consider only the feasibility of the schedule. It schedules one nurse at a time in such a way as to cover those days and nights with the most number of uncovered shifts.

This rule constructs solutions as follows. For each shift pattern in a nurse's feasible set, it calculates the total number of uncovered shifts that would be covered if the nurse worked that shift pattern. For instance, assume that a shift pattern covers Monday to Friday nights. Further assume that the current requirements for the nights from Monday to Sunday are as follows: $(-4, 0, +1, -3, -1, -2, 0)$, where a negative number means undercover and a positive overcover. The Monday to Friday shift pattern hence has a cover value of 8 as the sum of undercover is -8. In this example, a Tuesday to Saturday pattern would have a cover value of 6 as the sum of undercover is -6.

For nurses of grade s, only the shifts requiring grade s nurses are counted as long as there is a single uncovered shift for this grade. If all these are covered, shifts of the next lower grade are considered and once these are filled those of the next lower grade, etc. This operation is necessary as otherwise higher graded nurses might fill lower graded demand, whilst higher graded demand might not be met at all.

Due to the nature of this rule, nurses' preference costs p_{ij} are not taken into account. Therefore, the "Overall Cover" rule can be summarized as assigning the shift pattern with the largest amount of undercover for the nurse currently being scheduling.

(4) *"Contribution" Rule.*

The fourth "Contribution" rule is biased towards solution quality but includes some aspects of feasibility. It cycles through all shift patterns of a nurse, assigns each one a score, and chooses the one with the highest score. In case of more than one shift pattern with the best score, the first such shift pattern is chosen. Compared with the third "Overall Cover" rule, this "Contribution" rule is more complex because it considers the preferences of the nurses and also tries to look ahead a little.

The score of a shift pattern is calculated as the weighted sum of the nurse's p_{ij} value for that particular shift pattern and its contribution to the cover of all three grades. The latter is measured as a weighted sum of grade one, two and three uncovered shifts that would be covered if the nurse worked this shift pattern, i.e., the reduction in shortfall. Obviously, nurses can only contribute to uncovered demand of their own grade or below.

More precisely, using the same notation as before, the score S_{ij} of shift pattern j for nurse i is calculated as

$$S_{ij} = w_p(100 - P_{ij}) + \sum_{s=1}^{3} w_s q_{is} \left(\sum_{k=1}^{14} a_{jk} d_{ks} \right), \tag{14.6}$$

where parameter w_p is the weight of the nurse's, p_{ij} value for the shift pattern and parameter w_s is the weight of covering an uncovered shift of grade s. Variable a_{jk} is 1 if shift pattern j covers day k, 0 otherwise. Variable d_{ks} is 1 if there are still nurses needed on day k of grade s, 0 otherwise. Note that $(100 - p_{ij})$ must be used in the score, as higher p_{ij} values are worse and the maximum for p_{ij} is 100.

14.3.5 Fitness Function

Our nurse scheduling problem is complicated by the fact that higher qualified nurses can substitute less qualified nurses but not vice versa. Furthermore, the problem has a special day–night structure as most of the nurses are contracted to work either days or nights in one week but not both. These two characteristics make the finding and maintaining of feasible solutions in any heuristic search is extremely difficult. Therefore, a penalty function approach is needed while calculating the fitness of completed solutions. Since the chosen encoding automatically satisfies constraint set (14.2) of the integer programming formulation, we can use the following formula to calculate the fitness of solutions (the fitter the solution, the lower its fitness value):

$$\text{Min} \quad \sum_{i=1}^{n}\sum_{j=1}^{m} p_{ij}x_{ij} + w_{\text{demand}} \sum_{k=1}^{14}\sum_{s=1}^{p} \max\left(\left[R_{ks} - \sum_{i=1}^{n}\sum_{j=1}^{m} q_{is}a_{jk}x_{ij} \right], 0 \right). \tag{14.7}$$

Note that only undercovering is penalized not overcovering, hence the use of the max function. The parameter w_{demand} is the penalty weight used to adjust the penalty that a solution has added to its fitness, and this penalty is proportional to the number of uncovered shifts. For example, consider a solution with an objective function value of 22 that undercovers the Monday day shift by one shift and the Tuesday night shift by two shifts. If the penalty weight was set to 20, the fitness of this solution would be $22+(1+2)*20 = 82$.

14.4 Computational Results

This section describes the computational experiments that are used to test the proposed BOA. For all experiments, 52 real data sets as given to us by the hospital are available. Each data set consists of one week's requirements for all shift and grade combinations and a list of available nurses together with their preference costs pij and qualifications. The data was collected from three wards over a period of several months and covers a range of scheduling situations, e.g., some data instances have very few feasible solutions whilst others have multiple optima.

For all data instances, we used the following set of fixed parameters to implement our experiments. These parameters are based on our experience and intuition and thus not necessarily the best for each instance. We have kept them the same for consistency.

- Stopping criterion: number of generations = 200, or an optimal solution is found;
- Population size = 140;
- The number of solutions kept in each generation = 40;
- For the "k-Cheapest" rule, $k = 5$;
- Weight set in formula (14.6): $w_p = 1$, $w_1 = 8$, $w_2 = 2$ and $w_3 = 1$;
- Penalty weight in fitness function (14.7): $w_{\text{demand}} = 200$;
- Number of runs per data instance = 20.

The BOA was coded in Java 2, and all experiments were run on the same Pentium 4 2.0GHz PC with 512MB RAM under the Windows XP operating system. To test the robustness of the BOA, each data instance was run 20 times by fixing the above parameters and varying the pseudorandom number seed at the beginning. The executing time per run and per data instance varies from half second to half a minute depending on the difficulty of the individual data instance. For example, data instances 27, 29, 31 and 51 are harder to solve than others due to a shortage of nurses in these weeks.

The detailed computational results over these 52 instances are listed in Table 14.1, in which the last row (headed with "Av.") contains the mean values of all columns using following notations:

- IP: optimal solutions found with an integer programming software [13];
- Rd-1: random search, i.e., only the first "Random" rule is in use;
- Rd-2: random rule-selection, i.e., using four rules but every rule has an equal opportunity to be chosen all the time for all nurses;
- Best: best result out of 20 runs of the BOA;
- Mean: average result of 20 runs of the BOA;
- Fea: number of runs terminating with the best solution being feasible;
- #: number of runs terminating with the best solution being optimal;
- $\leq$3: number of runs terminating with the best solution being within three cost units of the optimum. The value of three units was chosen as it corresponds to the penalty cost of violating the least important level of requests in the original formulation. Thus, these solutions are still acceptable to the hospital.

Table 14.1. Comparison of results over 52 instances

Data	IP	RD1	RD2	Best	Mean	Fea	#	≤ 3
01	8	N/A	27	8	8.1	20	19	20
02	49	N/A	85	56	74.5	20	0	0
03	50	N/A	97	50	72.0	20	2	5
04	17	N/A	23	17	17.0	20	20	20
05	11	N/A	51	11	12.4	20	8	16
06	2	N/A	51	2	2.4	20	17	17
07	11	N/A	80	14	16.8	20	0	3
08	14	N/A	62	15	17.5	20	0	11
09	3	N/A	44	14	17.3	20	0	0
10	2	N/A	12	2	4.9	20	2	10
11	2	N/A	12	2	2.8	20	2	20
12	2	N/A	47	3	7.8	20	0	2
13	2	N/A	17	3	3.6	20	0	20
14	3	N/A	102	4	6.5	20	0	7
15	3	N/A	9	4	5.1	20	0	20
16	37	N/A	55	38	38.8	20	0	20
17	9	N/A	146	9	11.3	20	4	11
18	18	N/A	73	19	20.8	20	0	20
19	1	N/A	135	10	12.0	20	0	0
20	7	N/A	53	7	8.3	20	5	19
21	0	N/A	19	1	1.6	20	0	20
22	25	N/A	56	26	27.5	20	0	15
23	0	N/A	119	1	1.6	20	0	20
24	1	N/A	4	1	1.0	20	20	20
25	0	N/A	3	0	0.2	20	18	20
26	48	N/A	222	52	66.8	20	0	1
27	2	N/A	158	28	35.6	20	0	0
28	63	N/A	88	65	65.5	20	0	3
29	15	N/A	31	109	169.2	20	0	0
30	35	N/A	210	38	83.5	20	0	3
31	62	N/A	253	159	175.0	20	0	0
32	40	N/A	102	43	80.4	20	0	4
33	10	N/A	30	11	15.6	20	0	8
34	38	N/A	95	41	63.8	20	0	2
35	35	N/A	118	46	60.2	20	0	0
36	32	N/A	130	45	47.6	20	0	0
37	5	N/A	28	7	8.2	20	0	7
38	13	N/A	130	25	33.0	20	0	0
39	5	N/A	44	8	10.8	20	0	3
40	7	N/A	51	8	8.8	20	0	10
41	54	N/A	87	55	56.2	20	0	15
42	38	N/A	188	41	73.3	20	0	1
43	22	N/A	86	23	24.2	20	0	13
44	19	N/A	70	24	77.4	20	0	0
45	3	N/A	34	6	8.1	20	0	2
46	3	N/A	196	7	9.4	20	0	0
47	3	N/A	11	3	3.4	20	13	20
48	4	N/A	35	5	5.7	20	0	10
49	27	N/A	69	30	47.8	20	0	2
50	107	N/A	162	109	175.5	20	0	1
51	74	N/A	197	171	173.6	20	0	0
52	58	N/A	135	67	98.6	20	0	0
Av.	21.1	N/A	82.9	29.7	39.8	20	2.5	8.5

According to the computational results in Table 14.1, one can see that using the "Random" rule alone does not yield a single feasible solution, as the "Rd1" column shows. This underlines the difficulty of this problem. In addition, without learning, the results of randomly selecting one of the four rules at each move are much weaker, as the "Rd2" column shows. Thus, it is not simply enough to use the four rules to build schedules that are acceptable to the hospital for practical use. For the proposed BOA, 38 out of 52 data instances are solved to or near to optimality (i.e., within three cost units) and feasible solutions are always found for all instances.

The behavior of an individual run of the BOA is as expected. Figure 14.3 depicts the BOA search process for data instance 04. In this figure, the x-axis represents the number of generations and the y-axis represents the solution cost of the best individual in each generation. As shown in Fig. 14.3, the optimal solution, with a cost of 17, is produced at the generation of 57. The actual values may differ among various instances, but the characteristic shapes of the curves are similar for all seeds and data instances.

To understanding the results further, we have developed a graphical interface to visualize the BOA's learning process, instead of simply outputting the final numerical values from a "black box". The software package is developed as a Java applet, which enables the BOA to run real problems online at *http://www.cs.nott.ac.uk/~jpl/BOA/BOA_Applet.html.*

In our graphical interface, the conditional probability of choosing a rule for a specific nurse has been mapped into a 256-gray value and the causal relationship between two nodes (i.e., nurse/rule pairs) is denoted by drawing an edge in this grey value to connect the nodes. For instance, a probability of 50% would give a grey gradient of 128. The darker the edge, the higher the chance this edge (building block) will be used.

Please note that in our online demonstration two further rules can be selected. These rules are still experimental and currently under evaluation.

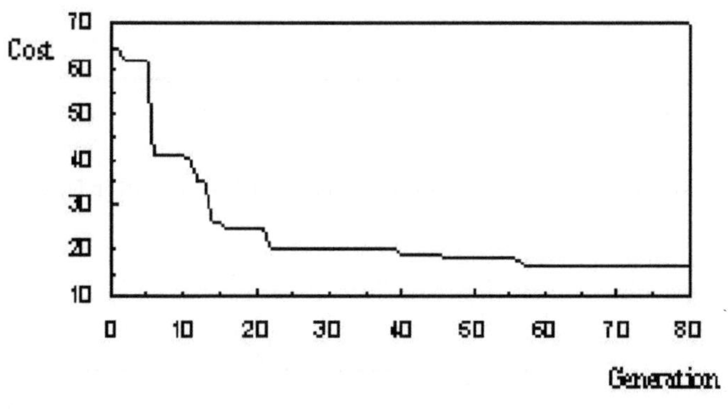

Fig. 14.3. a sample run of the BOA

The two additional rules are the "Highest Cover" rule and the "Enhanced Contribution" rule. The "Highest Cover" rule is very similar to the "Overall Cover" rule described in 3.4, with the difference in their ways of calculating the undercover: a "Highest Cover" rule is to find shift patterns which cover those days and nights with the highest number of uncover, while a "Overall Cover" rule is to find the largest total number of uncovered shifts. The second "Enhanced Contribution" rule is similar to the "Contribution" rule (described in Sect. 14.3.4), and also uses formula (14.6) to assign each feasible shift pattern a score. However, this rule uses a different definition of d_{ks}: d_{ks} is the actual number of required nurses if there are still nurses needed on day k of grade s, 0 otherwise.

Figures 14.4–14.6 show the graphic results equivalent to the experiments carried out in this paper, i.e., same parameter selection and 4 rules. The pictures show a single run for the "01" data instance which involves the scheduling of 26 nurses. Figure 14.4 shows the beginning, Fig. 14.5 the intermediate and Fig. 14.6 the final stages of the learning process respectively.

The graphic results in Figs. 14.4–14.6 are in accordance with our hypothesis. In the early process of the BOA, any edge (i.e., any building block) can be used to explore the solution space as much as possible. Thus, no particular edge stands out in Fig. 14.4. With the BOA in progress, some edges are

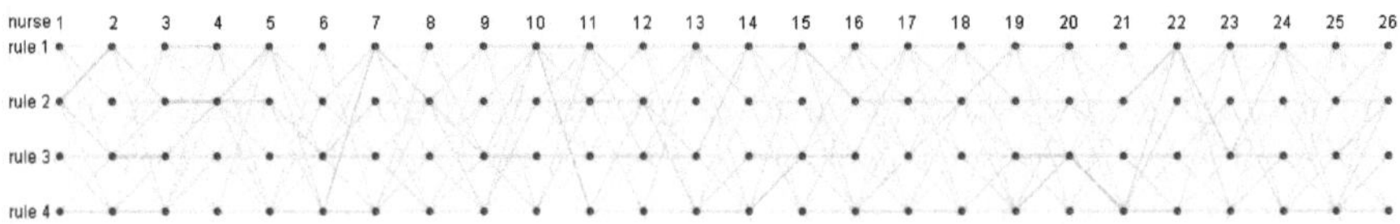

Fig. 14.4. Graphic display of the probability distributions at the initial generation

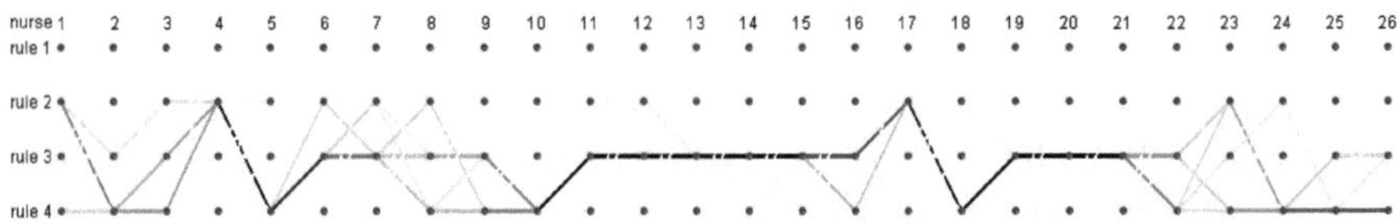

Fig. 14.5. Graphic display of the probability distributions after 50 generations

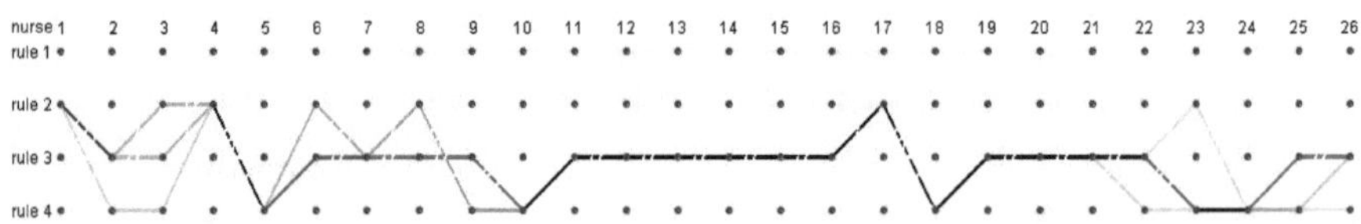

Fig. 14.6. Graphic display of the probability distributions after 100 generations

becoming more remarkable, while some ill-fitting paths become diminishing (shown in Fig. 14.5). Eventually, the BOA converges to several optimum solutions depicted as clear paths with some common parts, such as the segment of path from node 10 to node 21 (shown in Fig. 14.6). This is particularly true for our problem instances because we know from the hospital that the optimum schedules are often not unique.

Based on the summary results of 20 runs on the same 52 benchmark instances, Table 14.2 give a comparison of our BOA with some other approaches, which are briefly described as follows:

- LCS: a learning classifier system.
- Basic GA: a GA with standard genetic operators;
- Adaptive GA: the same as the basic GA, but the algorithm also tries to self-learn good parameters during the runtime;
- Multi-population GA: the same as the adaptive GA, but it also features competing sub-populations;
- Hill-climbing GA; the same as the multi-population GA, but it also includes a local search in the form of a hill-climber around the current best solution;
- Indirect GA: a novel GA-based approach that first maps the constrained solution space into an unconstrained space, then optimizes within that new space and eventually translates solutions back into the original space. Up to four different rules and hill-climbers are used in this algorithm.

Let us discuss the summary results in Table 14.2. Compared with the optimal results of the IP approach which can take more than 24 h runtime for each instances, the BOA's results are still more expensive on average but they are achieved within half minute. Compared with other meta-heuristic approaches which are all executed quickly, our BOA performs best in terms of feasibility. In terms of solution quality, in general our BOA performs better than the LCS, the basic GA, the adaptive GA and the multi-population GA, and performs slightly worse than the hill-climbing GA and the indirect GA. However, it is worth mentioning that the hill-climbing GA includes the features of

Table 14.2. Summary results of various approaches

Algorithm	References	Best	Mean	Feasibility (%)	Runtime
BOA	[16]	29.7	39.8	100	23.0 s
LCS	[17]	35.5	47.7	100	44.5 s
Optimal IP	[13]	21.4	21.4	100	>24 h
Basic GA	[3]	79.8	88.6	33	18.9 s
Adaptive GA	[3]	65.0	71.4	45	23.4 s
Multi-population GA	[3]	37.1	59.8	75	13.4 s
Hill-climbing GA	[3]	23.2	44.7	89	14.9 s
Indirect GA	[2]	22.0	25.5	95	11.9 s

multiple populations and elaborate local search which are not available in the proposed BOA, and the indirect GA uses the best nurses' ordering together with a local search to produce the final solution while our BOA only uses the ordering given in the original data sets throughout the search.

14.5 Conclusions

A BOA is presented for the nurse scheduling problem. Unlike most existing rule-based approaches, the proposed BOA has the ability to build schedules by using flexible, rather than fixed rules. Experimental results from real-world nurse scheduling problems have demonstrated the strength of the approach.

Although we have presented this work in terms of nurse scheduling, it is suggested that the main idea of the BOA could be applied to many other scheduling problems where the schedules will be built systematically according to specific rules. It is also hoped that this research will give some preliminary answers about how to include human-like learning into scheduling algorithms and thus may be of interest to practitioners and researchers in areas of scheduling and evolutionary computation.

Acknowledgments

The work was funded by the UK Government's major funding agency, the Engineering and Physical Sciences Research Council (EPSRC), under grants GR/R92899/02.

References

[1] Aickelin U, Dowsland K (2000) Exploiting problem structure in a genetic algorithm approach to a nurse rostering problem. Journal of Scheduling 3: 139–153
[2] Aickelin U, Dowsland K (2003) An indirect genetic algorithm for a nurse scheduling problem. Computers and Operations Research 31: 761–778
[3] Aickelin U, White P (2004) Building better nurse scheduling algorithms. Annals of Operations Research 128: 159–177
[4] Baker K (1976) Workforce allocation in cyclical scheduling problems: a survey. Operational Research 27: 155–167
[5] Bechtold SE, Brusco MJ, Showalter MJ (1991) A comparative evaluation of labor tour scheduling methods. Decision Sciences 22: 683–699
[6] Blau R, Sear A (1983) Nurse scheduling with a microcomputer. Journal of Ambulance Care Management 6: 1–13
[7] Bradley D, Martin J (1990) Continuous personnel scheduling algorithms: a literature review. Journal of the Society for Health Systems 2: 8–23

[8] Burke EK, Kendall G, Soubeiga E (2003) A tabu-search hyperheuristic for timetabling and rostering. Journal of Heuristics 9: 451–470

[9] Burke EK, De Causmaecker P, Vanden Berghe G, Van Landeghem H (2004) The state of the art of nurse rostering. Journal of Scheduling 7: 441–499

[10] Cheang B, Li H, Lim A, Rodrigues B (2003) Nurse rostering problems – a bibliographic survey. European Journal of Operational Research 151: 447–460

[11] Chen JG, Yeung T (1993) Hybrid expert system approach to nurse scheduling. Computers in Nursing 11: 183–192

[12] Dowsland K (1998) Nurse scheduling with tabu search and strategic oscillation. European Journal of Operational Research 106: 393–407

[13] Dowsland K, Thompson JM (2000) Nurse scheduling with knapsacks, networks and tabu search. Journal of Operational Research Society 51: 825–833

[14] Ernst AT, Jiang H, Krishnamoorthy M, Sier D (2004) Staff scheduling and rostering: a review of applications, methods and models. European Journal of Operational Research 153: 3–27

[15] Goldberg DE (1989) Genetic algorithms in search, optimization and machine leaning. Addison-Wesley, Reading, MA

[16] Li J, Aickelin U (2003) A Bayesian optimization algorithm for the nurse scheduling problem. In Proceedings of Congress on Evolutionary Computation. IEEE Press, pp. 2149–2156

[17] Li J, Aickelin U (2004) The application of Bayesian optimization and classifier systems in nurse scheduling. In Yao X et al. (eds) Proceedings of the 8th International Conference on Parallel Problem Solving from Nature. Springer, LNCS, Berlin Heidelberg, New York 3242: 581–590

[18] Miller HE, Pierskalla W, Rath G (1976) Nurse scheduling using mathematical programming. Operations Research 24: 857–870

[19] Mühlenbein H, Mahnig T (1999) FDA – a scalable evolutionary algorithm for the optimization of additively decomposed functions. Evolutionary Computation 7: 45–68

[20] Okada M, Okada M (1988) Prolog-based system for nursing staff scheduling implemented on a personal computer. Computers and Biomedical Research 21: 53–63

[21] Pearl J (1988) Probabilistic reasoning in intelligent systems: networks of plausible inference. Morgan Kaufmann, San Franscisco, CA

[22] Pelikan M (2005) Hierarchical Bayesian optimization algorithm – toward a new generation of evolutionary algorithms. Springer, Berlin Heidelberg New York

[23] Pelikan M, Goldberg D, Cantu-Paz E (1999) BOA: the Bayesian optimization algorithm. IlliGAL Report No. 99003, University of Illinois

[24] Scott S, Simpson RM (1998) Case-bases incorporating scheduling constraint dimensions: experiences in nurse rostering. In Smyth S, Cunningham P (eds) Advances in case-based reasoning, Springer, Berlin Heidelberg New York LNAI 1488: 392–401

[25] Sitompul D, Randhawa S (1990) Nurse scheduling models: a state-of-the-art review. Journal of the Society of Health Systems 2: 62–72
[26] Smith LD, Bird D, Wiggins A (1979) A computerized system to schedule nurses that recognizes staff preferences. Hospital Health Service Administration 24: 19–35
[27] Tien JM, Kamiyama A (1982) On manpower scheduling algorithms. Society for Industrial and Applied Mathematics 24: 275–287
[28] Trivedi VM, Warner M (1976) A branch and bound algorithm for optimum allocation of float nurses. Management Science 22: 972–981
[29] Warner M, Prawda J (1972) A mathematical programming model for scheduling nursing personnel in a hospital. Management Science 19: 411–422

15

Searching for Ground States of Ising Spin Glasses with Hierarchical BOA and Cluster Exact Approximation

Martin Pelikan and Alexander K. Hartmann

Summary. This chapter applies the hierarchical Bayesian optimization algorithm (hBOA) to the problem of finding ground states of Ising spin glasses with $\pm J$ and Gaussian couplings in two and three dimensions. The performance of hBOA is compared to that of the simple genetic algorithm (GA) and the univariate marginal distribution algorithm (UMDA). The performance of all tested algorithms is improved by incorporating a deterministic hill climber based on single-bit flips. The results show that hBOA significantly outperforms GA and UMDA on a broad spectrum of spin glass instances. Cluster exact approximation (CEA) is then described and incorporated into hBOA and GA to improve their efficiency. The results show that CEA enables all tested algorithms to solve larger spin glass instances and that hBOA significantly outperforms other compared algorithms even in this case.

Key words: Hierarchical Bayesian optimization algorithm, genetic algorithm, cluster exact approximation, spin glass, ground states, scalability

15.1 Introduction

Ising spin glasses are prototypical models for disordered systems and have played a central role in statistical physics during the last three decades [5, 9, 22, 42]. Examples of experimental realizations of spin glasses are metals with magnetic impurities, e.g., gold with a small fraction of iron added. The interaction between two magnetic spins, each sitting on an iron atom, is modulated by the conducting material leading to an effective interaction between these spins, which varies with the distance in magnitude *and* in sign. Due to the random distribution of the spins, competing interactions arise, where it is not possible to satisfy all interactions at the same time. This effect is called *frustration.* At low temperatures, this leads to an ordered low-temperature phase exhibiting the so-called spin-glass order, which is characterized by frozen local moments, but which does not exhibit a global magnetization. This spin glass phase poses a challenging, unsolved problem in theoretical physics because the nature of its phase-space structure is still not well understood [28]

M. Pelikan and A.K. Hartmann: *Searching for Ground States of Ising Spin Glasses with Hierarchical BOA and Cluster Exact Approximation*, Studies in Computational Intelligence (SCI) **33**, 333–349 (2006)
www.springerlink.com

despite the three decades of intensive research. It is widely believed that spin glasses exhibit a rough energy landscape leading to this intrinsic complexity.

Spin glasses represent also a large class of challenging problems for optimization algorithms [18–20] where the task is to minimize energy of a given spin glass instance [10, 17, 24, 27, 31, 40]. States with the lowest energy are called *ground states* and thus the problem of minimizing energy of spin glass instances can be formulated as the problem of finding ground states of these instances. There are two main challenges that must be tackled to find ground states of spin glasses efficiently and reliably (1) there are many local optima in the energy landscape (the number of local optima may grow exponentially with problem size) and (2) the local minima are often surrounded by high-energy configurations, which make it difficult for local operators to escape the local optimum once they get trapped in it.

The purpose of this chapter is to apply the hierarchical Bayesian optimization algorithm (hBOA) [29, 30] to a broad spectrum of instances of the problem of finding ground states of Ising spin glasses. The performance of hBOA is compared to that of the simple genetic algorithm (GA) and the univariate marginal distribution algorithm (UMDA). We also describe cluster exact approximation (CEA) [16], which provides an efficient method to perform large updates of spin glass configurations to decrease their energy. CEA is then incorporated into hBOA and GA to improve their performance, and the resulting hybrids are tested on a number of spin glass problem instances.

The chapter is organized as follows. Section 15.2 describes the problem of finding ground states of Ising spin glasses. Section 15.3 briefly describes algorithms hBOA, GA, and UMDA; additionally, the section describes the deterministic hill climber (DHC), which is incorporated into all tested algorithms to improve their performance. Section 15.4 presents initial experiments with hBOA, GA, and UMDA on several types of Ising spin glass instances. Section 15.5 describes CEA and discusses how CEA can be incorporated into evolutionary algorithms to improve their performance. Section 15.6 tests hBOA with CEA and GA with CEA on a number of spin glass instances. Finally, Sect. 15.7 summarizes and concludes the chapter.

15.2 Ising Spin Glass

A very simple model to describe a finite-dimensional Ising spin glass is typically arranged on a regular 2D or 3D grid where each node i corresponds to a spin s_i and each edge $\langle i, j\rangle$ corresponds to a coupling between two spins s_i and s_j. Each edge has a real value associated with it that defines the relationship between the two connected spins. To approximate the behavior of the large-scale system, periodic boundary conditions are often used that introduce a coupling between the first and the last element along each dimension.

For the classical Ising model, each spin s_i can be in one of two states: $s_i = +1$ or $s_i = -1$. Note that this simplification corresponds to highly anisotropic systems, which do indeed exist in some experimental situations. Nevertheless, the two-state Ising model comprises all basic effects also found in models with more degrees of freedom. A specific set of coupling constants define a spin glass instance. Each possible setting of the states of all spins is called a spin configuration.

Given a set of coupling constants $J_{i,j}$, and a configuration of spins $C = \{s_i\}$, the energy can be computed as

$$E(C) = \sum_{\langle i,j \rangle} s_i J_{i,j} s_j \ , \tag{15.1}$$

where $i, j \in \{0, 1, \ldots, n-1\}$ and $\langle i,j \rangle$ denote the nearest neighbors in the underlying grid (allowed edges).

Given a set of coupling constants, the usual task in statistical physics is to integrate a known function over all possible configurations of spins, where the configurations are distributed according to the Boltzmann distribution; that is, the probability of encountering a configuration C at temperature T is given by

$$p(C) = \frac{\exp\left(-E(C)/T\right)}{\sum_{\tilde{C}} \exp\left(-E(\tilde{C})/T\right)} \ , \tag{15.2}$$

where the sum in the denominator runs over all possible spin configurations.

From the physics point of view, it is interesting to know the ground states (configurations associated with the minimum possible energy). Finding extremal energies then corresponds to sampling the Boltzmann distribution with temperature approaching 0 and thus the problem of finding ground states is simpler *a priori* than integration over a wide range of temperatures. However, most of the conventional methods based on sampling the above Boltzmann distribution see (15.2) fail to find the ground states because they get often trapped in a local minimum.

The problem of finding ground states for a given set of coupling constants is a typical optimization problem, where the task is to find an optimal configuration of spins that minimizes energy (see Fig. 15.1). Although polynomial-time deterministic methods exist for 2D spin glasses [2, 4, 12, 13], most algorithms based on local search operators, including a (1+1) evolution strategy, conventional Monte Carlo simulations, and Monte Carlo simulations with Wang-Landau [41] or multicanonical sampling [3], scale exponentially [8] and are thus impractical for solving this class of problems. The origin for this slowdown is due to the suppressed relaxation times in the Monte Carlo simulations in the vicinity of the extremal energies because of the enormous number of local optima in the energy landscape. Recombination-based genetic algorithms succeed if recombination is performed in a way that interacting spins are located close to each other in the representation; k-point crossover with a rather small k can then be used so that the linkage between contiguous blocks of bits is preserved (unlike with uniform crossover, for instance). However, the behavior of such specialized representations and variation operators

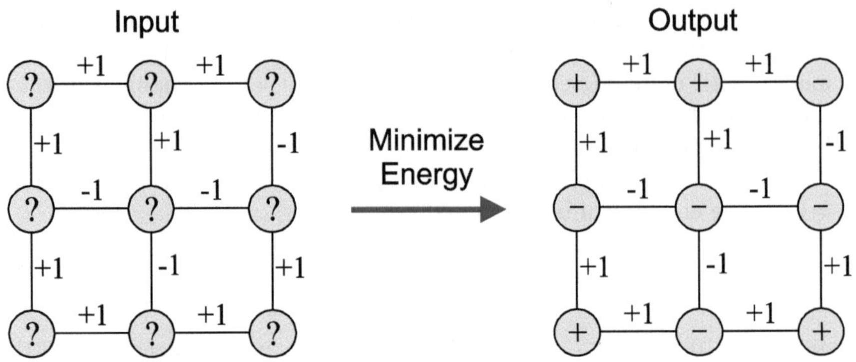

Fig. 15.1. Finding ground states of Ising spin glasses on a 2D system with no periodic boundary conditions. The left-hand side shows the input, which consists of a 2D grid with coupling constants between adjacent spins. The right-hand side shows the output, which consists of one of the ground states (the spin configurations with the minimum energy). The energy of the ground state shown in this figure is -10

cannot be generalized to similar slowly equilibrating problems that exhibit different energy landscapes, such as protein folding or polymer dynamics.

In order to obtain a quantitative understanding of the disorder in a spin glass system introduced by the random spin–spin couplings, one generally analyzes a large set of random spin glass instances for a given distribution of the spin–spin couplings. For each spin glass instance, the optimization algorithm is applied and the results are analyzed to obtain a measure of computational complexity. Here we consider two types of initial spin–spin coupling distributions, the $\pm J$ spin glass and the Gaussian spin glass.

15.2.1 $\pm J$ Spin Glass

In the $\pm J$ spin glass, each spin–spin coupling constant is set randomly to either $+1$ or -1 with equal probability. Energy minimization in this case can be transformed into a constraint satisfaction problem, where the constraints relate spins connected by a coupling constant. For any two connected neighbors i and j, if $J_{i,j} = +1$, then the constraint requires spins i and j to be different, whereas if $J_{i,j} = -1$, then the constraint requires spins i and j to be the same. Energy is minimized when the number of satisfied constraints is maximized.

15.2.2 Gaussian Spin Glass

In the Gaussian spin glass, coupling constants are generated according to a zero-mean Gaussian distribution with variance one. For real-valued couplings, energy minimization can be casted as a constraint satisfaction problem with weighted constraints. Gaussian spin glass instances are usually easier for methods based on local operators (such as MCMC) [8], whereas for recombination-based algorithms (such as hBOA), Gaussian spin glass instances tend to be slightly more difficult than $\pm J$ instances [32].

15.3 Compared Algorithms

This section briefly describes the algorithms compared in this chapter (1) The hierarchical Bayesian optimization algorithm (hBOA), (2) the genetic algorithm (GA), and (3) the univariate marginal distribution algorithm (UMDA). Additionally, the section describes the DHC that is used in all compared algorithms to locally improve candidate solutions.

15.3.1 Hierarchical Bayesian optimization algorithm (hBOA)

The hBOA [29, 30] evolves a population of candidate solutions. Each candidate solution is represented by an n-bit binary string where each bit specifies the value of one of the n spins (0 represents state -1 and 1 represents state $+1$). The population is initially generated at random according to a uniform distribution over all n-bit strings.

Each iteration starts by selecting a population of promising solutions using any common selection method of genetic and evolutionary algorithms, such as tournament and truncation selection. In this chapter, binary tournament selection is used. Binary tournament selection selects one solution at a time by first choosing two random candidate solutions from the current population and then selecting the best solution out of this subset. Random tournaments are repeated until the selected population has the same size as the original population and thus each candidate solution is expected to participate in two tournaments.

New solutions are generated by building a Bayesian network with decision trees [6, 11] for the selected solutions and sampling the built Bayesian network.

To ensure useful diversity maintenance, the new candidate solutions are incorporated into the original population using restricted tournament replacement (RTR) [15]. RTR incorporates the new candidate solutions into the original population one solution at a time. For each new candidate solution X, a random subset of candidate solutions in the original population is first selected. The solution that is closest to X is then selected from this subset. The new candidate solution replaces the selected closest solution if it is better. Here we measure the distance of two solutions by counting the number of bits where solutions differ. The size of the subsets selected in RTR is called window size.

The run is terminated when termination criteria are met; for example, the run can be terminated when the global optimum is found or when the number of iterations reaches some threshold.

15.3.2 Genetic algorithm (GA)

The GA [14, 21] also evolves a population of candidate solutions represented by fixed-length binary strings. The first population is generated at random.

Each iteration starts by selecting promising solutions from the current population. New solutions are created by applying variation operators to the population of selected solutions. Specifically, crossover is used to exchange bits and pieces between pairs of candidate solutions and mutation is used to perturb the resulting solutions. Here we use one-point crossover that exchanges the tails of two binary strings starting in a randomly chosen position. As a mutation operator, bit-flip mutation is used, which flips each bit with a specified probability p_{m} (p_{m} is usually small). The new candidate solutions are incorporated into the original population using RTR [15]. The run is terminated when termination criteria are met.

15.3.3 Univariate Marginal Distribution Algorithm (UMDA)

The UMDA [25] also evolves a population of candidate solutions represented by binary strings, starting with a random population.

Each iteration starts by selection. Then, the probability vector is learned that stores the proportion of 1s in each position of the selected population. Each bit of a new candidate solution is then set to 1 with the probability equal to the proportion of 1s in this position; otherwise, the bit is set to 0. Consequently, the variation operator of UMDA preserves the proportions of 1s in each position while decorrelating different string positions. The new candidate solutions are incorporated into the original population using restricted tournament replacement (RTR) [15]. The run is terminated when termination criteria are met.

The only difference between hBOA and the UMDA variant discussed in this paper is the type of the probabilistic model used to model promising candidate solutions and generate the new ones. The comparison between hBOA and UMDA should therefore indicate whether in this problem domain effective exploration necessitates complex probabilistic models that can efficiently encode large-order interactions between spins. For analogical reasons, the comparison between hBOA and GA will indicate whether it is important to use advanced variation operators that adapt to the problem at hand like in hBOA.

15.3.4 Deterministic Hill Climber

The DHC is incorporated to hBOA, GA, and UMDA to improve their performance. DHC takes a candidate solution represented by an n-bit binary string on input. Then, it performs one-bit changes on the solution that lead to the maximum improvement of solution quality (maximum decrease in energy). DHC is terminated when no single-bit flip improves solution quality and the solution is thus locally optimal. Here, DHC is used to improve every solution in the population before the evaluation is performed. The hybrids created by incorporating DHC into hBOA, GA, and UMDA are referred to as hBOA+DHC, GA+DHC, and UMDA+DHC, respectively.

DHC improves the performance of hBOA, GA, and UMDA by a constant factor [29]. In Sect. 15.5 we describe cluster exact approximation (CEA), which is capable of making large updates of spin glass configurations, providing a much more effective search for low-energy configurations than DHC.

15.4 Initial Experiments

This section presents the results of initial experiments on various instances of the problem of finding ground states of Ising spin glasses. Three algorithms are compared: hBOA+DHC, GA+DHC, and UMDA+DHC. The section starts by describing problem instances used in the tests. Next, experiments are described. Finally, the experimental results are presented and discussed.

15.4.1 Tested Spin Glass Instances

Three types of spin glass instances have been considered in the initial experiments:

1. Two-dimensional Ising spin glass with $\pm J$ couplings.
2. Two-dimensional Ising spin glass with Gaussian couplings.
3. Three-dimensional Ising spin glass with $\pm J$ couplings.

In all instances, periodic boundary conditions are used.

To analyze scalability, for 2D spin glasses, instances of size 6×6 ($n = 36$ spins) to 20×20 ($n = 400$ spins) have been considered; 1,000 random instances have been generated for each problem size. For 3D spin glasses, instances of size $4 \times 4 \times 4$ ($n = 64$ spins) to $7 \times 7 \times 7$ ($n = 343$ spins) have been considered; in this case, only eight random instances have been generated for each problem size because of the increased computational resources required to solve the 3D instances.

15.4.2 Description of Experiments

All compared algorithms use binary tournament selection to select promising solutions. As a replacement strategy, RTR is used where the window size w is set to the number of bits in solution strings but it is always ensured to be at most 5% of the population size, $w = \min(n, N/20)$. In GA+DHC, one-point crossover is used and the probability of crossover is $p_c = 0.6$. GA+DHC uses also bit-flip mutation, where the probability of flipping a bit is set to $1/n$.

For each problem instance, bisection is run to determine the minimum population size to ensure convergence in five independent runs (out of five runs total). Each run is terminated either when the algorithm has found the optimum or when the algorithm has failed to find the optimum for a large number of iterations. The optimum for most 2D instances was verified with the branch-and-cut algorithm provided at the Spin Glass Ground State Server at the University of Köln [35]. The remaining 2D instances with ground states were obtained from S. Sabhapandit and S. N. Coppersmith from the University of Wisconsin who identified the ground states using flat-histogram Markov chain Monte Carlo simulations [8]. All 3D instances with their ground states were obtained from previous simulations of one of the authors [17].

The upper bound on the number of iterations (generations) is determined empirically so that the number of iterations is sufficiently large for all tests. In general, the bound on the number of iterations for GA+DHC is larger than that for hBOA+DHC and UMDA+DHC because of the slower mixing with one-point crossover [34].

The performance of the compared algorithms is measured by the number of evaluated spin glass configurations until the optimum has been found. Additionally, we show the number of DHC updates until the optimum has been found.

15.4.3 Results

Figure 15.2 compares the performance of hBOA+DHC, UMDA+DHC, and GA+DHC on two-dimensional spin glasses with $\pm J$ couplings and periodic boundary conditions. The results indicate that the number of evaluations and the number of DHC flips for hBOA+DHC grows with a low-order polynomial of problem size, i.e., with $O(n^{1.63})$ and with $O(n^{2.46})$, respectively. Furthermore, the results show that hBOA significantly outperforms GA+DHC and UMDA+DHC in both the number of evaluations and the number of DHC flips.

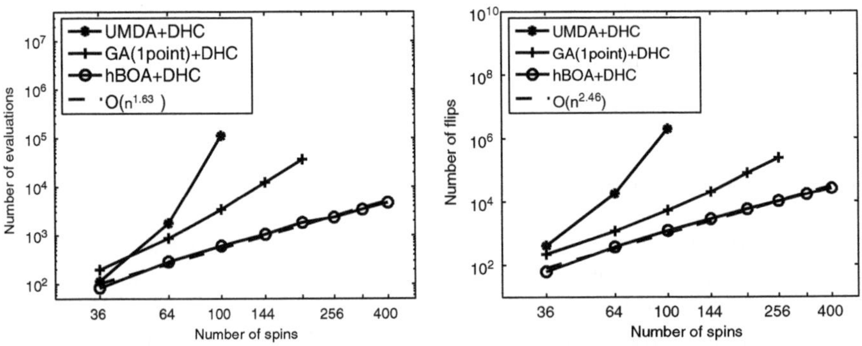

(a) Number of evaluations. (b) Number of flips.

Fig. 15.2. Performance of hBOA+DHC, UMDA+DHC, and GA+DHC on random 2D $\pm J$ Ising spin glasses

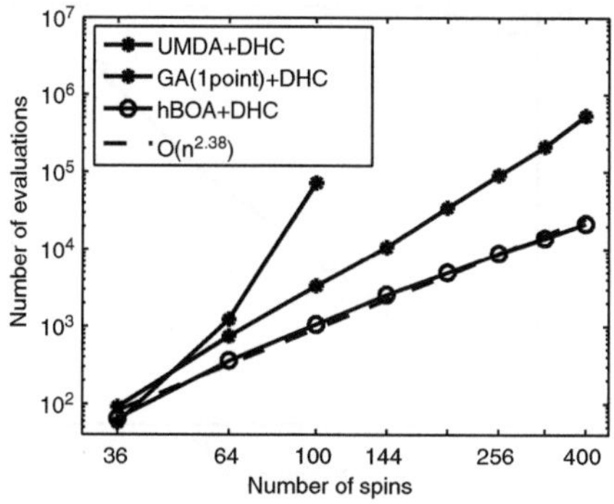

(a) Number of evaluations.

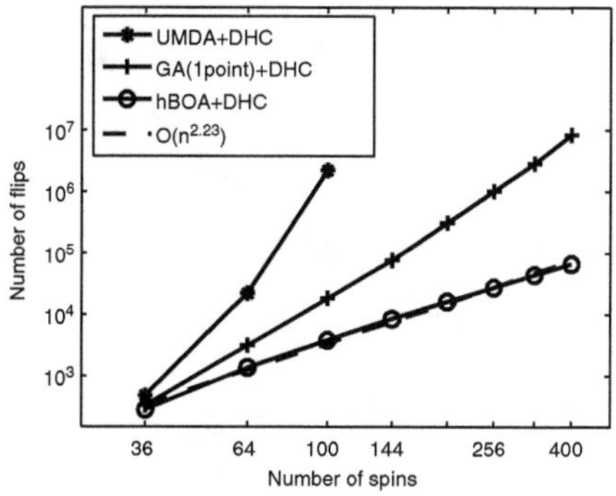

(b) Number of flips.

Fig. 15.3. Performance of hBOA+DHC and GA+DHC on random 2D Gaussian Ising spin glasses

The worst performance is achieved by UMDA+DHC, the time complexity of which grows faster than polynomially.

Recall that for spin glasses, one-point crossover performs relatively well because one-point crossover rarely breaks important interactions between spins due to the used representation. Nonetheless, this behavior cannot be generalized to other similar slowly equilibrating problems that exhibit different energy landscapes, such as protein folding or polymer dynamics.

Figure 15.3 compares the performance of hBOA+DHC and GA+DHC on 2D spin glasses with Gaussian couplings and periodic boundary conditions. The results indicate similar behavior as for $\pm J$ couplings in two dimensions; hBOA+DHC outperforms both GA+DHC and UMDA+DHC; UMDA+DHC performs worst and its time complexity grows faster than polynomially. The results also show that Gaussian couplings are harder for all tested algorithms. The reason for this is that in Gaussian spin glasses the constraints are scaled nonuniformly and it is known that most selectorecombinative evolutionary algorithms perform better on uniformly scaled problems than on nonuniformly scaled ones [26, 38]. Additionally, it seems that for Gaussian couplings, the number of flips grows slower than for $\pm J$ couplings, which indicates that for nonuniformly scaled spin glass couplings the impact of DHC on performance becomes much less significant.

For 2D Ising spin glasses, a polynomial algorithm [12, 13] with complexity $O(n^{3.5})$ exists that computes the number of states at each energy level, including the ground state. It was shown [32] that on 2D $\pm J$ Ising spin glasses, hBOA+DHC achieves asymptotic performance of the polynomial algorithm without any prior knowledge about spin glasses. The performance becomes slightly worse for the Gaussian spin glasses, although hBOA+DHC still retains low-order polynomial complexity.

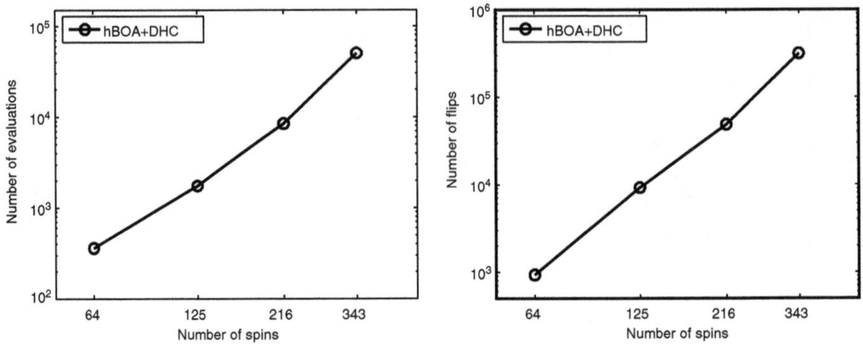

(a) Number of evaluations. (b) Number of flips.

Fig. 15.4. Performance of hBOA+DHC on random 3D $\pm J$ Ising spin glasses

Figure 15.4 shows the performance of hBOA+DHC on 3D $\pm J$ spin glasses. Since both GA+DHC and UMDA+DHC have not been capable of solving most 3D instances even with enormous computational resources, we only include the results for hBOA+DHC. The results show that the performance of hBOA+DHC appears to grow exponentially fast. This behavior is expected because the problem of finding ground states of 3D spin glasses is NP-complete [1] and it is thus unlikely that there exists an algorithm that can solve this class of problems in polynomial time. However, we see that hBOA+DHC is still capable of solving instances of several hundreds spins, which are intractable with most standard optimization algorithms, such as genetic algorithms and simulated annealing.

15.5 Cluster Exact Approximation (CEA)

Due to the complex structure of the energy landscape of spin glasses, many local minima exist, which have energies very close to the ground-state energy. Usually these minima differ from the true ground states by flips of large domains. Hence, as already mentioned, the minima are surrounded by high energy barriers from the viewpoint of single-spin-flip dynamics. This leads to a very small efficiency of algorithms which apply single-bit (spin) changes as DHC. For this reason, we also consider *cluster exact approximation* [16], which provides an efficient method that can change many spins at the same time optimally (assuming that the remaining spins remain fixed).

CEA constructs iteratively and randomly a nonfrustrated cluster of spins. During the construction of the cluster a local gauge-transformation of the spin variables $s'_i = h_i s_i$ ($h_i = \pm 1$) is applied so that all interactions between cluster spins become ferromagnetic ($J'_{i,j} = h_i J_{i,j} h_j < 0$), which is always possible for nonfrustrated systems by definition. For each spin, which has not

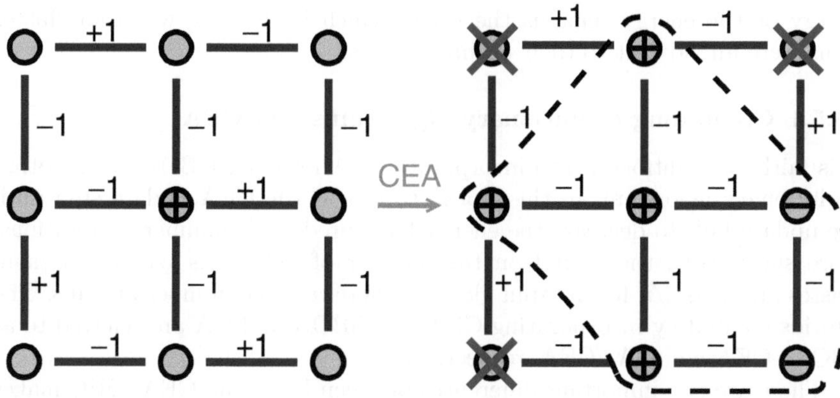

Fig. 15.5. Example of the cluster-exact approximation method. A part of a spin glass is shown, ferromagnetic bonds are denoted by −1, while antiferromagnetic bonds are denoted by +1, see (15.1). The construction starts with the spin i_0 at the center, the gauge field is set to $h_{i_0} = +1$. The right part displays the final situation. The spins which belong to the cluster carry a plus or minus sign which indicates how each spin is gauge-transformed, i.e., the gauge-field h_i, such that only ferromagnetic interactions remain inside the cluster. All other spins cannot be added to the cluster because it is not possible to multiply them by ±1 to turn all adjacent bonds ferromagnetic. Please note that many other combinations of spins can be found to build a cluster without frustration

been considered so far, it is tested whether a value $h_i = \pm 1$ can be found, which makes all adjacent bonds ferromagnetic. If this is possible, the spin is added to the cluster, if not, it becomes a noncluster spin. Figure 15.5 shows an example of how the construction of the cluster works for a small spin-glass system. The order in which the spins are considered can be random or governed by some heuristics; here pure random order was used. Note that the clusters constructed in this way are typically very large, e.g., For 3D $\pm J$ spin glasses, each cluster contains typically 55% of all spins.

The noncluster spins remain fixed during the following calculation; they act like local magnetic fields on the cluster spins. Consequently, the ground state of the gauge-transformed cluster is not trivial, although all interactions inside the cluster are now ferromagnetic. Since the cluster exhibits no bond-frustration, an energetic minimum state for its spins can be calculated in polynomial time by an algorithm which has been applied successfully for ferromagnetic random-field Ising models [18]. This algorithm is based on graph-theoretical methods [7, 36]: an equivalent network is constructed [33], the maximum flow is calculated [37, 39] and the spins of the cluster are set to orientations leading to a minimum in energy.

The CEA update step ensures that the spins in the cluster minimize energy of the considered spin glass assuming that the remaining (noncluster) spins remain fixed to their current values. Each CEA iteration either decreases the energy or the energy remains the same, which is the case when all cluster spins were already set to their optimal values.

15.5.1 Combining evolutionary algorithms and CEA

It is fairly straightforward to incorporate CEA into GA, hBOA or any other evolutionary algorithm. In this work, we repeat the CEA update step until the update fails to decrease the energy for a predefined number of iterations; more specifically, the bound on the number of failures is $\sqrt{n}$ for 2D spin glasses and it is $\sqrt[3]{n}$ for 3D spin glasses, where n is the number of spins. The hybrids created by incorporating CEA into hBOA and GA are referred to as hBOA+CEA and GA+CEA, respectively.

There are two important differences between DHC and CEA. DHC makes only single-bit updates and that is why it is able to reach the local optimum (with respect to one-bit flips) quickly but it is often trapped in a not-so-good local minimum. CEA makes much bigger updates, leading to a much more significant decrease in the energy, but also requiring more computational resources for each update; for example, the time complexity of each CEA step on 3D spin glasses can be expected to be about $O(n^{4/3})$ [23], whereas each single-bit update using DHC can be done in only $O(\log n)$ time.

Section 15.6 shows that CEA significantly improves the performance of both hBOA and GA, allowing these algorithms to solve much bigger problem instances efficiently.

15.6 Experiments

This section tests hBOA+CEA and GA+CEA on a broad range of 2D and 3D $\pm J$ Ising spin glass instances. Since the initial experiments indicated that UMDA can be expected to perform significantly worse than both hBOA and GA, we do not include UMDA in the comparison.

15.6.1 Tested Spin Glass Instances

Two types of spin glass instances have been considered in this set of experiments:

1. 2D Ising spin glass with $\pm J$ couplings.
2. 3D Ising spin glass with $\pm J$ couplings.

In all instances, periodic boundary conditions are used.

In the 2D case, instances of size 15×15 ($n = 225$ spins) to 50×50 ($n = 2,500$ spins) have been considered; $10,00$ random instances have been generated for each problem size. In the 3D case, instances of size $6 \times 6 \times 6$ ($n = 216$ spins), $8 \times 8 \times 8$ ($n = 512$ spins), and $10 \times 10 \times 10$ ($n = 1,000$ spins) have been considered; $1,000$ random instances have been generated for each problem size.

15.6.2 Description of Experiments

All parameters have been set like in the initial set of experiments in Sect. 15.4 of this chapter. The only difference from the set-up of the initial experiments is that here CEA is used to improve all candidate solutions instead of DHC.

Since one update step of CEA is usually more computationally expensive than the entire evaluation [23], the time complexity of the compared algorithms is measured by the number of iterations of CEA as opposed to the number of evaluations.

15.6.3 Results

Figure 15.6a shows the performance of hBOA+CEA and GA+CEA on 2D Ising spin glasses with $\pm J$ couplings. The results indicate that hBOA+CEA significantly outperforms GA+CEA and thus hBOA retains superior performance even with CEA. The results also show that incorporating CEA leads to a somewhat faster asymptotic growth of time complexity with problem size; on the other hand, the use of CEA provides a significant decrease of running time for the tested range of problems and, consequently, much larger problem sizes can be treated currently as compared to hBOA+DHC. Nonetheless, based on these results, it can be hypothesized that hBOA+DHC will become faster than hBOA+CEA for much larger spin glass instances.

Figure 15.6b shows the performance of hBOA+CEA and GA+CEA on 3D Ising spin glasses with $\pm J$ couplings. The results indicate that the performance of both algorithms grows faster than polynomially even with the use of CEA as is expected from the NP-completeness of this problem. However, CEA improves the performance of GA significantly and makes the difficult 3D instances tractable even with GA. Nonetheless, hBOA+CEA still retains superior performance.

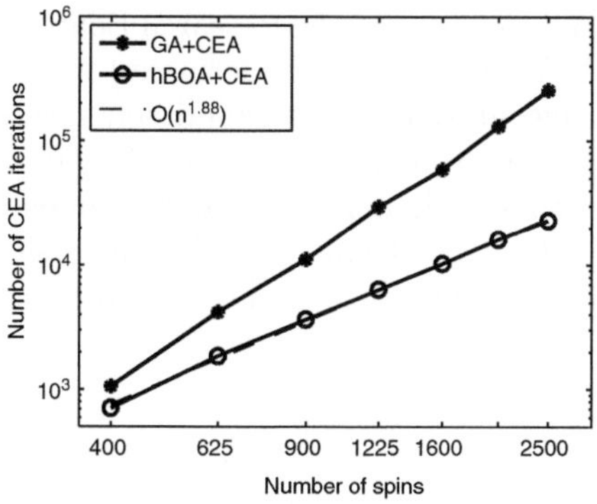

(a) 2D $\pm J$ Ising spin glass.

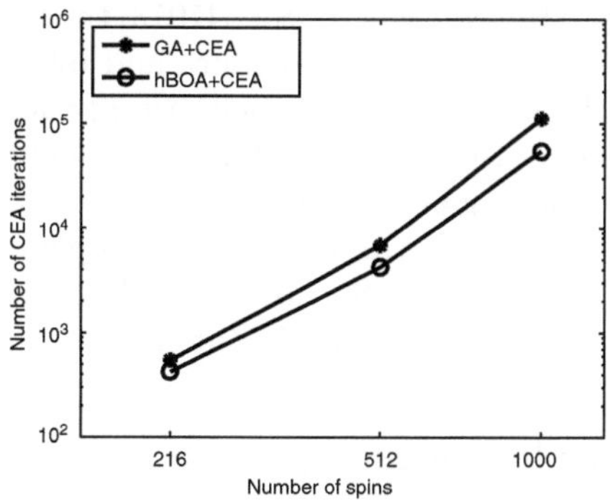

(b) 3D $\pm J$ Ising spin glass.

Fig. 15.6. Performance of hBOA+CEA and GA+CEA on $\pm J$ Ising spin glasses

15.7 Summary and Conclusions

This paper tested the hBOA on a large number of instances of the problem of finding ground states of Ising spin glasses with random couplings in two and three dimensions. The performance of hBOA was compared to that of the simple GA and the UMDA. All algorithms were hybridized by using either a simple DHC or the CEA. The results showed that hBOA significantly outperforms all other compared methods in all cases and that in some classes of Ising spin glasses, it achieves the performance of best polynomial-time analytical methods.

The problem of finding ground states of Ising spin glasses can be formulated as a constraint satisfaction problem and it has many features that can be found in other difficult constraint satisfaction problems, such as MAXSAT or graph coloring. Furthermore, this problem shares many challenges with other important classes of problems, such as protein folding. The most important source of problem difficulty is the enormous number of local optima that may grow exponentially with problem size. Additionally, to ensure effective exploration of the space of low-energy configurations, large-order interactions between spins must be considered.

The results presented in this paper thus confirm that using hierarchical decomposition for solving difficult optimization problems with little problem-specific knowledge holds a big promise and that this line of computational-optimization research is likely to advance a number of areas that crucially depend on solving difficult optimization problems scalably.

Acknowledgments

The authors would like to thank Fabien Alet, Martin V. Butz, David E. Goldberg, Jiri Ocenasek, Kumara Sastry, Simon Trebst, and Matthias Troyer for comments and help with the design, presentation and analysis of experiments.

This work was supported by the National Science Foundation under NSF CAREER grant ECS-0547013, the Research Award and the Research Board at the University of Missouri, and by the *VolkswagenStiftung* (Germany) within the program "Nachwuchsgruppen an Universitäten." Most experiments were done using the hBOA software developed by Martin Pelikan and David E. Goldberg at the University of Illinois at Urbana-Champaign. The experiments presented in this paper were completed at the Beowulf cluster at the University of Missouri at St. Louis and the Asgard cluster at the Swiss Federal Institute of Technology (ETH) Zürich.

References

[1] Barahona, F. (1982). On the computational complexity of Ising spin glass models. *Journal of Physics A: Mathematical, Nuclear and General*, 15(10):3241–3253

[2] Barahona, F., Maynard, R., Rammal, R., and Uhry, J. (1982). Morphology of ground states of a two dimensional frustration model. *Journal of Physics A*, 15:673

[3] Berg, B. A. and Neuhaus, T. (1992). Multicanonical ensemble - a new approach to simulate first order phase-transition. *Physical Review Letters*, 68(9)

[4] Bieche, I., Maynard, R., Rammal, R., and Uhry, J. (1980). On the ground states of the frustration model of a spin glass by a matching method of graph theory. *Journal of Physics A*, 13:2553

[5] Binder, K. and Young, A. (1986). Spin-glasses: Experimental facts, theoretical concepts and open questions. *Review of Modern Physics*, 58:801

[6] Chickering, D. M., Heckerman, D., and Meek, C. (1997). A Bayesian approach to learning Bayesian networks with local structure. Technical Report MSR-TR-97-07, Microsoft Research, Redmond, WA

[7] Claiborne, J. (1990). *Mathematical Preliminaries for Computer Networking*. Wiley, New York

[8] Dayal, P., Trebst, S., Wessel, S., ürtz, D., Troyer, M., Sabhapandit, S., and Coppersmith, S. (2004). Performance limitations of flat histogram methods and optimality of Wang-Langdau sampling. *Physical Review Letters*, 92(9):097201

[9] Fischer, K. and Hertz, J. (1991). *Spin Glasses*. Cambridge University Press, Cambridge

[10] Fischer, S. and Wegener, I. (2004). The Ising model on the ring: Mutation versus recombination. *Proceedings of the Genetic and Evolutionary Computation Conference (GECCO-2004)*, pages 1113–1124

[11] Friedman, N. and Goldszmidt, M. (1999). Learning Bayesian networks with local structure. In Jordan, M. I., editor, *Graphical models*, pages 421–459. MIT Press, Cambridge, MA

[12] Galluccio, A. and Loebl, M. (1999a). A theory of Pfaffian orientations. I. Perfect matchings and permanents. *Electronic Journal of Combinatorics*, 6(1). Research Paper 6

[13] Galluccio, A. and Loebl, M. (1999b). A theory of Pfaffian orientations. II. T-joins, k-cuts, and duality of enumeration. *Electronic Journal of Combinatorics*, 6(1). Research Paper 7

[14] Goldberg, D. E. (1989). *Genetic algorithms in search, optimization, and machine learning*. Addison-Wesley, Reading, MA

[15] Harik, G. R. (1995). Finding multimodal solutions using restricted tournament selection. *Proceedings of the International Conference on Genetic Algorithms (ICGA-95)*, pages 24–31

[16] Hartmann, A. K. (1996). Cluster-exact approximation of spin glass ground states. *Physica A*, 224:480

[17] Hartmann, A. K. (2001). Ground-state clusters of two, three and four-dimensional +/-J Ising spin glasses. *Physical Review E*, 63:016106

[18] Hartmann, A. K. and Rieger, H. (2001). *Optimization Algorithms in Physics*. Wiley-VCH, Weinheim

[19] Hartmann, A. K. and Rieger, H., editors (2004). *New Optimization Algorithms in Physics*. Wiley-VCH, Weinheim

[20] Hartmann, A. K. and Weigt, M. (2005). *Phase Transitions in Combinatorial Optimization Problems*. Wiley-VCH, Weinheim

[21] Holland, J. H. (1975). *Adaptation in natural and artificial systems*. University of Michigan Press, Ann Arbor, MI

[22] Mezard, M., Parisi, G., and Virasoro, M. (1987). *Spin glass theory and beyond*. World Scientific, Singapore

[23] Middleton, A. and Fisher, D. S. (2002). The three-dimensional random field Ising magnet: Interfaces, scaling, and the nature of states. *Physical Review B*, 65:134411

[24] Mühlenbein, H. and Mahnig, T. (1999). Convergence theory and applications of the factorized distribution algorithm. *Journal of Computing and Information Technology*, 7(1):19–32

[25] Mühlenbein, H. and Paaß, G. (1996). From recombination of genes to the estimation of distributions I. Binary parameters. *Parallel Problem Solving from Nature*, pages 178–187

[26] Mühlenbein, H. and Schlierkamp-Voosen, D. (1993). Predictive models for the breeder genetic algorithm: I. Continuous parameter optimization. *Evolutionary Computation*, 1(1):25–49

[27] Naudts, B. and Naudts, J. (1998). The effect of spin-flip symmetry on the performance of the simple GA. *Parallel Problem Solving from Nature*, pages 67–76

[28] Newman, C. and Stein, D. (2003). Finite-dimensional spin glasses: states, excitations and interfaces. preprint cond-mat/0301022

[29] Pelikan, M. (2005). *Hierarchical Bayesian optimization algorithm: Toward a new generation of evolutionary algorithms*. Springer, Berlin Heidelberg New York

[30] Pelikan, M. and Goldberg, D. E. (2001). Escaping hierarchical traps with competent genetic algorithms. *Proceedings of the Genetic and Evolutionary Computation Conference (GECCO-2001)*, pages 511–518

[31] Pelikan, M. and Goldberg, D. E. (2003). Hierarchical BOA solves Ising spin glasses and MAXSAT. *Proceedings of the Genetic and Evolutionary Computation Conference (GECCO-2003)*, II:1275–1286

[32] Pelikan, M., Ocenasek, J., Trebst, S., Troyer, M., and Alet, F. (2004). Computational complexity and simulation of rare events of Ising spin glasses. *Proceedings of the Genetic and Evolutionary Computation Conference (GECCO-2004)*, 2:36–47

[33] Picard, J.-C. and Ratliff, H. (1975). Minimum cuts and related problems. *Networks*, 5:357

[34] Sastry, K. and Goldberg, D. E. (2002). Analysis of mixing in genetic algorithms: A survey. IlliGAL Report No. 2002012, University of Illinois at Urbana-Champaign, Illinois Genetic Algorithms Laboratory, Urbana, IL

[35] Spin Glass Ground State Server (2004). `http://www.informatik.uni-koeln.de/ls_juenger/research/sgs/sgs.html`. University of Köln, Germany
[36] Swamy, M. and Thulasiraman, K. (1991). *Graphs, Networks and Algorithms.* Wiley, New York
[37] Tarjan, R. (1983). *Data Structures and Network Algorithms.* Society for industrial and applied mathematics, Philadelphia
[38] Thierens, D., Goldberg, D. E., and Pereira, A. G. (1998). Domino convergence, drift, and the temporal-salience structure of problems. *Proceedings of the International Conference on Evolutionary Computation (ICEC-98)*, pages 535–540
[39] Träff, J. (1996). A heuristic for blocking flow algorithms. *European Journal of Operations Research*, 89:564
[40] Van Hoyweghen, C. (2001). Detecting spin-flip symmetry in optimization problems. In Kallel, L. et al., editors, *Theoretical Aspects of Evolutionary Computing*, pages 423–437. Springer, Berlin Heidelberg New York
[41] Wang, F. and Landau, D. P. (2001). Efficient, multiple-range random walk algorithm to calculate the density of states. *Physical Review Letters*, 86(10):2050–2053
[42] Young, A., editor (1998). *Spin glasses and random fields.* World Scientific, Singapore

MIX
Papier aus verantwortungsvollen Quellen
Paper from responsible sources
FSC® C105338

If you have any concerns about our products,
you can contact us on
ProductSafety@springernature.com

In case Publisher is established outside the EU,
the EU authorized representative is:
Springer Nature Customer Service Center GmbH
Europaplatz 3, 69115 Heidelberg, Germany

Printed by Libri Plureos GmbH
in Hamburg, Germany